THE HANDLING OF CHEMICAL DATA

The Handling of Chemical Data

BY

P. D. LARK, B. R. CRAVEN

School of Chemistry, University of New South Wales

AND

the late R. C. L. BOSWORTH

THE QUEEN'S AWARD TO INDUSTRY 1966

PERGAMON PRESS

OXFORD · EDINBURGH · LONDON · NEW YORK

TORONTO · SYDNEY · PARIS · BRAUNSCHWEIG

Pergamon Press Ltd., Headington Hill Hall, Oxford
4 & 5 Fitzroy Square, London W. 1

Pergamon Press (Scotland) Ltd., 2 & 3 Teviot Place, Edinburgh 1

Pergamon Press Inc., Maxwell House, Fairview Park, Elmsford, New York 10523

Pergamon of Canada Ltd., 207 Queen's Quay West, Toronto 1

Pergamon Press (Aust.) Pty. Ltd., 19a Boundary Street,
Rushcutters Bay, N.S.W. 2011, Australia

Pergamon Press S.A.R.L., 24 rue des Écoles, Paris 5^e^

Vieweg & Sohn GmbH, Burgplatz 1, Braunschweig

First edition 1968
Reprinted 1969

Library of Congress Catalog Card No. 66-17264

Printed in Great Britain by A. Wheaton & Co., Exeter

08 011849 6

CONTENTS

PREFACE

THE proper handling or management of experimentally determined data requires a knowledge of their fundamental nature, of the patterns they follow and of the influence of experimental errors upon them. Given such knowledge, the experimenter can attempt to extract the essential information his observations contain or to present them in such a way that others can do so.

This book is concerned with the handling of measurements such as are obtained in chemical experimentation. A general account of the nature of measurements and of their specifically dimensional character is contained in the first parts of Chapters I and VII. The remainder of the book is largely devoted to the ways in which particular kinds of observations may be treated and described by graphical and numerical methods. Observations relating to single variables are discussed in Chapters II and III and those involving two or more variables in Chapters IV, V and VI.

The selection of topics is based to some extent on the authors' own experience and appreciation of the difficulties of colleagues and students in academic and industrial practice. Solutions to problems have been collated from the recent chemical literature as well as from specialized works. Where possible, simple methods of treatment are presented with some emphasis on graphical procedures and in many cases alternative procedures are described to give the user the opportunity of selection in accordance with the means of computation at his disposal.

The treatment aimed at is essentially elementary and the discussion discursive but, in spite of this, some attempt has been made to include material found mainly in more advanced monographs. We have done this because problems encountered in practice are often complicated but have still to be solved by research workers who generally lack the time and sometimes the training in mathematical and statistical analysis to find and apply methods of solution given in rigorous works written in an unfamiliar idiom. It is to such people that this work is addressed and to those, in particular, who prefer or are compelled to treat their own data with the aid of simple calculating devices.

The borderline between topics proper to the theme of the book and others closely related is not easy to define. Two of several which might have been included are the solution of equations, especially sets of linear equations, and nomography. The first is basic to some of the procedures discussed but only becomes of special importance in connection with rather more elaborate problems than we have thought it useful to include. The second is a means of graphically representing experimental data but its primary function lies in calculation rather than in representation as such. Treatment of these topics and more thorough treatment of those dealt with is to be found in many books on statistical and numerical methods, nomography and dimensional analysis. Some are listed for guidance in further reading and as a record of our indebtedness in the general bibliography.

The senior author, R. C. L. Bosworth, was the author of several books and many papers of theoretical and industrial interest in the fields of physical chemistry and chemical engineering, his *Physics in the Chemical Industry* and *Transport Processes in Applied Chemistry* being the most notable. After his appointment as Associate Professor of Physical Chemistry in the University of New South Wales in 1957 he became interested in the teaching of the applications of mathematics and statistics in a specifically chemical context. This book, which is partly the result of Bosworth's own interest and enthusiasm, was incomplete at the time of his sadly early death in 1964. His co-authors have attempted to preserve the spirit if not always the exact letter of his contribution in their final editing.

The authors wish to record their gratitude for help received from many colleagues, in particular Professor R. L. Werner, Dr. E. S. Swinbourne and Mr. P. Souter of the University of New South Wales, Professor J. R. Anderson of Flinders University of South Australia, and Dr. C. G. Barraclough of the University of Melbourne, to the holders of copyright who have given permission for the reproduction of tables and graphs and to the Publishers for their assistance in overcoming many difficulties in the production of the book.

P. D. LARK and B. R. CRAVEN

School of Chemistry,
University of New South Wales

NOTATION

SINCE this book deals with physico-chemical data in particular and with their treatment in general the notation is of some complexity. Italic type has been used in the main for general mathematical and statistical symbols, Greek or Roman bold type for physico-chemical properties and ordinary Greek type symbols for statistical parameters.

We have not distinguished symbolically between variates and variables and have avoided the use of subscripts to distinguish individual values of a variable from the variable itself except in a few sections where the more rigorous notation helps to clarify the argument. As far as possible double letter symbols are reserved for abbreviations (for example, SS for sum of squares of deviations) and the dimensionless ratios of Chapter VII.

To assist the reader lists containing the relevant symbols are given at the end of each chapter after the first.

CHAPTER I

THE ORIGIN AND PRESENTATION OF CHEMICAL DATA

DATA, which are the raw material of science, and in particular of physical science, attain their greatest usefulness when they are in the form of a quantitative measure. Indeed, in the view of Lord Kelvin: "I often say that when you can measure what you are speaking about, and express it in numbers, you know something about it, but when you cannot express it in numbers, your knowledge is of a meagre and unsatisfactory kind; it may be the beginning of knowledge but you have scarcely in your thoughts advanced to the stage of a Science, whatever the matter may be."

This has always been the attitude of the physical scientist. If one wishes to know anything about any subject whatsoever, one should first devise an appropriate scale of measurement, and then obtain pointer readings on that scale. The pointer readings will contain all the real information on the subject, and the extension of scientific principles will involve an extension of mensuration to include the new subject. This attitude in the physical sciences contrasts sharply with that of the descriptive sciences, where the classical developments have followed an organization of material rather than mensuration. Knowledge concerning the external world comes to us through sense perceptions. Data concerned with exact measurement, and therefore of value in scientific analysis, come to us only through the senses of seeing and hearing. Data coming through the senses of taste, touch or smell may express aesthetic values but cannot be given numerical form. Even among the visual and aural senses there are whole regions of sensations which cannot be utilized for numerical measurement. Senses of tonal value or musical appreciation cannot, or at least hitherto have not been made the basis of a numerical scale, and neither has the sense of colour harmony. Peripheral vision, vital as a warning of danger to primitive man in the jungle and equally so to motorists in a modern city, plays no part in setting up any form of mensurational scale. Indeed Eddington has claimed that the ideal observer is blind in one eye, colour-blind in the other and endowed only with intense foveal visual acuity.

While data on the behaviour of the physical world may come to us, as scientists, through all senses and all aspects of these senses, it is only a very narrow range of experience which can provide raw material for scientific organization. In order to understand those aspects of data of value in the exploration of chemical facts and chemical hypotheses, we must first examine the various ways in which data can arise from a chemical investigation, and then go on to consider the methods of presentation of this information so that it is available in a convenient form for subsequent treatment and interpretation.

1.1. The Origin and Nature of Quantitative Data

Chemistry is closely related to the other physical sciences in which progress has been based on the measurement of specific properties. This metrical approach is very prominent in physical chemistry, whereas some other branches of the discipline, for example organic chemistry, are related to the descriptive sciences in that emphasis is laid more strongly on organization than on measurement. Thus, spectroscopic data, molecular weights, boiling points, etc., are purely quantitative; relations between structures, however, are entirely organizational in nature.

Other aspects of chemistry point to its close relationship to certain pre-scientific technologies. Assessments in such systems can be carried out by a far wider range of sense impressions than those envisaged by Eddington. Some chemical properties, for example, can be gauged through senses of taste or smell. The nature and roughly the amount of certain contaminants in an alcoholic beverage can be judged far more quickly and accurately by an "expert" depending on these senses than by a chemist using the methods of quantitative analysis based on visual inspection. The "expert" does not use a scale, and probably cannot communicate the basis of his assessment, but nevertheless he is able to arrange a group of beverages in *order* according to the notions of greater or less. The sense of feeling can be used in grading the rheological properties of composite mixtures such as butter and paint in much the same way.

To understand the ways in which the chemical properties of natural systems can be investigated quantitatively we must first look at the ways in which numbers can be applied to such systems. The fact that information may be communicated or stored most conveniently in the form of numbers is perhaps partly responsible for the fact that we devise all types of indices of performance for assessment of qualities which are not normally or cannot be considered as metric quantities. Thus, purely qualitative attributes may be identified by numerals to establish a *nominal scale* (e.g., 0, 1 to indicate absence or presence of an attribute).

To return to ordering: we judge the performance of students in terms of their examination marks. Suppose four students obtain the following marks in two different examinations:

Mulcahy	72	95	Smith	52	68
Fletcher	54	91	Lowe	46	57

The marks at the two different examinations are quite different and we may well ask if the students were more competent in the subject at the time of the second examination than they were at the time of the first or if the second examination were less stringent. Can we get any absolute measure of the performance of any one of the four candidates from these data? Ideally, of course, the examiner has in mind an average grade passable student, and each actual student should be compared with this average student. The examination then simply becomes the mechanism whereby the thing being tested for a property is compared with a standard, real or ideal, of that property. The fact that the standard can be an idealized concept with no real constant physical counterpart (in which case the mechanism of comparison is purely mental) is in the main responsible for the poor reproducibility obtained in situations like that of the two examinations given above.

There is, however, a means of attaining a real standard which is at least more constant than any individual student. If we rank the students in decreasing order of marks we find in this case the same order in the two examinations, namely

Mulcahy	1	Smith	3
Fletcher	2	Lowe	4

The number 2 applied to Fletcher indicates his position in this group but does not indicate how far he may be below the standard of Mulcahy or above that of Smith. Indeed, in one case he appears to be almost equivalent to the former and in the other almost equivalent to the latter. The number 2 refers only to the rank or order of greater or less, and gives no indication of its bearing to other neighbours on the examination sheet. The quantity 2 is in effect an ordinal number and the *ordinal scale* (1 ... 4) is an example of the simplest application of numerical measure to a property which changes progressively. Such scales are of little use in deductive processes although they are still to be found in some of the more primitive concepts of science. Thus Moh's scale of hardness of minerals is ordinal. A Moh's hardness of 6·5 simply means that the mineral tested is harder than orthoclase (hardness 6) but softer than quartz (hardness 7). This may be helpful in classifying an unknown mineral but the number itself would be useless for deducing mechanical strength.

Science based on measured and interrelated quantities requires that each property examined be compared by a graduated scale for which there is a one-to-one correspondence between the marks on the scale and the ensemble of unknown quantities. Scale marks relating to a particular property must be compared with manifestations of that property. This comparison may be effected by means of a transducer, which transforms one particular property into another, followed by the use of an appropriate standard for the second property. The apparatus with which the comparison is made may be called the detector. A suitable detector must respond to slight mismatching of the standard and unknown. In the ultimate analysis the detectors are the human senses but often these are supplemented by instrumental methods.

1.2. Operation of a Scale of Measurement

The basis of the operation of a scale measurement lies in the attainment of a one-to-one correspondence between the scaled (known) property and that to be measured (unknown). There is a limited number of physical properties which may be compared directly by the method of superposition in which the unknown is placed over the known and shown to be equivalent. Thus, one length can be placed over another and the two compared directly. They may be found to be equal or one longer than the other. Replication of a standard length is easily obtained and checked by repeated superposition and multiples and submultiples fabricated. In terms of these a graduated scale can be built up. The act of measurement is that of estimating the position of the pointer with respect to the graduations of the scale. Here we may use visual aids or mechanical detectors for amplifying the differences between the pointer and the graduation marks.

There are, of course, properties other than length which are superposable; for instance volumes, angles, forces—both mechanical and electrical—and intensities of light and sound. Many of these superposable properties can be measured by using a balancing mechanism in which the unknown is caused to act in opposition to a measured quantity of the known and the zero point indicated by suitable detectors. Examples of such devices are the chemical balance, the potentiometer and the photometer. Because the detector is only required to operate for small differences between the unknown and the known it may be given a high degree of sensitivity without being particularly robust. A still greater degree of sensitivity can be obtained by combining balancing systems with pointer movement so that near balance is reached and the out-of-balance component measured by pointer displacement as, for example, in the micro-chemical balance in the use of which the last figures for the mass are obtained from pointer deflection and the rest by counting the number of weights used. The recording Wheatstone bridge operates on the same principle.

While there is only a limited class of physical properties which can be treated by the pointer or balance methods, there is always the possibility of transforming the change in any particular property into a change in a completely different property by a transducer. The mercury-in-glass thermometer transforms a change in temperature into a change in volume, a watch transforms changes in time into changes in angle, a katharometer transforms changes in thermal conductivity into changes in electrical resistance, the second mentioned properties in each case being superposable. Provided that the transduced property is monotonically related to the property to be changed and that the transduced property is superposable, the transducer provides a means of measurement of the primary property even if there is no direct way of assigning a number to the primary property itself. There is also a direct corollary to this: any effect which produces a uniform real change in the external world can be measured and units of measurement devised.

If the scale is to be usable, the units and subunits in which it is composed must refer to the primary property and not the secondary or transduced property. Therefore, although the peculiar properties of the transducer may affect the sensitivity and accuracy of measurement, they should not in any way alter the scales of measurement of the primary property. In other words, scales of temperature may be obtained through the operation of the mercury-in-glass transducer, the platinum resistance transducer, the constant volume nitrogen transducer, the thermocouple transducer or the colour or brightness radiator transducer, but in any of these systems temperature is being compared with temperature and not resistance with pressure or volume with light intensity. Different transducers may offer different degrees of experimental difficulty or may require different techniques but there is no reason why measurements on one scale should not be the same as those on another, provided only that the various methods have been standardized with respect to the same fixed points.

This is an aspect of transducer mensuration which is too often ignored in technical or non-scientific applications. A measurement of one quantity must be referred to other quantities of the same nature and not to the qualities of the transducer. For example, filterability may be measured in terms of the pressure drop across a filter cake but the units of filterability are certainly not those of pressure.

1.3. Measuring and Counting

So far we have considered the assignment of effectively continuous numbers on the scale of a variable by superposition but there is another and often much simpler way of assigning numbers to the performance of an unknown with respect to the standard. Suppose, for example, that the unknown is the aperture of a batch of ring spanners from a production line. Individuals differ slightly one from another but a spanner is acceptable if it fits a standard bolt-head. The results of the test against the standard would divide the batch into acceptable and unacceptable articles. This is an example of counting attributes rather than measuring variables. The fraction accepted is a measure of the efficiency of the manufacturing operation. A standard exists, namely the test article, or the bolt-head in the present example. The values obtained, however, are not continuously variable, but are the results of a counting and not a measuring operation. This counting procedure is used widely with respect to manufactured articles and also in quantum or atomic manifestations, for example in the use of photoelectric multipliers to measure very low light intensities where single quanta actuate a counting mechanism.

1.4. Units, Fixed Points and Relationship between Scales

The object of any communication between a sender and a receiver is to transfer information with a minimum of loss, an object which can only be achieved if the numerical measure of a quantity of communicated data has the same significance to sender and receiver. This is done by referring the measure to a common standard known to both. The relevant standard whose nature is common knowledge is in fact the unit of measurement. A unit quantity for the measurement of a property—say temperature, time, viscosity, etc.—must, as we have said, be a quantity of the same physical nature as the property being measured. Thus we do not measure quantities of temperature in units of volume of mercury or pressure of nitrogen and the transducer which measures the unknown quantity must be calibrated in terms of the reactions of other transducers carrying out the same operation. The measure of the unknown in terms of the performance of the agreed unit cannot, however, be ascertained until there are at least two fixed points or, alternatively, an arbitrary zero point and an arbitrary unit point on the scale. The difference between these points may be used to estimate how far the unknown diverges from the zero point (or from one of the fixed points) and this estimation may be given in terms of an arithmetic relation. If the transducer shows a value x as the deflection from the zero for a superposable quantity and has a value y for the difference in superposable quantity between the two states defined as the zero and unit points, then the measure of the unknown in terms of the unit is x/y units.

This method of setting up a scale may be seen from the operation of the mercury-in-glass thermometer where units of temperature are defined in terms of equal volumetric changes between fixed points on the thermometer. The fixed points are in turn defined by reproducible states of temperature to which are given arbitrary measures. However, there is in general no justification for accepting the scale as linear, although the smaller the difference between the two fixed points the more nearly linear becomes the scale so defined. The use of other subsidiary fixed points in addition to the two primary fixed points will improve the linearity of the scale but this method of improvement is only applicable if the measures of the subsidiary fixed points can be assessed independently with respect to the behaviour of the transducer. This means that we must have in hand a second more accurately known transducer, possibly one which can only be used under special conditions or, alternatively, we must have plausible theoretical relationships between the measures of the primary and secondary fixed points. Taking a temperature scale as an example, we may supplement the primary fixed ice and steam points by such secondary points as the boiling point of oxygen, the boiling point of sulphur and the melting points of antimony, silver or gold. However, these secondary points may be fixed accurately with respect to the primary points by means of the constant pressure nitrogen thermometer for which there is a plausible relationship between volume and temperature based on the known thermodynamic properties of nitrogen gas.

Summarizing, we may conclude that any two fixed points, which can be obtained independently and to which can be assigned arbitrary values, can be used to define an *interval scale* of measurement provided only that their use gives rise to a real and superposable effect. The scale so defined may not be linear with respect to another scale but, if we have independent knowledge of a property which fixes the size of the effects, we may find relationships between a whole series of fixed points which can be used to define as linear a scale as is required. A multiple set of fixed points may define a usable and reproducible scale, but unless the measures of the fixed points have been related by independent measurements, the meaning assigned to different intervals from various parts of the scales may not be the same. In Moh's scale we use ten fixed points for hardness but there are no independent methods used to interrelate the points. If we could do this, say by using Brinnell or Rockwell tests to assign definite values to the ten fixed points, then we could use the system of these points (no longer valued 1 to 10) to fix a reasonably linear scale of hardness from below that of talc to that of diamond and beyond.

If a property is measured by two different types of transducer the numerical scores for each may be quite different although, if the scales are normalized by the use of a number of fixed points, they will necessarily be the same at and in the immediate vicinity of these points. Thus, if there is a sufficiently large number of fixed points chosen, the difference between the scores may be negligible over the whole range of measurement. Temperature, for example, can be measured in a number of different ways—by the differential volumetric expansion of mercury and glass, by the change in the electrical resistance of a platinum wire, by the change in the e.m.f. generated by a platinum–platinum rhodium thermocouple or by the pressure of nitrogen in an enclosed vessel within the system whose temperature is to be measured. These four transducers will necessarily give the same temperature score at and about 0 °C and 100 °C. However, they need not, and in practice do not, give the same scores for temperatures in the middle of the scale, indeed differences between the scores obtained by the various transducers in this range may be as much as 4 °C (Paul, 1951). At first sight, therefore, it appears that temperature measurement in the 50 °C region could involve errors of this order. However, there exists independent evidence to the effect that the ideal gas temperature scale is a true measure of the thermodynamic scale in which the property "temperature" is defined in terms of idealized, and therefore theoretical, concepts. The platinum scales are non-linear, but can be normalized, as parabolic functions for example, by using a third fixed point.

Multiple fixed points necessarily reduce the differences between scores on various scales and if we use a sufficiently large number and can define some form of measurement as canonical (for example, that of temperature by the ideal gas thermometer) we can make all the scales identical. Canonical scales have uniquely fixed zero points and may be termed *ratio scales* to distinguish them from the arbitrarily based interval scales defined only by fixed points (Stevens, 1955).

1.5. Dimensions

It is a matter of experience in the physical sciences that the various units as defined by accepted scales are not wholly independent, nor yet completely dependent but a limited number of relationships hold between different units. The unit of length, the metre, may be regarded as fundamental. The unit of time, the mean solar second, is independent of the metre, but derived units, such as those of velocity, acceleration, etc., are related to both the metre and the second. The unit of electric charge may be regarded as either fundamental or derived; the unit of electric current is directly related to the unit of charge and the unit of time. The relationship which exists between the unit of any quantity and the fundamental units is known as the *dimensions* of the quantity. For example, if the unit of a particular property had been shown to be directly proportional to the unit of mass and proportional to the square of a linear measurement, the dimensions of that property would be $[ml^2]$.

Let us apply the concept of dimensions to the measurement of a property with a transducer. The method may, in practice, consist in measuring a property of the transducer, e.g. temperature in terms of volume of mercury, but this does not mean that the dimensions of the property being investigated are those of the transducer itself. However we have seen that the measurement is referred to that of one or more fixed points. Dimensions, therefore, are defined by the properties of the fixed points and these do not alter when the transducer is changed. For example, it has sometimes been claimed that the dimensions of velocity as measured by a Doppler shift of wavelength should be those of a frequency change over the original frequency, i.e. of $\delta\boldsymbol{v}/\boldsymbol{v}$ (dimensionless) and not $[lt^{-1}]$ as defined otherwise. The Doppler shift is only a transducer of velocity and the fixed point defining the scale is the velocity of the radiation concerned, e.g. light or sound. The dimension of the velocity as measured by the device of the Doppler shift is, therefore, the dimension of the fixed point, a velocity of radiation.

1.6. The Presentation of Quantitative Data

The raw products of laboratory experiments in the physical sciences consist, as we have seen in the previous sections, in numbers. These numbers may refer to isolated facts, for example physical properties of an organic compound or the number of impulses from a radiation counter in a given time, or they may refer to measures of two or more quantities which are suspected of being related, for example the refractive index of a liquid and temperature.

In the first case, that of successive isolated measurements, the investigator has a certain amount of scope in presenting his raw data and the various methods available for this purpose, e.g. tables, frequency distribution graphs of various sorts, probability paper plots, etc., are discussed in some detail in Chapter II. They are usually the initial step in an examination of the data by statistical methods.

The aim of the present section is to set out the various ways in which the second type of numerical data can be presented, that is, where there is a suspected relationship between two or more of the measured quantities. If the measurements are concerned with the variation of, say, density or refractive index with temperature, the data are said to be singly associated. If the solubility of a gas in a liquid is studied jointly at various temperatures and pressures (both of which influence solubility), the data are said to be doubly associated. If we are studying the yield of a chemical process in which the reaction mixture is passed at different space velocities and pressures over catalysts of different activities and at different temperatures, we are concerned with multiply associated data.

There are three main ways in which associated data may be presented: by table, graph or empirical equation. We shall now consider each of these methods of data presentation in greater detail.

1.7. Tabular Presentation

Data are almost invariably arranged in tabular form in an investigation, either as the final, publishable, form or as the first step in the use of other methods. Its main advantage lies in the fact that any desired number of significant figures can be reported, irrespective of the range of the data included.

When prepared for publication a table should be, as far as possible, self-explanatory, a legend under the heading giving both the nature of the data and an indication of its source. The data to be presented should be set out in either columns or rows, with appropriate headings or side headings giving the name or symbol of the quantity recorded and the unit in which it has been measured. Top headings with the data in columns are easier to follow than side headings with the data in rows, although if only singly associated data are to be presented the use of columns may not be justifiable on account of its expense.

The data should be tabulated in order of ascending or descending values of the independent variable, that is the variable which the experimenter has controlled or attempted to control to a series of chosen values. Any additional information on the individual measurements, such as origin or order of determination, could conveniently be recorded in a separate column or row. For presentation in manuscript or typescript it is desirable for reasons of neatness to box in all headings and tabulations by ruled horizontal and vertical lines. An example is given in Table 1.1 in which the ratios of heat capacities and the number of atoms per molecule are given for various gases with an idea of their size and shape implied by the chemical formulae. Note that for easy reading of data tabulated in columns, decimal points should be vertically aligned in each column.

TABLE 1.1. RATIO OF THE HEAT CAPACITIES OF GASES ($C_p/C_v = \gamma$)

Gas	Molecule	Atoms per molecule	γ
Mercury	Hg	1	1·67
Argon	A	1	1·65
Hydrogen	H_2	2	1·41
Oxygen	O_2	2	1·40
Chlorine	Cl_2	2	1·32
Carbon dioxide	CO_2	3	1·31
Ammonia	NH_3	4	1·30
Methane	CH_4	5	1·27

The methods for carrying out mathematical operations such as smoothing, integration, etc., of tabulated data will be considered in some detail in Chapter VI.

1.8. GRAPHICAL PRESENTATION

This method of presentation is very commonly used when it is desired to clarify trends in data and it also gives a reasonable qualitative idea of the scatter displayed by the information. The number of significant figures which can be included in a graph is limited by the size of the graph and the range of the data. Data represented graphically can be differentiated, integrated, interpolated and extrapolated very rapidly, although not with great accuracy.

The association between two sets of experimental data, and particularly those of limited accuracy, are more speedily communicated by graphical means than by any other. For this reason, graphs are widely used for representing all kinds of relationships, varying from such trivial instances as those pictorial representations showing that ice manufacturers effect a greater turnover in the summer months to detailed and complicated curves of the relationships between current density and applied voltage in a polarograph, in which the position and magnitude of each discontinuity in the curve may be strikingly relevant. While there are ways, which we shall consider later in this volume, of handling relationships between more than two variables, we shall at present restrict ourselves to the case of only two variables, values of one being plotted as ordinates and of the other as abscissae to give the required graph.

The primary representation of two sets of interdependent variables will thus consist of a series of isolated points on ruled paper. The investigator then usually makes the assumption that if he were able to take a sufficiently large number of observations under sufficiently well controlled conditions, the experimental points would tend to join up to form a line or curve which would be the experimental functional relationship between the variables. In practice, of course, replication does nothing of the sort, but rather, as the number of the observations is increased, denser and denser clusters of experimental points form a cloud which surrounds the hypothetical "experimental" curve, the form of which may be "induced" by statistical methods as the locus about which the experimental points cluster most closely.

If the relationship connecting two variables is to be anything more than a purely qualitative or indicative one, we must develop accurate methods for estimating the position of the experimental curve and, having done this, develop other methods for summarizing the information it contains. Both of these objectives are most readily reached if the curve can be reduced to a straight line. While this is not always possible, there are many types of relationships for which it can be done.

Normal linear graph paper has both axes ruled arithmetically. Any functional relationship between variables x and y of the type

$$y = ax + b \tag{1.1}$$

will give a straight line if plotted on this paper. Thus the line

$$2x + y + 1 = 0$$

is shown in Fig. 1.1. If we have any functional relationship between the variables of such a nature that we can transpose x into X and y into Y in a way which will give an equation of the form of eqn. (1.1) above, we may then represent the x–y data as a straight-line relationship.

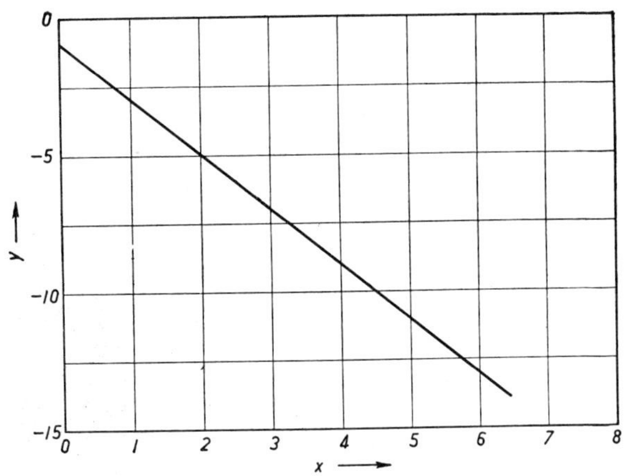

FIG. 1.1. A Graph of the Relationship: $2x + y + 1 = 0$.

Suppose for example the equation between x and y which best fits their empirical relationship is of the form

$$x = ab^{cy} \tag{1.2a}$$

where a, b and c are constant. Transposition of variables in eqn. (1.2a) to

$$X = \log x, \quad Y = y$$

would give the expression

$$X = \log a + (c \log b)\, Y \tag{1.2b}$$

which is now in the form of eqn. (1.1) and therefore a plot of X against Y would give a straight line.

Again, a relationship of the type

$$x^a = by^c \tag{1.3a}$$

could be transposed to

$$aX = \log b + cY \tag{1.3b}$$

by the transformations

$$X = \log x, \quad Y = \log y.$$

Further, suppose x and y are connected by a parabolic relationship, thus

$$y = a + bx + cx^2. \tag{1.4a}$$

Select (x_0, y_0) as any point on this curve and transpose

$$X = x, \quad Y = (y - y_0)/(x - x_0).$$

Then X and Y are related by the expression

$$Y = (b + cx_0) + cX \tag{1.4b}$$

which again represents a linear relationship.

As a last example, take the relationship between x and y as being

$$y = x/(a + bx). \tag{1.5a}$$

Now let

$$X = 1/x, \quad Y = 1/y.$$

We then obtain the linear relationship

$$Y = aX + b. \tag{1.5b}$$

These are examples of relatively simple transformations; the topic is further discussed in Chapter V.

The plotting of the experimental data to yield straight lines can in some cases be carried out by using a suitable non-linear graph paper. For handling data such as might be represented by eqn. (1.2) the most useful type of paper is *semi-logarithmic*, i.e. paper ruled logarithmically along one axis and arithmetically along the other. The scale rulings on the log-axis are crowded together at the higher values. Excessive crowding on a long scale can be avoided by periodically changing the value of the intercept between the rulings. A common logarithmic ruling is that in which the first ten rulings from 1 to 10 are shown, the next nine from 20 to 100, the next nine from 200 to 1000, and so on. The number of replications of the rulings is frequently referred to as the number of cycles.

For handling the data represented by eqn. (1.3) a *double logarithmic* or *log–log* paper is used on which both axes are ruled with logarithmic scales. Data connected by the function

$$y = Ax^n \tag{1.6}$$

will also give a straight line when plotted on log–log paper, but it must be remembered that the slope of this line will be equal to the exponent n of eqn. (1.6) while its intercept on the y-axis will be equal to the logarithm of A in the same equation.

For special purposes there are many other types of graph paper rulings available, for example *probability* paper, on which the plotted error distribution becomes linear, *log-reciprocal* paper on which the equation for an activated process

$$y = Ae^{-b/x} \tag{1.7}$$

becomes a straight line (x here is usually the temperature) and *triangular* paper, on which three component phase diagrams may be conveniently constructed.

Graphical representation of data has an advantage over other methods in that it provides a quick and easily read indication of the type of relationship holding between the two variables as well as a relative measure of the precision of their relationship. In order to use the method to greatest advantage and avoid misinterpretations certain rules for the construction of graphs must be followed.

First, the nature and source of the data represented should be made perfectly clear on the diagram with a suitable heading and, where necessary, legend, with both the nature and magnitude of the variables clearly indicated on the axes. The scale of an axis should be selected in such a way that the general trend of the curve will lie at roughly 45° to each axis. Attention should also be given to the choice of a scale factor which will facilitate reading of scale values. Thus scale factors of 2 units or 5 units of the variable per linear ruled unit of the axis are easily read, whereas 3 to 1 or 4 to 1 are more difficult.

If the data come from different sources or have been obtained under different conditions it is convenient, indeed essential in the preliminary drawing, to represent points referring to different conditions by different marks, for example ⊙, ⊡, △. Even a casual inspection may then show whether or not different sets of data can or cannot be used to define a single line and thus yield a single relationship between the variables.

A preliminary plot of experimental data may indicate whether the expected functional relationship between variables is to be a smooth curve or whether discontinuities or cusps are to be expected. No departure from a smooth curve should be accepted unless there are several neighbouring points supporting the existence of the discontinuity. Preliminary plotting may therefore be used as an indication that further data are required before proper completion of the curve is attempted. If there is no definite evidence for the existence of discontinuities, a smooth curve with as few inflections as possible should be drawn through the cluster of points in such a way that, in any moderately sized group, one half of the points lie above and one half below the curve. The shape of the curve may then indicate some transformation of the variables by which it might be rectified.

The actual drawing of the line representing the estimated functional relationship between the variables is the final task to be accomplished and the reliability of the line drawn depends on a number of factors. Of course, greatest reliability can be most easily attained when the data have been transformed into variables which are linearly related and we shall consider first the drawing of such a graph.

The data which are to be represented by the line may be subject to all the ordinary sources of experimental error. Such errors (see Chapter III) may be subdivided into two types, systematic or constant errors and random errors. The former are the more difficult to eliminate, and will be dealt with first.

The essential problem involved in drawing a graph is one of finding a functional relationship connecting x and y, that is

$$y = \varphi(x). \tag{1.8}$$

If the variable x is measured with a constant but unknown error, x_0, then the actual relationship holding between the variables is

$$y = \varphi(x' - x_0) \tag{1.9}$$

where x' is the measured value of x.

If this function is linear, namely

$$\varphi(x) = ax + b \tag{1.10}$$

then the presence of the constant error will simply displace the line (eqn. 1.8) by a value of the ordinate equal to x_0, whereas random errors in the variable x will simply give a random scattering of the data about the line (eqn. 1.8) with points above and below it.

However, if the function $\varphi(x)$ is a strongly curved one the relationship between y and $(x' - x_0)$, where x_0 is unknown, may be a very difficult one to find. Suppose, for example, x and y are connected by the simple relationship

$$y = \tfrac{1}{2} \log x \tag{1.11}$$

holding over the range $1 \leqslant x \leqslant 100$, and suppose that in an experiment, the measured values of x, that is x', are subject to a constant error of 10. These together with the supposed recorded values of y are given in Table 1.2.

TABLE 1.2. LOGARITHMIC RELATIONSHIP BETWEEN x AND y INVOLVING A CONSTANT ERROR, x_0

"True" value, x	y	Measured value, $x' = x + x_0$	$\log x'$
1	0·00	11	1·04
2	0·15	12	1·08
5	0·35	15	1·18
10	0·50	20	1·30
20	0·65	30	1·48
50	0·85	60	1·78
100	1·00	110	2·04

In the fourth column we have a transposed value of a variable which, but for the presence of the systematic error of 10, would be a linear function of y. Instead, as is shown in Fig. 1.2, the relationship between $\log x'$ and y is of a curved nature, as of course is the relationship x and y itself. If now, instead of plotting $\log x'$ against y, we plot x' versus 10^{2y}, we get a true linear relationship (Fig. 1.3) in which the error x_0 appears as a constant displacement of all of the points involved.

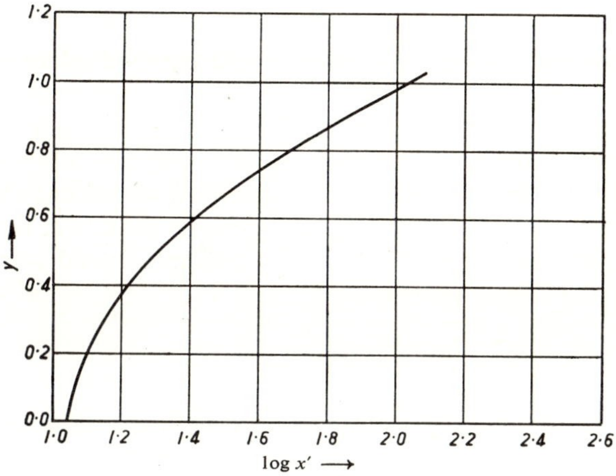

FIG. 1.2. A Graph of y versus $\log x'$.

x' is the measured value of x and is subject to a constant error of $+10$.

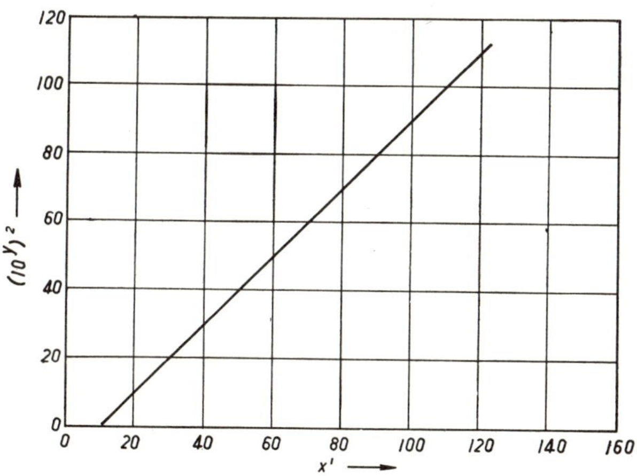

FIG. 1.3. A Graph of $(10^y)^2$ versus x'.

The graph shows the error in the measurement of x, x_0, as a constant displacement of the line.

We are led therefore to a most important rule for the handling of data in which one of the variables is suspected of being subject to systematic error. This variable should be plotted only on a linear scale and any non-linearity in its relationship with the other variable should be taken up by plotting appropriate functions of the second variable. Let us consider a study of reaction kinetics in which some measure of extent of reaction, χ, is determined at times $t_1, t_2, t_3, t_4, \ldots$ etc. Suppose the actual time of commencement of the reactions is uncertain due to the existence of induction periods arising from the experimental establishment of the desired reaction conditions, e.g. temperature. The true time variable present in the function of χ will not be the measured t, but $(t - t_0)$ where t_0 is a constant for any given run but varies from run to run. If the reaction is of the second order, it follows that

$$\frac{1}{\chi_0} - \frac{1}{\chi} = k(t - t_0). \qquad (1.12)$$

It would, in this instance, be quite impossible to obtain the value of χ_0 (the initial degree of advancement) by plotting χ against $1/t$. Unless the measured values of t were very much greater than t_0 the resulting points would describe a curved relationship. Thus, unless the value of χ_0 is known it is impossible to prepare the data for a linear plot by any method other than a series of successive approximations.

We now come to a consideration of random error in the variables and how it is taken into account when drawing a graph. (Numerical and statistical methods for handling this problem will be dealt with in later chapters.) The random error in the variables will appear as a scatter of points about the most likely estimate of the true line. The aim when constructing the graph is to obtain the line of "best fit" which passes "most evenly" through the cloud of points. Two criteria of this property of "evenness" are in general use, and curve fittings which are based on them are called first and second moment fittings respectively. In practice the two criteria are usually equivalent. In first moment fitting the algebraic sum of the perpendicular distances drawn from all points to the given line must be equal to zero, whereas in second moment fitting the sum of squares of all perpendicular distances drawn from the points to the line is adjusted to a minimum value. This is the principle of "least squares" largely adopted in Chapter IV. As long as the scatter of the experimental points is not too great, however, the first of these conditions may be quite accurately satisfied by eye. A transparent ruler placed over the points can be adjusted so that half the points lie above and half below its edge with some allowance made for the magnitude of the perpendicular distances. This condition should be independently satisfied for both right hand and left hand sides of the diagram.†

1.9. Graphical Representation of more than Two Variables

When a system involves many variables it is preferable to keep all but two constant, to vary one of those left and to study the change in the other under these restricted conditions. Having thus studied the effect of one variable on another under stated conditions for all other variables, a third variable is permitted to change and, in turn, its effect on the relationship between the first two is systematically studied. In many instances, particularly in the observational as opposed to the experimental sciences, this condition cannot be obtained and two or more independent variables must be allowed to change simultaneously. In these instances certain specialized statistical techniques elaborated in Chapter V are available.

In the simpler situation, typical presentation of data concerning three variables gives a relationship between two of them, say x and y, while the third, say z, is kept at a certain fixed value. Graphically, this relationship will be represented by a curve in an x–y plane, and will have the generalized form

$$y = \varphi_1(x). \tag{1.13}$$

At a different value of z, y will be a different function of x and may be a line, probably curved, represented by

$$y = \varphi_2(x). \tag{1.14}$$

† Where the scatter of the individual points is too great for the method given above, others, such as the method of averages and least squares adjustment, can be used. These are described in Chapter IV (linear) and Chapter V (curvilinear relationships).

For a series of selected z values there will be a corresponding series of y versus x curves, and from these it is quite easy to construct a series of "x versus y at constant z" or "y versus z at constant x" curves.

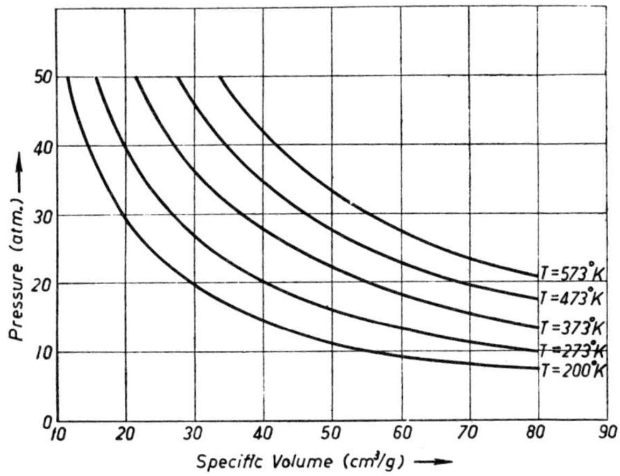

FIG. 1.4. The Variation of the Specific Volume of Nitrogen with Pressure at Various Temperatures (assuming ideal behaviour).

The following is a simple example of the technique. The pressure–volume relationships at a series of different temperatures for nitrogen are given in Fig. 1.4. It is required to obtain a set of "pressure versus temperature at constant volume" curves. Lines of constant specific volume (*isochores*) are drawn parallel to the pressure axis. These will cut the "pressure versus volume at constant temperature" curves at different values of pressure corresponding to each of the fixed isotherms. Thus at a specific volume of 50 cm^3/g the following pressures can be read off;

Temperature (°K):	200	273	373	473	573
Pressure (atm):	11·0	16·0	22·2	27·8	33·5

These points can be graphed to give one isochore and, by reading off pressure–temperature data along a series of constant specific volume lines, the required pressure–temperature diagram can be constructed (Fig. 1.5). In a similar manner "volume versus temperature at constant pressure" curves can be constructed (Fig. 1.6).

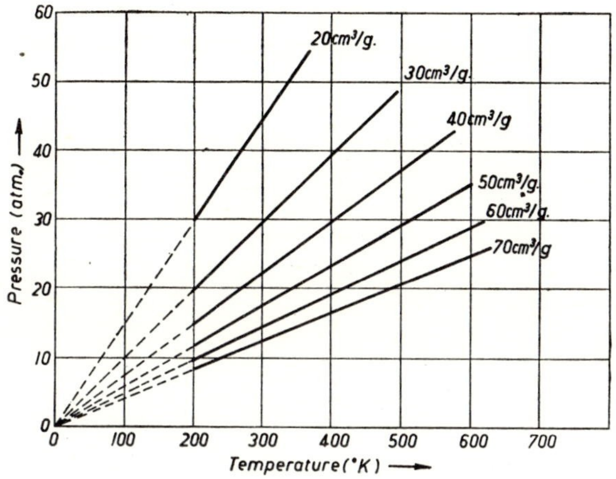

FIG. 1.5. The Pressure–Temperature Relationships of Nitrogen at Various Specific Volumes (assuming ideal behaviour).

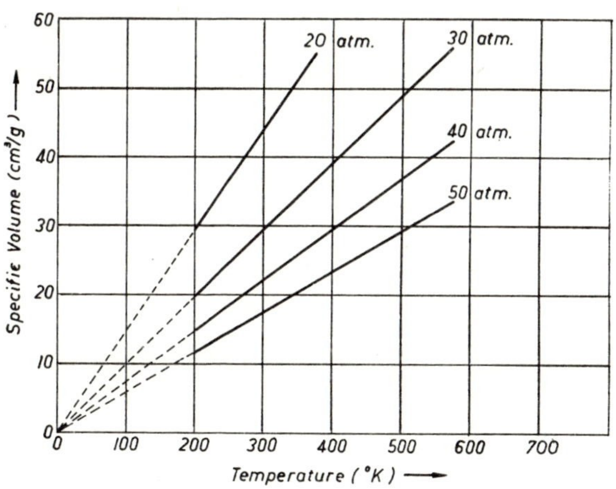

FIG. 1.6. The Variation of Specific Volume of Nitrogen with Temperature at Various Pressures (assuming ideal behaviour).

Data involving more than two variables also occur in a whole range of phase and mixture diagrams. Here the variables representing all the fractions of the components in a mixture are necessarily connected by the relationship

$$\text{Sum of all fractions} \equiv \text{Unity}$$

so that all the fractions are not mathematically independent. This fact can often most conveniently be taken into account by the use of triangular graph paper ruled in the form of equilateral triangles with the sum of the perpendicular distances from the three sides constant.

1.10. Alignment Charts

We have seen that in graphical representation the connection between two variables becomes a line on a chart. Often this is curved, but at least in principle it can be transformed into a straight line by suitable manipulation of the form of representation of one or both of the coordinates.

Now the geometry of a straight line is completely fixed by stating the positions of two points which lie on that line. On the other hand, any two straight lines lying in the same plane will intersect at one and only one point, so that the geometry of a point is completely fixed by stating the positions of two straight lines which intersect at that point. The apparently exceptional case where the two straight lines are parallel defines a point at infinity.

It is thus possible to set up two completely analogous systems of geometry in one of which lines are defined in terms of points and the other of which points are defined in terms of lines. It is the first system of geometry which we use in graphical representation. However, by using the other system of geometry we can arrive at a representation in which a whole set of bivariate quantities can be represented by a single point and in which specific examples can be read off by drawing lines through that point instead of selecting points on a line as in ordinary graphical representation.

Systems in which the reading of data is accomplished by drawing lines through a point are in general referred to as nomographic. Nomography can be used in all applications for which graphical methods are also available, i.e. the presentation of data and the rapid numerical solution of equations. Details of these systems are described in detail elsewhere.† Here, we mention only that nomographs are convenient for the presentation of numerical data.

1.11. Empirical Equations

Empirical equations are frequently used to represent data in the physical sciences and, in addition to being the most compact method of presentation, they have the advantage that certain mathematical operations (differentiation, integration, etc.) can be carried out with speed and accuracy. In some cases the form of an empirical equation leads directly to theoretical development. A statement of the range of the variables over which such an equation holds should always be made.

To return to the matter of Section 1.8. If a suitable operation on variables results in the production of a straight line then the transformed variables, say X and Y, are connected by the equation

$$Y = a + bX \tag{1.15}$$

where b is the measured slope of the line and a the measured intercept on the Y axis. The transformed variables X and Y need not, of course, be the measured variables x and y. For example X may be $\log x$ and Y may be $\log y$: then the indicated relationship between the variables takes the form

$$\log y = a + b \log x \tag{1.16a}$$

or

$$y = Ax^b \tag{1.16b}$$

where

$$a = \log A.$$

In other instances eqn. (1.15) may be valid with X and Y equal to the untransformed variables x and y. In still other instances data may only lie approximately on the line defined by eqn. (1.15) with a divergence which becomes progressively greater as we approach the edge of the diagram. By plotting the divergence from linearity against y and repeating this operation as many times as necessary,

† See, for example, Davis (1962).

it is always possible to get equations of the type

$$y = a + bx + cx^2 + \ldots. \tag{1.17}$$

Generally speaking, equations of this type are a useful way of representing data when the terms on the right hand side become successively smaller and smaller.

Thermodynamic data are commonly tabulated as empirical equations of the type represented by eqn. (1.17). Thus, the molar heat capacity of a gas, C_p, usually varies slightly and generally non-linearly with the temperature

$$C_p = a + bT + cT^2 + \ldots. \tag{1.18}$$

Also the density, ϱ, of aqueous solutions varies approximately linearly with the concentration, m, of the solution, and empirical data can frequently be given in the form

$$\varrho = \varrho_0 + am + bm^2 + \ldots. \tag{1.19a}$$

In this instance ϱ_0 is the density of the solvent, a known constant, and the equation can be reduced in degree to

$$(\varrho - \varrho_0)/m = a + bm + \ldots. \tag{1.19b}$$

So far in our discussion of empirical equations we have not considered any way of indicating the level of accuracy of the relationship. This has to be done by statistical methods which will be applied to this particular problem in Chapters IV and V.

CHAPTER II

MEASUREMENTS OF A SINGLE VARIABLE

MOST of the data dealt with by chemists is "associated" in the sense that it consists of measurements in related pairs. As a preliminary to the problems involved in the treatment of such data it is necessary to consider the situation in which we have measurements on a single scale—for example, the results of repeated determinations of a particular physical constant or measurements of some dimension of a number of similar articles. In such cases we have "unassociated" or "frequency" data whose quantitative aspect arises out of the scale values and number of times each occurs in the collection of values of the measurements in hand. Our concern in such cases is with general location and spread, and perhaps with the manner of spread, along the scale of the single variable. Frequency data also arise when the scale of the variable is replaced by a classification of attributes which need have no quantitative significance and in this situation the quantitative aspect of the data resides in counting alone.

2.1. THE PROVENANCE OF FREQUENCY DATA

The data of Table 2.1 are values resulting from the measurement of density of each of a collection of eighty-one blocks of solid carbon dioxide drawn from a day's production. The eighty-one blocks constitute a sample of a much larger number which we call a *population* or *universe*. The values of the measurements themselves constitute the sample array and our interest in them lies in the light they throw on the properties of all the values the universe itself could provide were we able to extract this totality of information. It is the universe and not the sample, only one of many which could have been obtained, which is the true object of our interest.

In the same way determinations of some property of a physico-chemical system constitute a sample from a much larger number of determinations which could have been obtained under the same conditions. In this case we normally regard the universe as being of infinite size (adopting the fiction that an infinite number of determinations could conceivably be made by the same observer using the same apparatus and without change in the system itself) while the universe of a day's production of carbon dioxide blocks is finite. It is the universe whose characteristics it is desired to study but even finite universes are generally so large that the process of sampling is inescapably necessary if information is to be obtained at all, and even if it were not so large as to preclude complete investigation (the array would then include every possible measurement) sampling saves both time and money. It follows, however, that the sample, being in effect a smaller edition of the universe from which it has come, must provide a fair picture of it if we are to arrive at fair conclusions as to its nature.

The problem of the relationship of the sample in hand to the universe we wish to study is the problem of sampling error. This has two main aspects—relative size and the manner of selection. That this error will be reduced by increasing the size of the sample relative to that of the universe is obvious enough. If an almost completely accurate picture is required, as for example in the political opinion sampling of a general election, then the sample must as nearly as possible coincide with the universe. In attempting to achieve accuracy in the case of smaller samples, these are commonly drawn at *random*, i.e. so that every item in the universe has an equal chance of being drawn into the sample. On the other hand, a universe divided into strata or regions or intervals of time can be sampled so that its various parts are represented in the sample in proportion to their size. Thus, the resulting sample is *stratified* in the same way as the universe itself and may be called a *proportional* sample. A few determinations of a physico-chemical property like the surface tension of an organic liquid constitute an unrestricted random sample from a hypothetically infinite universe: the eighty-one blocks of carbon dioxide of our previous example could be an unrestricted random sample from a day's production or a stratified sample which contains so many blocks per shift or per hour into which the production period is divisible but, if these in turn have been drawn at random, the whole is said to be a *stratified random* sample. The taking of ore from the bottom, middle and top of a mound followed by coning and quartering is an example of a carefully stratified procedure. On the other hand, the serial selection of, say, every tenth item from an enumerable set might result in a random sample if the items were originally quite randomly ordered or in a completely stratified sample under other conditions. Serial (one-by-one) selection is in fact the usual practice leading to random samples of determinations of physical properties but careful randomizing procedures (as described, for example, by Nalimov (1963)) may be needed in dealing with finite universes of discrete items.

In all these cases, failure to draw the sample fairly may result in *bias* or *systematic error*. This can occur if items are excluded from the sample on subjective grounds (as by the rejection of suspiciously high or low values from the array) or when some parts of a heterogeneous universe are better represented than others, always a risk in political and social investigations. Allied to this is failure to draw the sample from the universe which it is desired to study, or failure to recognize that from which it has actually been drawn. Thus, a number of determinations of the density of nitrogen from air less oxygen do not come from the universe of determinations on the chemically prepared and pure gas, as Lord Rayleigh was able to show.

Errors of the systematic kind can be reduced by careful experimental technique but, however good a sample may be from this point of view, its relative smallness raises a problem of uncertainty which it is the central task of statistical methods to solve. The characteristic features of the universe which are of interest can only be estimated from the available sample data and the estimates are always subject in greater or less degree to *random error*. If, for example, the mean or central point of the universe is the characteristic (*parameter*) of interest, the means of successive samples as estimates of this parameter will show some degree of scatter from it. Thus, the universe of experimental determinations of the surface tension of water at 20 °C (given the observer, instrument, temperature control and pure substance) has as its mean the value 72·8 dyn/cm, let us say, but a random sample of six observations may give values averaging 72·4 dyn/cm, another five may average 73·3 dyn/cm, a further eight 73·0 dyn/cm and so on and the individual values (which can be regarded as samples of unit size) are still more widespread. The likely extent of such familiar discrepancies can be estimated, and limits can then be put to the uncertainty of particular sample estimates. However, the smaller the sample the vaguer is the picture of the universe it presents, the wider the likely limits of such random error and the less reliable will be the conclusions that can be drawn.

2.2. Classification of Data—The Frequency Distribution

The amount of work to be done on an array of sample data depends on their number and the purposes to which they are to be applied or the features of the universe which it is desired to reveal. If the sample is large and a qualitative idea of the shape of the universe is all that is required, the array is reduced by classification to a *frequency distribution* from which a tabular or graphical picture can be obtained. If, alternatively or in addition, certain of the universe parameters are of interest, the data of the array must be condensed to estimates of these and the process of classification, via the frequency distribution, is of considerable assistance in carrying out the computations required. If, finally, it is desired to give mathematical expression to the distribution of values by fitting an equation, classification of the data is an essential preliminary. On the other hand, experimental values of a physical quantity are seldom classified (however numerous) unless for purposes of computation because the shape of the universe is generally well known and is not of particular interest in itself. Small samples are not so treated because of the paucity of data.

The operation is simple and largely mechanical though there is room for taste in the matter of representation of the frequency distribution. In the first instance this may take the form of a table in which each value of the variable has set against it the number of times it occurs in the array, its *frequency*. When the range of values of the variable covered by the sample is large it is usual and convenient to divide this into between five and fifteen narrower *class intervals* or groups. These are normally made equal in size for convenience in graphical representation and subsequent calculation and, where possible, their width is made to coincide with a natural interval on the scale. For purposes of representation alone a small number of class intervals is generally adequate but a larger number is best for accuracy in computation and these should be positioned in some rational way in relation to the density scale. (A more precise guide as to number is given at the end of Section 2.6 below.)

TABLE 2.1. DENSITIES OF COMPRESSED CO_2 BLOCKS, g/cm³

1·498	1·510	1·496	1·498	1·512	1·491	1·497	1·487	1·493
1·482	1·474	1·504	1·467	1·503	1·490	1·493	1·497	1·513
1·517	1·481	1·491	1·482	1·494	1·489	1·494	1·504	1·527
1·516	1·493	1·506	1·460	1·490	1·491	1·479	1·507	1·506
1·508	1·475	1·492	1·498	1·488	1·497	1·489	1·522	1·497
1·487	1·526	1·488	1·506	1·493	1·506	1·498	1·504	1·504
1·516	1·480	1·508	1·505	1·491	1·489	1·477	1·514	1·508
1·503	1·511	1·500	1·504	1·490	1·483	1·487	1·499	1·484
1·496	1·521	1·494	1·516	1·494	1·503	1·480	1·508	1·498

For example, the frequency distribution of the density values in Table 2.1 may be found as follows:

(a) *Select a Suitable Class Interval Width*

Here, the range of values in the array, the difference between the highest and the lowest, is 1·527 − 1·460 or 0·067. Division by 5 and 15 gives 0·013 and 0·004 between which may be found a convenient width. For tabular or graphical representation 0·010 is not unduly wide.

(b) *Position the Intervals Conveniently*

If possible, the extreme values in the array should be close to the mid-points of the intervals in which they fall and by which they will be represented in subsequent graphing and calculation. If we start at 1·455, the first interval will be 1·455 to 1·464 and its centre is close to the only value which falls within it. A check on the last interval on this scheme, 1·525 to 1·534, reveals it to be fairly suitable from the same point of view.

(c) *Set up a Table and Tally the Values*

It is tedious and conducive to error to hunt through the array for values belonging to a particular class interval—the entry of a tally mark for each successive value and the grouping of these in fives as shown in Table 2.2 is the commonly recommended procedure.

TABLE 2.2. FREQUENCIES OF COMPRESSED CO_2 BLOCK DENSITIES

Class int. limits	Tally	Frequency		Cumulative frequency, No. and %					
		No.	%	Above lower boundary			Below upper boundary		
1·455–1·464	/	1	1·2	81	100·0	1·4545	1	1·2	1·4645
1·465–1·474	//	2	2·5	80	98·8	1·4645	3	3·7	1·4745
1·475–1·484	卌 卌	10	12·4	78	96·3	1·4745	13	16·1	1·4845
1·485–1·494	卌 卌 卌 卌 ////	24	29·6	68	83·9	1·4845	37	45·7	1·4945
1·495–1·504	卌 卌 卌 卌 /	21	25·9	44	54·3	1·4945	58	71·6	1·5045
1·505–1·514	卌 卌 卌	15	18·5	23	28·4	1·5045	73	90·1	1·5145
1·515–1·524	卌 /	6	7·4	8	9·9	1·5145	79	97·5	1·5245
1·525–1·534	//	2	2·5	2	2·5	1·5245	81	100·0	1·5345
		81	100·0						

The intervals chosen in (b) are so demarcated as to embrace any of ten possible values on the basis of the number of significant figures to which these are expressed. Thus, the first interval will include such of the values 1·455, 1·456 ... 1·464 as occur in the array. Its mid-point is, properly, 1·4595—this being the mean of all ten and also of the terminal values alone. It is unaffected by the consideration that the true scale is continuous, recorded measurements of 1·455 and 1·464 meaning 1·455 $\pm$ 0·005 and 1·464 $\pm$ 0·005. As a consequence, however, the class intervals can be delimited in two ways, (i) according to the recorded measurements (e.g. 1·455–1·464) and (ii) according to the possible extremes of true values represented by the measurements (e.g. 1·4545–1·4645). Mode (1961) makes a useful distinction in calling the first set the class interval "limits" and the second the "boundaries" and, although the former remain the basis of tallying, it is the latter which are the more relevant in graphical representation. It is quite permissible, and was once the practice, to select such boundaries as 1·455–1·465 with mid-point 1·460, but in this case a value like 1·465 would be counted half into each of two adjacent intervals and lead to awkward fractional frequencies, whereas in the scheme adopted here it goes into a particular interval. These distinctions become trivial when the class intervals are wide with respect to the difference between values on the scale of measurement.

For purposes of representation there are many alternatives. Class frequencies can be expressed as such or relative to the total frequency or as percentages and the cumulative frequencies, either above or below selected scale values, likewise. The latter are obtained by accumulating the class counts either upwards or downwards as illustrated in Table 2.2. For example, eighty of the eighty-one values in the array are equal to or greater than the lower limit of the second interval, 1·465, and one is equal to or less than the upper limit of the first, 1·464, and they are designated as being "above the lower boundary" or "below the upper boundary" of their appropriate intervals, both 1·4645 in this instance. In tabulation, the percentage class interval frequency is most commonly used and is generally quoted against the lower boundary or limit of the interval (the upper being implied in the table) but mid-points are sometimes found. Thus, the frequency distribution of the CO_2 block densities might be reported as in Table 2.3.

TABLE 2.3. COMPRESSED CO_2 BLOCK DENSITIES

Density (g/cm^3)	Percentage
1·455–	1·2
1·465–	2·5
1·475–	12·4
1·485–	29·6
1·495–	25·9
1·505–	18·5
1·515–	7·4
1·525–	2·5

The conventional graphical forms are the *histogram* and the *frequency polygon.* When, as is generally to be preferred, the class intervals are of equal width, the first takes the form of a set of rectangles (erected on the class intervals as base) whose height is proportional to the observed frequency. The histogram corresponding to the data of Table 2.3 is shown in Fig. 2.7 where it is made the basis of some graphical calculation. When the class intervals are of unequal width, the area of the rectangles is made proportional to the frequency and by this means the eye is not misled as to the actual way in which the values are distributed. The frequency polygon, in the case of equal intervals, is made by joining points having frequency as ordinate and class interval mid-point as abscissa. The result is shown in Fig. 2.1. The frequeney values in this case may be regarded as being particular to the intervals above whose mid-points they are plotted or as frequency densities (number or percentage of values per given interval of variable scale) as observed at the scale values in question. If unequal class intervals have been used in tallying the data, the frequency polygon will only give a valid representation of the form of the distribution when frequency density is plotted.

In the case of very large samples and vanishingly small class intervals both histogram and polygon approach a smooth *frequency curve* for the data. This is a smoothed representation of frequency density and it conveys the form of a distribution rather better than other graphical means. In principle, such a curve could be obtained graphically by (a) differentiation of the cumulative frequency curve (see below), (b) ordinary item difference smoothing methods (see Chapter VI) or the approximate method of Thurstone (described by

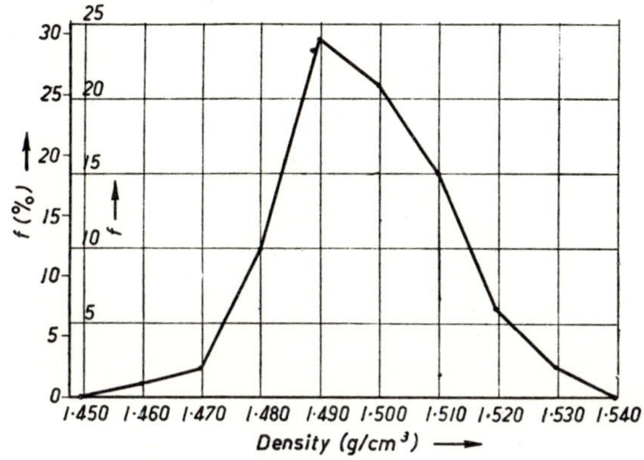

FIG. 2.1. Densities of Solid CO_2 Blocks. The Frequency Polygon.

The frequencies are plotted against the mid-points of intervals whose boundaries are 1·4445–1·4545, 1·4545–1·4645, etc.

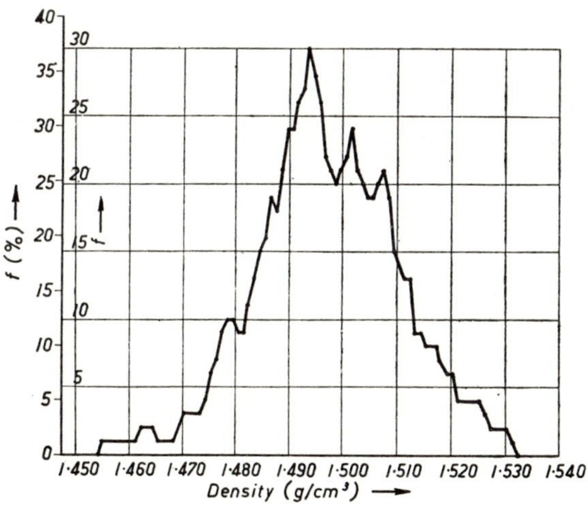

FIG. 2.2. Densities of Solid CO_2 Blocks. Frequency Density.

The numbers of observations per 0·01 g/cm³ as found from Table 2.1 are plotted against the mid-points of intervals having lower limits 1·451, 1·452, 1·453, etc.

Schmid, 1954) or (c) drawing by eye through the points of a polygon or the tops of histogram rectangles. None of these methods is satisfactory when applied to relatively sparse or coarsely grouped data. Part of the difficulty lies in the commonly encountered bell shape of the curves which makes them hard to draw (or their item differences hard to smooth) through an inadequate number of points. In addition, frequencies and the shapes of curves derived from them are somewhat dependent on the accidental location of class intervals when the sample is not of very large size. Some idea of the differences to be expected may be gained from Fig. 2.2 which shows, in effect, the superposition of a number of frequency polygons as well as the actual variation of frequency density revealed by the data of Table 2.1. In this figure the frequency density per 0·01 units of scale is plotted at 0·001 unit spacing; e.g. the number of values between the limits 1·450–1·459 is plotted against mid-point 1·4545, values between 1·451–1·460 against 1·4555 etc. The resulting impression is rather different from that conveyed by Fig. 2.1 but even with all the points now available a smooth curve would still be difficult to fit by eye.

Plots of cumulative data on the other hand have the advantage that smooth curves can, if desired, be readily fitted by eye and that several distributions can be presented on the same graph without the curves crossing. They have the disadvantage that they do not show the actual form of the distribution in a way which is easy to appreciate. Which of the two types of cumulative frequency in Table 2.2 is made the basis of graphical representation is generally immaterial—both are plotted in Fig. 2.3 but only to show the shape of the curves, the manner of plotting (cumulative frequency above a lower boundary against that boundary and vice versa) and that they intersect on the 50% line. The percentage cumulative frequencies below the upper boundaries are also given in Fig. 2.4 on "probability paper" which uses a scale having the effect of straightening out the tails of the S-shaped curves (*ogives*) resulting from cartesian coordinates. Complete rectification can only be expected in the case of certain symmetrically distributed data but any cumulative curve is straightened to some extent in such a plot.

There are a number of other types of diagrams for illustrating frequency density or cumulative frequency. These are described in many books on statistical methods (especially those having economic bias) and a very extensive account is given by Schmid (1954).

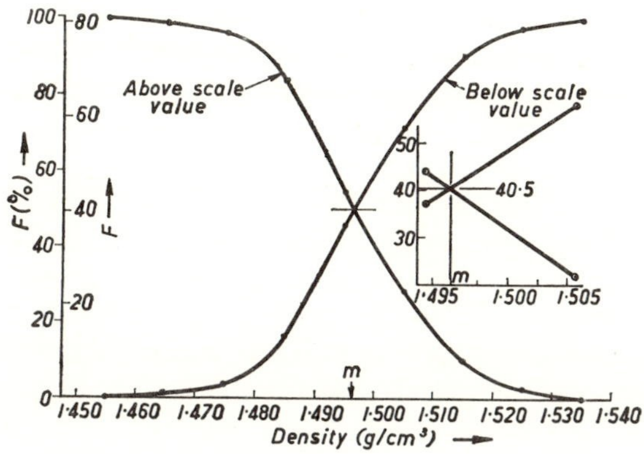

FIG. 2.3. Densities of Solid CO_2 Blocks. Cumulative Frequency Curves.

The curves are constructed from the data of Table 2.2. They intersect on the $F = 50\%$ line and the median. The linear central portions are enlarged in the inset to show the determination of the median.

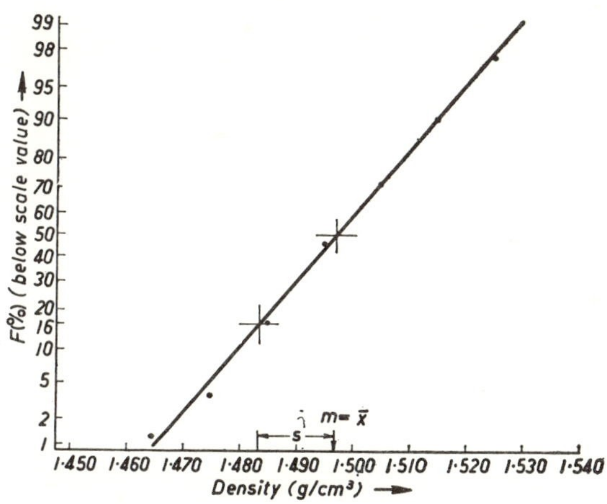

FIG. 2.4. Densities of Solid CO_2 Blocks. A Cumulative Frequency Curve on Probability Paper.

The straight line through the cumulative (below scale value) frequencies from Table 2.2 has been fitted by eye. Its intersection with the $F = 50\%$ line gives the median (which is approximately the arithmetic mean in this instance) and the difference between this and the $F = 16\%$ or 84% line is the standard deviation. The values, 1·4967 and 0·0135, should be compared with those calculated in Table 2.9.

A more complete description of a distribution can be given by a *frequency function* which relates frequency to particular values of the variable or to intervals on its scale or by a *distribution function* which relates cumulative frequency to particular values of the variable. The former is the derivative of the latter. It will be seen that if F_i is the cumulative frequency to x_i and F_j to a higher value x_j, then $(F_j - F_i)$ is the area of a histogram rectangle based on the interval $x_i - x_j$. $(F_j - F_i)/(x_j - x_i)$ is therefore the height of the rectangle and is the frequency per unit of x or average frequency density over the interval in question. In the limit as x_j approaches x_i, the frequency becomes $\mathrm{d}F$ in the interval $x_i + \mathrm{d}x$ and the frequency density at this point on the scale becomes $\mathrm{d}F/\mathrm{d}x$, which we denote by f_i. The frequency distribution curve may be regarded as being made up of a series of infinitesimally thin histogram rectangles whose height is $\mathrm{d}F/\mathrm{d}x$, so we have in general

$$f = \mathrm{d}F/\mathrm{d}x \quad \text{or} \quad \mathrm{d}F = f\,.\,\mathrm{d}x$$

and, on integration,

$$\int_{\alpha}^{x_i} \mathrm{d}F = F_i = \int_{\alpha}^{x_i} f\,.\,\mathrm{d}x$$

where α is the lower limit of population on the scale of the variable. α is $-\infty$ in some of the special expressions (often exponential) which are fitted to frequency density and cumulative frequency. A brief account of the methods used will be found in Chapter V.

2.3. Types of Frequency Distribution

Distributions of frequency data encountered in practice vary considerably in shape but can be sorted into a few broad classes which we designate as follows.

(1) "*Normal*" *Distributions*

The term is used to describe distributions which fall away symmetrically from a central value of the variable and in doing so follow closely the theoretical normal frequency function or Gaussian distribution which is discussed in Chapter III. The real distributions which follow this most nearly are those in which a multiplicity of small sources of variation operate this way and that on the value of a variable and in the majority of instances tend to cancel. Thus, repeated determinations of a physico-chemical quantity like the surface tension of a liquid tend to distribute themselves symmetrically about a "true" value with relatively few (subject to unfortunate combinations of errors operating in the same direction) in the tails. Biometric measurements, for example the heights of individuals in a homogeneous population group, are similarly subject to a multitude of factors, hereditary and environmental, and are often distributed in the same way. Samples of given size drawn at random from a universe are unlikely to contain a concentration of members from either end of the scale of the universe variable but, rather, individuals spread throughout the universe in accordance with its own shape. Hence, the arithmetic means of the samples are normally distributed as shown in Fig. 3.3 and, if the samples are large enough, this will apply to other properties also.

The universe which gave rise to the data of Table 2.1 is one to which the term "normal" as used here would be fairly applied and the curves given by its sample in Figs. 2.1–2.4 compare well with those for the theoretical function in Fig. 3.2. In what follows the term "Gaussian" will be used to denote the theoretical function and "normal" to imply a practical approximation to it.

(2) "*Non-normal Unimodal*" *Distributions*

The normal distribution is not very common but its shape is familiar and a convenient standard by which to judge others. There are many distributions in which the individual values follow a broadly similar pattern, i.e. concentrated so as to give a peak or *mode* and falling away in number on each side, except that the fall off is unsymmetrical and/or differs in rate from the normal. Departures from symmetry are referred to as *skewness*, this being counted positive if the distribution tails off to the right and negative if to the left. Departures in rate of falling away, i.e. of curvature of the frequency function or curve, imply differing relative concentrations of values in different parts of the distribution and are described under the term *kurtosis*. The normal standard is called *mesokurtic*. A distribution having relatively more values round the modal region and in the extreme tails and whose frequency curve therefore looks more than normally peaked is *leptokurtic* while one

TABLE 2.4. MISCELLANEOUS DISTRIBUTIONS

(a) Sulphur sol particle sizes		(b) Differences between first and second of pairs of titrations		(c) α-particle counts		(d) Guessed final digits	
Diameter (microns)	Percentage of particles	Difference (ml)	No. found (%)	Counts in 1/8 minute period	No. of periods	Digit	No. of times found
0·5–	1	−0·6	2	0	57	0	8
1·0–	9	−0·5	1	1	203	1	7
1·5–	16	−0·4	4	2	383	2	5
2·0–	22	−0·3	7	3	525	3	9
2·5–	16	−0·2	9	4	532	4	12
3·0–	12	−0·1	12	5	408	5	2
3·5–	10	0·0	31	6	273	6	10
4·0–	5	+0·1	12	7	139	7	12
4·5–	3	+0·2	7	8	45	8	11
5·0–	2	+0·3	7	9	27	9	5
5·5–	2	+0·4	4	10	10		81
6·0–	1	+0·5	2	11	4		
6·5–	1	+0·6	2	12	0		
	100		100	13	1		
				14	1		
					2608		

showing the reverse characteristics is *platykurtic*. In terms of departure from the normal this property is conveniently referred to as *excess of kurtosis*, this being positive for a leptokurtic and negative for a platykurtic distribution.

Distributions of this general type will arise if the sources of variation tend to be influenced by the value of the variable or by some particular limitation. A familiar physico-chemical example is Maxwell's distribution of the speeds of gas molecules at a given temperature. These cannot be less than zero but may, subject to enough favourable molecular collisions, be quite considerable and the distribution, therefore, tails off in the direction of high values of the variable. Similarly, particle sizes in powders produced by precipitation or comminution tend to be generally small but the growth or grinding processes result in a few particles of much larger than average dimensions. The effect is illustrated by the data of Table 2.4(a) for the particle size distribution of precipitated sulphur globules. The vast majority of these distributions are positively skewed and particle size distributions are often markedly leptokurtic as well. In many cases a quite simple transformation of the variable results in a normal distribution—in the case of the sulphur globules this is true of the logarithms of the diameters.

Skewness is easy to recognize in any kind of frequency diagram but this is not so for excess of kurtosis. Differences are, however, recognizable in plots on probability paper. The values in Table 2.4(b) are differences between duplicates of analytical titrations where the analyst, knowing his first burette reading, is subconsciously influenced to record his second reading rather closer to the first than it might actually be. The result is a leptokurtic distribution. Were he not to know the first reading the differences should be normally distributed. The data of Table 2.4(b) are artificial but the case is discussed and an interesting example given by Hald (1952). Table 2.4(d) presents a special case of a platykurtic distribution and it is instructive to compare its cumulative curve with that for the differences between duplicates in Fig. 2.5. In using probability paper (as in Fig. 2.5) it must be remembered that the cumulative frequency scale is derived from the unbounded Gaussian function—the values in all real distributions are confined within some limits on the scale, however far apart these may be, and their cumulative frequency curves will eventually turn perpendicular to the scale axis if carried far enough.

The effect of skewness will also be produced (or enhanced) if the sample is not or cannot be drawn from the whole of the universe or if the sample array is incomplete. When the limitation is imposed by the nature of the system at some particular value of the variable

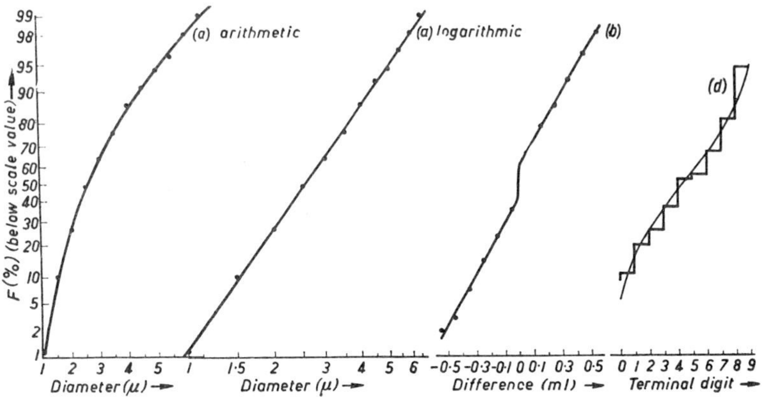

FIG. 2.5. Various Distributions from Table 2.4. on Probability Paper.
(a) Sulphur sol particle diameters (microns, arithmetic and logarithmic scales). The distribution of the diameters is positively skewed but that of the logarithms of the diameters is normal; (b) Differences between first and second titrations (millilitres). The distribution is symmetrical but leptokurtic; (d) Guessed final digits. The distribution is symmetrical (rectangular) but platykurtic. In this case the variable is discrete and a step diagram is the result but the smooth curve is that for the rectangular distribution of a variable continuous over the range $-0{\cdot}5$ to $+9{\cdot}5$.

the sample is said to be *truncated* and when the values of the variable are unknown for some part of the sample items it is said to be *censored* but the distinction is not always clear. Thus, measurements of the size of particles on a microscope slide are truncated by the resolution of the microscope but the data might be further censored by the observer if he decides to discard the unreliable values obtained for a proportion of the smaller particles or does not bother to measure them.

(3) "*J-shaped*", "*U-shaped*" *and* "*Rectangular*" *Distributions*

The first in this miscellaneous category may be regarded as the final form of the skewed type and results when values at one end of the variable scale are most favoured; for example, when the variable is a discrete number of rare events. In addition the shape can be due to conditions of observation. The data of Rutherford and Geiger

(1910) as presented in Table 2.4(c) provide an example of a moderately skewed distribution. They counted the number of α-particles causing scintillations on a fluorescent screen in 2608 periods of 1/8 min each, the highest number or value of the variable here being 14 and the lowest zero. It will be appreciated that if the length of the period of observation had been much shortened many more zero counts would have occurred and the resulting distribution would have been J-shaped but if it had been greatly lengthened all the counts would have been higher and the distribution more symmetrical. Particle size distributions, within and because of limits of measurement, are sometimes found to be of this type.

The rare U-shaped distribution arises when values of the variable tend to be either high or low and the rectangular when all values over the range are equally likely. If an observer records many measurements with a guessed final figure from 0 to 9 he should show no preference for any particular digit and all should be roughly equally represented in a sample as befits one drawn from a rectangular universe. Table 2.4(d) provides a check on the observer of the densities of CO_2 blocks in Table 2.1, supposing the third decimals to be the results of guesses to within one-tenth of a scale interval. It is obvious that although there is no tailing off in this instance there are grounds for suspicion that the observer is prejudiced against the middle and ends of the series and that the universe from which these data come is not fact rectangular. A test to see if this suspicion is justified is described in the next chapter. When a scale is marked so that an observer has to guess a final digit in the series 0 to 5 a tendency towards a U-shaped distribution is sometimes found, because digits 2 and 3 are subconsciously avoided.

(4) "*Irregular*", "*Heterogeneous*", "*Compound*" *or* "*Multimodal*" *Distributions*

These descriptive terms are used synonymously here to describe situations in which values are irregularly concentrated about two or more points on the scale of the variable. Samples showing this effect can generally be regarded as having been drawn from a combination of universes whose different characteristics may well be of interest. For example, a metal sol formed by an a.c. arc consists in part of coarse particles formed by the action of the liquid on the molten metal during interruptions in the arc and in part of fine particles formed by condensation of the metal gas due to vaporization by the arc, and the overall size distribution of the sol appears to be composed of a highly positively skewed (nearly J-shaped) distribution in the smaller sizes and a more normal distribution in the larger (Burton, 1938). A very similar effect in the case of bending strengths of glass rods freshly drawn from the melt has been demonstrated by Cornelissen *et al.* (1962) and is explained by the existence of different kinds of flaws in the glass. An adaptation of a set of their data appears in Table 2.5 and is graphed in Fig. 5.7.

TABLE 2.5. BENDING STRENGTHS OF GLASS RODS

Bending strength (kg/mm^2)	No. of rods	Bending strength (kg/mm^2)	No. of rods
20–40	4	200–220	5
40–60	7	220–240	8
60–80	7	240–260	18
80–100	9	260–280	30
100–120	11	280–300	51
120–140	18	300–320	59
140–160	10	320–340	36
160–180	12	340–360	17
180–200	6	360–380	6

The question of the decomposition of compound distributions into hypothetical components is discussed in Chapter V. The problem concerns not only frequency data of the kind we have here but also the resolution of overlapping chromatographic and spectrographic peaks which in themselves often have the shapes, normal and skewed, of the distributions described above.

2.4. CONDENSATION OF DATA—THE SAMPLE ESTIMATES

It will be apparent from a study of the preceding section that most distributions can be characterized by a few measures, viz. a measure of *central tendency* or *location* to describe broadly the position of the individual values with respect to the scale of the variable, a measure of *spread* or *dispersion* to describe the distribution of values about the position of central tendency, a measure of *skewness* to describe departures from symmetry, and a measure of *kurtosis* (or of *excess of kurtosis* from the normal) to describe the relative concentration of values in the central and tail parts of the distribution. In many cases measures of central tendency and dispersion are sufficient for practical purposes. In the matter of skewness, symmetry is the standard and a measure is not required unless there is an obvious departure from this. When the data are in the form of small random samples measures of skewness and of kurtosis are very unreliable, i.e. because of sampling accidents they may be far from the true universe value, and are not worth calculating. For each of the properties, but especially for central tendency and dispersion, there is a wide choice of measures and in the main only the principal ones and their relationships are discussed here.

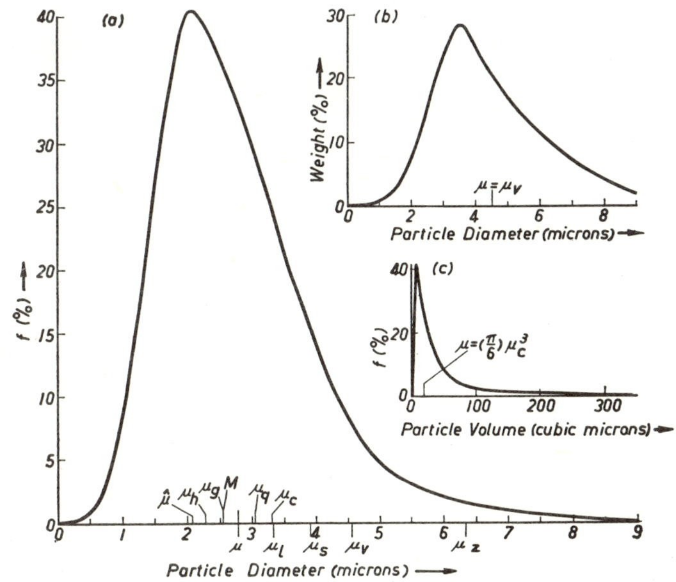

FIG. 2.6. Sulphur Sol Particle Size and Derived Distributions.

(a) is the number distribution, an idealized log-normal version of the data of Table 2.4(a). The various positions of the measures of central tendency of Table 2.6 are shown; (b) is the weight-percent distribution whose arithmetic mean is the volume mean of (a); (c) is the distribution of particle volumes whose arithmetic mean is the volume of a sphere of diameter equal to the cubic mean of (a).

The symbols are for the most part those in common use although in the matter of central tendency some discrepancies arise which are due to an attempt to reconcile the notation for the few ordinary measures of statistical practice in general and the many of particle size and molecular weight distributions in particular. Wherever possible the convention of using Greek letters for universe parameters and small Roman letters for their sample estimates is followed, but sometimes the parameters are indicated by bold capitals.† The following have the same meaning in all the expressions in which they occur:

x_i = an individual value in the sample array (x_1 = lowest, x_n = highest),
X_i = its deviation from the sample arithmetic mean, $\bar{x}$,
f_i = its frequency in the universe or sample,
N = the number of individuals in the universe (if finite), and
n = the number in the sample or sample size.

When the variable is continuous, x_i is to be regarded as the centre of an interval containing f_i values but the subscript is dropped except where (in certain derivations) it is needed for clarity, and f itself is dropped in writing expressions for small samples whose values are taken one by one. Summation limits are only indicated if their use resolves ambiguity. When the variable is measured on a continuous scale and the universe is infinite the parameters of the latter should be defined not by summation but by integral expressions. For example, the equation for the arithmetic mean in such a case may be written

$$\mu = \int_{\alpha}^{\beta} fx \,.\, \mathrm{d}x \bigg/ \int_{\alpha}^{\beta} f \,.\, \mathrm{d}x \tag{2.1}$$

the denominator being the total area under the frequency curve which lies between α and β, the lower and upper bounds of x. Such expressions are not used in computation and will not be given but in the ordinary summation formulae f_i/N is to be regarded as a relative frequency in the case of infinite universes.

(1) *Measures of Central Tendency*

The principal measures of central tendency, and a number besides, are given in Table 2.6. Of these, the *mode*, being the most

† A table of symbols is to be found at the end of this chapter.

TABLE 2.6. MEASURES OF CENTRAL TENDENCY

Measure	Symbol for universe, sample	Definition or defining equation	
Mode	$\hat{\mu}$ $\hat{m}$	Value of x for a peak on the frequency curve.	
Median	M m	Value of x corresponding to 50% cumulative frequency.	
		For a small sample having values in order $x_1 \cdots x_n$,	
		$m = x_{\frac{1}{2}n+\frac{1}{2}}$ (n odd), $m = (x_{\frac{1}{2}n} + x_{\frac{1}{2}n+1})/2$ (n even)	(2.4a, b)
Arithmetic mean	μ $\bar{x}$	$\mu = \Sigma f_i x_i/N$, $\bar{x} = \Sigma f_i x_i/n$ (Σ = summation)	(2.5a, b)
Harmonic mean	μ_h m_h	$\mu_h = N/\Sigma(f_i/x_i)$, $m_h = n/(\Sigma f_i/x_i) = 1/n_{-1}$	(2.6a, b)
Geometric mean	μ_g m_g	$\mu_g = \sqrt[N]{(\Pi x_i^{f_i})}$, $m_g = \sqrt[n]{(\Pi x_i^{f_i})}$ (Π = multiplication)	(2.7a, b)
		$\log \mu_g = \Sigma f_i \log x_i/N$, $\log m_g = \Sigma f_i \log x_i/n$	(2.8a, b)
Quadratic mean	μ_q m_q	$\mu_q = \sqrt{[(\Sigma f_i x_i^2)/N]}$, $m_q = \sqrt{[(\Sigma f_i x_i^2)/n]} = \sqrt{n_2}$	(2.9a, b)
Cubic mean	μ_c m_c	$\mu_c = \sqrt[3]{[(\Sigma f_i x_i^3)/N]}$, $m_c = \sqrt[3]{[(\Sigma f_i x_i^3)/n]} = \sqrt[3]{n_3}$	(2.10a, b)
Linear average	μ_l m_l	$\mu_l = \sum^N f_i x_i^2/\sum^N f_i x_i$, $m_l = \sum^n f_i x_i^2/\sum^n f_i x_i = n_2/n_1$	(2.11a, b)
Surface average	μ_s m_s	$\mu_s = \sum^N f_i x_i^3/\sum^N f_i x_i^2$, $m_s = \sum^n f_i x_i^3/\sum^n f_i x_i^2 = n_3/n_2$	(2.12a, b)
Volume / Weight } average	μ_v m_v	$\mu_v = \sum^N f_i x_i^4/\sum^N f_i x_i^3$, $m_v = \sum^n f_i x_i^4/\sum^n f_i x_i^3 = n_4/n_3$	(2.13a, b)
z-average	μ_z m_z	$\mu_z = \sum^N f_i x_i^7/\sum^N f_i x_i^6$, $m_z = \sum^n f_i x_i^7/\sum^n f_i x_i^6 = n_7/n_6$	(2.14a, b)

(For notation, see the table of symbols at the end of this chapter. n_{-1}, n_1, n_2 etc., are sample moments about the origin defined in general by eqn. (2.25b) below.)

frequent value or that corresponding to the peak of the frequency curve, is perhaps the most obvious measure but is difficult to calculate. If the variable is continuous and a frequency function has been fitted to the sample data then the value corresponding to its maximum turning point can be found by differentiation but such an estimating procedure involves remote values in the tails of the distribution. For practical purposes approximate rules are used:

$$\hat{m} = \bar{x} - 3(\bar{x} - m) \tag{2.2a}$$

or

$$\hat{m} = l_{\hat{m}} + h\{(f_2 - f_1)/[(f_2 - f_1) + (f_2 - f_3)]\} \tag{2.2b}$$

where $l_{\hat{m}}$ is the lower boundary of the most populated class interval containing f_2 values, h the class interval width, and f_1 and f_3 the frequencies in the intervals below and above the modal interval.

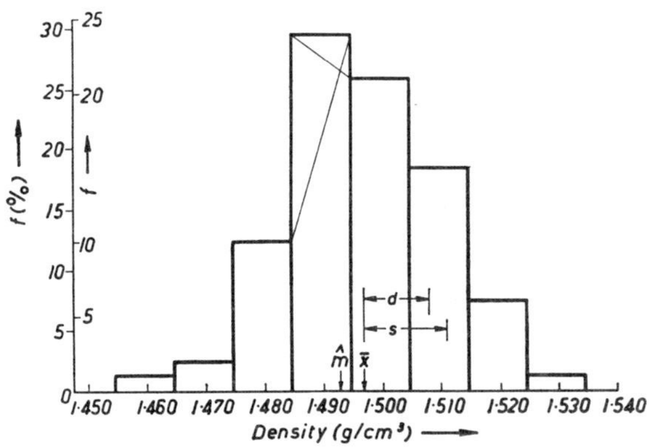

FIG. 2.7. Densities of Solid CO_2 Blocks. The Histogram.

The construction used to determine the mode is based on eqn. (2.2b). The arithmetic mean and the standard and average deviations as computed in Table 2.9 are also shown.

Equation (2.2b) is the basis of a simple graphical construction illustrated in Fig. 2.7 which amounts to the fitting of a parabola, having the mode as maximum, through the tops of the histogram rectangles. Kelly (1947) describes a more elaborate method involving the four neighbouring most populated classes:

$$\hat{m} = l_{\hat{m}} + h[(-0{\cdot}6f_1 - 0{\cdot}2f_2 + 0{\cdot}2f_3 + 0{\cdot}6f_4)/(-f_1 + f_2 + f_3 - f_4)] \tag{2.2c}$$

f_1, f_2, f_3 and f_4 are the frequencies in the contiguous intervals and $l_{\hat{m}}$ the boundary between the second and the third, one of these two being the most populated. The mode is sometimes taken as the mid-point of the most populated class interval but this is a rule of no great utility.

The *median* is the central value which divides the universe into two equal parts by frequency. It is best estimated from the intersection of a distribution curve and the abscissa at 50% cumulative frequency or, relying on the linearity of such curves in their middle regions, from the cumulative frequencies above or below neighbouring class interval boundaries by proportion. This is illustrated graphically in Fig. 2.3 and can be represented analytically by the equation

$$m = l_m + h[(n/2 - F_m)/f] \tag{2.3}$$

where l_m is the lower limit of the class interval containing the median, F_m the cumulative frequency up to that point, f the frequency within the interval and $n/2$ half the sample size. In the case of very small samples ($n \leqq 12$ in practice) the sample median is specially defined by eqns. (2.4a) and (2.4b) in Table 2.6 as the middle value in order of magnitude of an odd-numbered array or the arithmetic mean of the two middle values of an even-numbered array.

The *arithmetic mean* (or, simply, the mean) is defined for the universe by eqns. (2.1) or (2.5a) and for the sample by eqn. (2.5b). It is the point on the scale the deviations from which sum to zero, i.e. $\Sigma f_i(x_i - \bar{x}) = \Sigma f_i X_i = 0$ for the sample and similarly for the universe. It is also the point from which the sum of squares of the deviations is minimized for either universe or sample, the "least squares" value. This is easily demonstrated. Consider c to be that scale value from which the sum of squares of the deviations, $X_i = (x_i - c)$, is to be a minimum. For any individual squared deviation we have

$$X_i^2 = (x_i - c)^2 = (x_i^2 - 2x_i c + c^2)$$

If we add the n such for the sample (taking the values severally so all $f_i = 1$) we have

$$\Sigma X_i^2 = \Sigma x_i^2 - 2c\,\Sigma x_i + nc^2$$

and differentiating with respect to c and equating to zero to obtain the position of the minimum

$$\mathrm{d}\Sigma X_i^2/\mathrm{d}c = -2\Sigma x_i + 2nc = 0$$

so $c = \Sigma x_i/n$ which is the arithmetic mean as defined. Its position corresponds to that of the centre of gravity of a uniform slab having the shape of the frequency distribution—a property which also arises out of the nature of the sum of the deviations from it.

The mean, median and mode of a symmetrical (unimodal) universe coincide and their estimates from a sample may be expected to be close. If the distribution is skewed the relationship between the three is expressed approximately by eqn. (2.2a) and it is useful to remember that they lie in alphabetical (or reverse) order, the median being between the mean and the mode but closer to the mean. The latter, being subject to the effect of extreme values, lies on the side of the long tail.

Two general classes of average, i.e. measures of location having the same units as the variable x itself, are given by expressions of the type $\sqrt[r]{[(f_i x_i^r)/n]}$ (1) and also $\sum^{n} f_i x_i^r / \sum^{n} f_i x_i^{r-1}$ (2). The arithmetic mean is in fact a member of both of these, being that for which r is unity. Expression (1), called by Kelly (1947) the "generalized mean", gives rise to the *harmonic mean* ($r = -1$), the *geometric mean* (Lt $r \to 0$), the *quadratic mean* ($r = 2$) and the *cubic mean* ($r = 3$). These are sometimes encountered in connection with physico-chemical variables; for example, the quadratic mean as the result of the simplest equation for averaging the speeds of gaseous molecules (the *root mean square* value of Maxwell's distribution), the geometric mean as the activity coefficient of an electrolyte and the cubic mean in particle size studies. All these measures are obviously inappropriate when the range of the variable overlaps zero but when its values are all positive their order along the scale is the order of r as shown in Fig. 2.6. The second expression gives rise to averages which are normally used only in connection with particle size and polymer molecular weight distributions. The formulae in Table 2.6 may be altered according to whether (a) the variable is particle size (i.e. a "diameter") or particle mass (molecular weight in the case of polymer molecules) and (b) the "weighting factor" is number of particles (i.e. frequency as in eqns. (2.11)–(2.14)), length, surface or weight per interval. Thus m_z for a sample of particles characterized by linear size is given by eqn. (2.14b) but if x_i = molecular weight of a fraction or class of molecules

$$m_z = \Sigma f_i x_i^3 / \Sigma f_i x_i^2 \tag{2.14c}$$

and if w_i = mass of molecules possessing that molecular weight

$$m_z = \Sigma w_i x_i^2 / \Sigma w_i x_i. \tag{2.14d}$$

This quantity is given directly by the sedimentation equilibrium technique. Accounts of the interrelationships between these various averages and how they arise in direct experimental measurement are given by Bailey *et al.* (1962) and by Herdan (1960).

The question of choice of a measure of central tendency turns on a number of factors. Reliability (in the sense of the sample estimate being generally close to the universe parameter) is the principal criterion and is discussed quantitatively in the next chapter. The mean is on the whole the most desirable measure from this point of view and those based on higher powers of the variable or which do not utilize all the information available (said to be *insufficient* estimators, e.g. the small-sample median which uses only the order of values) are less satisfactory. If the sample is small, however, the median (as defined by eqns. (2.4a) and (2.4b)) is nearly as reliable as the mean and might be chosen for its ease of calculation and if the distribution is markedly leptokurtic it is in fact more reliable. If the distribution is markedly platykurtic another insufficient estimator is superior to the mean—this is the mean of the highest and lowest values, generally called the *mid-range*,

$$c = (x_1 + x_n)/2. \tag{2.15}$$

The nature of the scale also influences the choice of measure. The mode alone is appropriate if the scale is *nominal*, i.e. one whose "values" simply identify types or classes. The median as well as the mode is appropriate if the scale is ordinal like Moh's scale of hardness, and the mean in addition if the scale is interval like the temperature scales of Fahrenheit and Celsius. Members of the series of generalized means, other than the arithmetic, are appropriate only with ratio scales (length, surface, volume, energy, speed etc.) whose zero point is not arbitrary as is the case with interval scales. A third factor is the form or nature of the universe. If the sample is large and the universe from which it comes is one of a family having related characteristics (length, surface and weight of particles or energy and speed of molecules) more than one average might have special significance. For the comparison of several samples from skewed distributions it is generally desirable to choose a measure which is the mean of a symmetrically distributed transformation of the variable, thus $\sqrt[r]{[(f_i x_i^r)/n]}$ or the geometric mean if x^r or $\log x$ is symmetrical. Such a value will approximate the median on the transformed scale since this measure is not affected by scale transformation. These questions are discussed and an instructive example given by Stevens (1955).

(2) *Measures of Dispersion or Variability*

The principal measures of dispersion are listed with their notation and formulae in Table 2.7. It is usual and convenient to base such measures on differences from the arithmetic mean but these are both positive and negative and sum to zero so their own arithmetic mean cannot be used for measuring spread. The choice is the quadratic mean of the differences, called the *standard deviation*, and its square, the *variance*, is the arithmetic mean of the squares of the differences. These quantities are defined for the universe by eqns. (2.17a) and (2.16a). Analogous formulae, (2.16c) and (2.17c), in which the denominator is the sample size n, give the so-called "sample" values but these are biased estimators of their parameters and better estimates are obtained by manipulating the denominator or applying correction factors.

In the case of the variance we have the situation that a random collection of squared differences from the true mean, $\Sigma f_i(x_i - \mu)^2$, would average to give an unbiased estimate of σ^2 but, with μ unknown and $\bar{x}$ of necessity used in its place, we have $\Sigma f_i(x_i - \bar{x})^2$ which is minimized in the estimation of $\bar{x}$ and thus is always less than the former quantity. Hence n as denominator would tend to underestimate σ^2 and we must divide by something less, in fact by $n - 1$, to counteract this. The $n - 1$ is the number of *degrees of freedom* or independent differences on which the estimate is based. This is so because all the n individual sample values have a share in the value of $\bar{x}$ and if $\bar{x}$ is known only $n - 1$ differences need be known for the last to be found.

The variance is unsuitable as a descriptive measure of spread since its units are not those of the variable itself and the standard deviation is to be preferred. Commonly, the s of eqn. (2.17b) is used as the estimator (for example in the tests of significance described in the next chapter) but again bias arises, this time because we have taken the square root. If a large number of independent sample s^2 values have as their arithmetic mean the universe σ^2 (i.e. s^2

TABLE 2.7. MEASURES OF DISPERSION

Measure	Universe value	Sample value	
Variance	$\sigma^2 = \Sigma f_i(x_i - \mu)^2/N$	$s^2 = \Sigma f_i X_i^2/(n-1)$	(2.16a, b)
Sample variance		$\acute{\sigma}^2 = \Sigma f_i X_i^2/n$	(2.16c)
Standard deviation	$\sigma = \sqrt{\sigma^2}$	$s = \sqrt{[\Sigma f_i X_i^2/(n-1)]}$	(2.17a, b)
Sample standard deviation		$\acute{\sigma} = \sqrt{(\Sigma f_i X_i^2/n)}$	(2.17c)
		$\mathrm{s} = \sqrt{[\Sigma f_i X_i^2/(n - 1\cdot 63)]}$	(2.17d)
		$s' = \sqrt{[\Sigma(x_{i+1} - x_i)^2/2(n-1)]}$	(2·17e)
Coefficient of variation	$V = 100\sigma/\mu$	$v = 100s/\bar{x}$	(2.18a, b)
Average deviation	$\delta = \Sigma f_i \|x_i - \mu\|/N$	$d = \Sigma f_i \|X_i\|/n$	(2.19a, b)
Range	$W = \beta - \alpha$	$w = x_n - x_1$	(2.20a, b)
Quartile deviation	$Q = (\mathrm{F}_{\cdot 75} - \mathrm{F}_{\cdot 25})/2$	$q = (F_{\cdot 75} - F_{\cdot 25})/2$	(2.21a, b)
Geometric deviation	$\log \sigma_g = \sqrt{[\Sigma f_i(\log x_i - \log \mu_g)^2/N]}$	$\log s_g = \sqrt{[\Sigma f_i(\log x_i - \log m_g)^2/n]}$	(2.22a, b)

(For notation, see the table of symbols at the end of this chapter.)

is unbiased) then the $\sqrt{s^2}$ values cannot have σ as their arithmetic mean (see remarks on means above). For this reason, when an unbiased estimate of σ is required, it is immaterial whether s or $\mathscr{s}$ is used—both must be corrected by an appropriate factor. On the other hand, for samples of size greater than three, the s of eqn. (2.17d) provides a good estimate (Bosworth, 1944). To summarize and to show the degree of bias; suppose a corrected or "best" estimate of σ is given by

$$\hat{\sigma} = s/c_1 = \mathscr{s}/c_2 = \mathrm{s} \quad (n > 3) \qquad (2.23\,\mathrm{a, b, c})$$

then $1/c_1$ and $1/c_2$ vary as follows with sample size:

n	2	4	6	8	10	20	50	100
$1/c_1$	1·253	1·085	1·051	1·036	1·028	1·013	1·005	1·003
$1/c_2$	1·772	1·253	1·151	1·108	1·068	1·040	1·015	1·008

Formulae for c_1 and c_2, as the correcting factors are usually denoted, are given by Cowden (1957), values of c_2 in the A.S.T.M. Manual (1951) and of $1/c_1$ by Dixon and Massey (1957). So long as n is not very small the latter is approximately $1 + 1/4(n - 1)$.

A less reliable estimate of σ than the foregoing can be obtained from the differences between successive values; second minus first, third minus second, and so on. The s' of eqn. (2.17e) is in effect the quadratic mean of these successive differences divided by $\sqrt{2}$. It is useful when the values of the variable are known to be drifting during the progress of measurement because of some secular influence and it is desired to remove the effect of this so as to characterize the dispersion of the universe in the absence of such an influence. The average deviation is also of interest mainly because of its relationship to σ. The sample value, d, is a biased estimator of δ and better estimates are obtained by replacing the denominator n in eqn. (2.19b) by $\sqrt{[n(n-1)]}$ or $\sqrt{[n(n-1{\cdot}28)]}$ (Bosworth, 1950). Although found in scientific papers, the use of this measure is to be avoided except perhaps as a quick means of estimating σ. For this purpose

$$\sigma \doteqdot 1{\cdot}25\,d \qquad (2.24\,\mathrm{a})$$

may be taken as a fair approximation even in the case of moderately skewed distributions. The geometric deviation, defined in terms of its logarithm by eqns. (2.22a) and (2.22b), has application to some particle size distributions.

Measures of dispersion can also be devised from frequency partitioning values of a distribution called, generally, *quantiles* or *fractiles*. The median, which partitions the distribution into two equal parts by frequency, is one such. The three *quartiles*, of which the median is one, divide the distribution into four equal parts and differences between such values are obviously related to spread. One measure, not much used now, is the *quartile deviation* or *semi-interquartile range* which is defined by eqn. (2.21) as half the difference between the first and third quartiles. In the case of more or less normal distributions it is about two-thirds of the standard deviation. An analogous measure, very simply found from the sample array, is the *range*, w. It is not of interest as an estimator of the universe range (being always too low) but its ease of calculation and its close relationship to σ make it useful for estimating that parameter and for the comparison of the variability of small samples ($n \leqq 12$). The relationship between w and σ depends on sample size, since the larger the sample the more chance there is of it containing individuals nearer the bounds of the universe. In the usual notation

$$\sigma = w/d_2 \qquad (2.24\text{b})$$

The actual values of d_2 for small samples are to be found in Table V of the Appendix but, very approximately, it can be taken as 3 for $n = 10$, 4 for $n = 25$, 5 for $n = 100$ and 6 for larger samples from a normal universe. With the aid of these factors, eqn. (2.24b) is useful in providing an approximate check on a computed value of s.

Apart from the variance, the measures of dispersion described here are in the units of the variable scale but all could be reduced to a relative or dimensionless measure if divided by a measure of central tendency. The *coefficient of variation*, v, which expresses the standard deviation as a fraction or more generally as a percentage of the arithmetic mean, is the preferred measure of this type. It is only appropriate when the variable values derive from a ratio scale since the arbitrary location of the origin on an interval scale affects the mean. Thus, v would be useless for characterizing the dispersion of temperature readings about 0 °C for comparison with some about 25 °C but it might be of value if the readings were expressed in °K. It is sometimes of use in comparing the variability of samples from distributions whose dispersion is related to location or whose values are measured in different units.

(3) *Measures of Skewness and Kurtosis*

Differences in the manner of spread of values in a distribution reveal themselves and can be characterized in a number of different ways. Skewness affects the relative positions of averages and of quantiles on the scale of the variable and there are several simple measures based on these; for example, $(\bar{x} - \hat{m})/s$ and $[(F_{.75} - m) - (m - F_{.25})]/q$. Kurtosis affects measures of spread and the ratio of the average to the standard deviation provides a test of this property. The most used measures of both skewness and kurtosis, however, are based on the *moments* of the distribution or related properties called *cumulants*.

The term *moment* in this connection has a significance similar to that in physics, and the general formula, that of the rth moment about a value on the scale of the variable, A, is $\Sigma f_i(x_i - A)^r/N$ for the universe and it is estimated by $\Sigma f_i(x_i - A)^r/n$ from the sample data. From this simple beginning parameters, estimators and symbols proliferate. We can define

(i) moments about the origin

$$\nu_r = \Sigma f_i x_i^r/N \quad \text{estimated by} \quad n_r = \Sigma f_i x_i^r/\text{n} \qquad (2.25\,\text{a, b})$$

(ii) moments about the mean or *central moments*

$$\mu_r = \Sigma f_i(x_i - \mu)^r/N \quad \text{estimated by} \quad m_r = \Sigma f_i X_i^r/n \qquad (2.26\,\text{a, b})$$

and (iii) dimensionless moments or α *coefficients*

$$\alpha_r = \mu_r/\sigma^r \quad \text{estimated by} \quad a_r = m_r/\hat{\sigma}^r. \qquad (2.27\,\text{a, b})$$

In addition there are the cumulants or $\varkappa$ *coefficients* of Fisher. These are defined, and their relationships to the moments given in great detail, by Kendall and Stuart (1958). They have the advantage that, whereas the sample moments given by eqns. (2.26b) and (2.27b) are biased estimators, the sample "k statistics" defined as follows are not:

$$k_1 = \Sigma f_i x_i/n = n_1 \qquad (2.28\,\text{a})$$

$$k_2 = \Sigma f_i X_i^2/(n-1) \doteqdot m_2 \qquad (2.28\,\text{b})$$

$$k_3 = n\,\Sigma f_i X_i^3/(n-1)(n-2) \doteqdot m_3 \qquad (2.28\,\text{c})$$

$$k_4 = n[(n+1)\Sigma f_i X_i^4 - 3(n-1)(\Sigma f_i X_i^2)^2/n]/[(n-1)(n-2)(n-3)]$$
$$\doteqdot m_4 - 3m_2^2. \qquad (2.28\,\text{d})$$

It can be seen from these expressions that there is a close relationship between moments and cumulants (the first three universe values being identical) and that several have been encountered already. Thus, ν_1 is the arithmetic mean μ, μ_2 is σ^2, k_2 is s^2, m_2 is ς^2 and so on. Many theoretical distributions (like the Gaussian treated in the next chapter) have zero values for most of the higher moments and cumulants and samples drawn from universes which approximate such theoretical distributions should yield higher sample values round about zero. It should be noted that central moments are always positive if r is even, but can take either sign if r is odd.

For our present purposes we are concerned with the third and fourth central moments and cumulants which are related to skewness and to kurtosis or excess of kurtosis respectively. Dimensionless ratios are used: in terms of the notation above we have
(i) for the universe skewness

$$\gamma_1 = \varkappa_3/\varkappa_2^{3/2} = \mu_3/\mu_2^{3/2} = \mu_3/\sigma^3 = \alpha_3 = \pm\sqrt{\beta_1} \quad (2.29\,\text{a})$$

and the sample estimator is

$$g_1 = k_3/k_2^{3/2} \doteqdot m_3/m_2^{3/2} = m_3/\varsigma^3 = a_3 = \pm\sqrt{b_1} \quad (2.29\,\text{b})$$

(ii) for the universe excess of kurtosis

$$\gamma_2 = \varkappa_4/\varkappa_2^2 = \mu_4/\mu_2^2 - 3 = \mu_4/\sigma^4 - 3 = \alpha_4 - 3 = \beta_2 - 3 \quad (2.30\,\text{a})$$

and the sample estimator is

$$g_2 = k_4/k_2^2 \doteqdot m_4/m_2^2 - 3 = m_4/\varsigma^4 - 3 = a_4 - 3 = b_2 - 3 \quad (2.30\,\text{b})$$

The additional symbols introduced here, β_1 and β_2 estimated by b_1 and b_2, are Pearson's measures of skewness and kurtosis and are used to characterize a system of frequency curves referred to in Chapter V. The kurtosis as such of the Gaussian distribution is 3 so this becomes in effect the standard for excess of kurtosis measured by the appropriate ratio of the cumulants.

2.5. Computation of the Estimates from Sample Data

The method of computation depends on the size of the array and the number of estimates required.

(1) *Small Sample Methods*

Consider first the computation of the arithmetic mean and standard deviation of a very small sample, say of the following values of the atomic weight of carbon: 12·0095, 12·0101, 12·0106, 12·0118. In directly applying formulae (2.5 b) and (2.17 b) one might proceed as in Table 2.8 according to the long method illustrated therein.

In practice, the labour of calculation is reduced by *coding* the values of the variable, say by subtracting twelve and removing the decimal point in this case so that the relationship between the coded value and the original is $\not{x} = 10^4(x - 12)$. (In what follows coded values and quantities computed in terms of them are denoted by a stroke through the symbol. Thus $\bar{\not{x}}$, $\not{s}^2$ and $\not{s}$ are the mean, variance and standard deviation on the new scale and $\not{X}$ is the deviation of a coded value from the coded mean.) For a small array of close values the process of coding and of decoding would be done mentally. $\not{s}$ is decoded simply by multiplying by the ratio of the sizes of the units on the two scales (10^4 in the example) but in the case of $\bar{\not{x}}$ allowance must be made for any change of origin as well. The calculations are illustrated in the central columns of Table 2.8.

In coding, the closer the subtracted figure (the new origin) to the mean, the smaller will be the coded values and the simpler the calculations. Also, the deviations between the values and the mean need not be found individually but use made of the relationship between the sum of the squares of the deviations and the sum of the squares of the original values. This is simply derived. For n values of x (taken separately so that $f_i = 1$):

$$\Sigma X_i^2 = \Sigma(x_i - \bar{x})^2 = \Sigma(x_i^2 - 2\bar{x}x_i + \bar{x}^2) = \Sigma x_i^2 - 2\bar{x}\Sigma x_i + n\bar{x}^2$$

$$= \Sigma x_i^2 - 2n\bar{x}\Sigma x_i/n + n\bar{x}^2 = \Sigma x_i^2 - n\bar{x}^2 \qquad (2.31\,\text{a})$$

This can be expressed in various other ways:

$$\Sigma X_i^2 = \Sigma x_i^2 - (\Sigma x_i)^2/n \qquad (2.31\,\text{b})$$

$$n\Sigma X_i^2 = n\Sigma x_i^2 - (\Sigma x_i)^2 \qquad (2.31\,\text{c})$$

$$m_2 = \Sigma X_i^2/n = n_2 - n_1^2 \qquad (2.31\,\text{d})$$

the last being in terms of the notation of eqns. (2.25 b) and (2.26 b). Equation (2.31 a) has been applied to the coded values in the final columns of Table 2.8 but, as ΣX_i^2 is found as the difference between

TABLE 2.8. SMALL SAMPLE COMPUTATIONS

Long method			Coding method			Short coding method	
x	X	X^2	$\mathfrak{x}$	$\lvert\mathfrak{X}\rvert$	$\mathfrak{X}^2$	$\mathfrak{x}$	$\mathfrak{x}^2$
12·0095	−0·0010	0·00000100	95	10	100	−5	25
12·0101	−0·0004	0·00000016	101	4	16	1	1
12·0106	+0·0001	0·00000001	106	1	1	6	36
12·0118	+0·0013	0·00000169	118	13	169	18	324
4) 48·0420		3) 0·00000286	4) 420		3) 286	4) 20	386
$\bar{x}$ = 12·0105		s^2 = 0·00000095	$\bar{\mathfrak{x}}$ = 105		$\mathfrak{s}^2$ = 95	$\bar{\mathfrak{x}}$ = 5	$-n\bar{\mathfrak{x}}^2$ = −100
		s = 0·00097			$\mathfrak{s}$ = 9·7		3) 286 = $\Sigma\mathfrak{X}^2$
			$\mathfrak{x} = 10^4(x - 12)$				$\mathfrak{s}^2$ = 95
			$\bar{x}$ = 12·0105				$\mathfrak{s}$ = 9·7
			s = 0·00097			$\mathfrak{x} = 10^4(x - 12{\cdot}0100)$	
						$\bar{x}$ = 12·0105	
						s = ·00097	

two quantities which can be very large and similar, these need to be found accurately and eqns. (2.31 b) and (2.31 c) are better. This preference, especially for eqn. (2.31 c), is partly based on the proposition that "approximate operations should be deferred as long as possible" but it is particularly apposite when a calculating machine is available and applied to large arrays of uncoded data.†

(2) *Grouping Methods*

When the sample is large and classification has already been resorted to for purpose of graphical representation, the calculations are conveniently performed on the grouped values after coding. In the method now to be described they are simplified, first by letting the mid-point of a class interval represent all the values in that interval, and secondly by coding in such a way that the units of the variable scale are changed to the interval width. It follows that the method is approximate and its accuracy depends on the number of class intervals rather than on refinements of computational procedure.

In coding the values, a central and preferably well populated class interval is selected and its mid-point designated the arbitrary mean, A. This is made the origin of the coded scale and the other mid-points are allocated values which change by unity according to their distance from A. Some care is needed in locating A on the original scale—in the CO_2 densities example, it may be taken as the mid-point of the class interval 1·495– into which values of the variable equal to or greater than 1·495 but less than 1·505 have been collected (see Table 2.2). If it is considered that a value only infinitesimally lower than 1·505 could occur (in which case the variable would be strictly continuous), the mid-point would be 1·500. In the example it appears that the variable can only take the values 1·495, 1·496 ... 1·504 in this interval as defined (see Section 2.2 above), i.e. the density cannot be measured or has not been recorded to better than the third decimal place, and the mid-point is taken as 1·4995.

In addition to coding in this way, short methods of computation are used. If x_i is now taken as the mid-point of any individual interval on the coded scale and f_i as the number of values falling within it, we may write for second central moment in class interval units

† See Dwyer (1951) for a valuable discussion of the principles of computational design and methods in general.

Class interval	f	x	fx	fx^2	fx^3	fx^4	$x+1$	$(x+1)^4$	X	$f\|X\|$
1·455–	1	−3	−3	9	−27	81	−2	16	−3·741	3·741
1·465–	2	−2	−4	8	−16	32	−1	2	−2·741	5·482
1·475–	10	−1	−10	10	−10	10	0	0	−1·741	17·410
1·485–	24	0	0	0	0	0	1	24	−0·741	17·784
1·495–	21	+1	21	21	21	21	2	336	+0·259	5·439
1·505–	15	+2	30	60	120	240	3	1215	+1·259	18·885
1·515–	6	+3	18	54	162	486	4	1536	+2·259	13·554
1·525–	2	+4	8	32	128	512	5	1250	+3·259	6·518
	$n = 81$		$\Sigma fx = 60$	$\Sigma fx^2 = 194$	$\Sigma fx^3 = 378$	$\Sigma fx^4 = 1382$		$\Sigma f(x+1)^4 = 4379$		$\Sigma f\|X\| = 88{\cdot}813$

$A = 1{\cdot}4895$ $n_1 = 0{\cdot}7407$; $n_2 = 2{\cdot}3951$; $n_3 = 4{\cdot}6667$; $n_4 = 17{\cdot}0617$; $\Sigma f(\bar{x}+1)^4/n = 54{\cdot}0617$; $d = 1{\cdot}096$

$h = 0{\cdot}01$ $n_1^2 = 0{\cdot}5487$; $n_1^3 = 0{\cdot}4064$; $n_1^4 = 0{\cdot}3011$; $n_1 n_2 = 1{\cdot}7741$; $n_1 n_3 = 3{\cdot}4568$; $n_1^2 n_2 = 1{\cdot}3142$

Check, eqn. (2.35d): $1382 + 4(378) + 6(194) + 4(60) + 81 = 4379$, cf. $\Sigma f(x+1)^4$

or, eqn. (2.36): $17{\cdot}0617 + 4(4{\cdot}6667) + 6(2{\cdot}3951) + 4(0{\cdot}7407) + 1 = 54{\cdot}0619$, cf. $\Sigma f(x+1)^4/n$

Arithmetic mean (eqn. 2.33), $\bar{x} = 1{\cdot}4895 + 0{\cdot}01(0{\cdot}7407) = 1{\cdot}49691$

Median (eqn. 2.3), $m = 1{\cdot}4945 + 0{\cdot}01(40{\cdot}5 - 37)/21 = 1{\cdot}4962$

Mode (eqn. 2.2b), $\hat{m} = 1{\cdot}4845 + 0{\cdot}01[(24 - 10)/(24 - 10) + (24 - 21)] = 1{\cdot}4927$ (Fig. 2.7)

(eqn. 2.2c), $\hat{m} = 1{\cdot}4945 + 0{\cdot}01[(-6{\cdot}0 - 4{\cdot}8 + 4{\cdot}2 + 9{\cdot}0)/(-10 + 24 + 21 - 15)] \doteq 1{\cdot}4957$

Variance (eqns. (2.32a), (2.37)), $s^2 = m_2' = 2{\cdot}3951 - 0{\cdot}5487 - 0{\cdot}0833 = 1{\cdot}7631$; $s^2 = h^2 s^2 = 1{\cdot}7631 \times 10^{-4}$

(from eqn. (2.28b)), $s^2 = k_2' = n m_2'/(n-1) = 81(1{\cdot}7631)/80 = 1{\cdot}7851$; $s^2 = 1{\cdot}7851 \times 10^{-4}$

Standard deviation (eqn. 2.17c), $s = \sqrt{1{\cdot}7631} = 1{\cdot}328$; $s = hs = 0{\cdot}01328$

(eqn. 2.17b), $s = \sqrt{1{\cdot}7851} = 1{\cdot}336$; $s = hs = 0{\cdot}01336$

Eqns. (2.23a), (2.17d), $\hat{\sigma} = (1 + 1/4 \times 80)\,0{\cdot}01336 = 0{\cdot}01340$; $\doteq s = \sqrt{(81/79{\cdot}37)}\,0{\cdot}01328 = 0{\cdot}01342$

Check, eqn. (2.24a): $s \doteq 1{\cdot}25d \doteq 0{\cdot}0137$; Check eqn. (2.24b): $s \doteq w/5 \doteq (1{\cdot}527 - 1{\cdot}460)/5 = 0{\cdot}0134$

Average deviation (eqn. 2·19b), $d = 1{\cdot}096$ (above); $d = hd = 0{\cdot}0109$

3rd moment (eqn. 2.32b), $m_3 = 4{\cdot}6667 - 3(1{\cdot}7741) + 2(0{\cdot}4064) = +0{\cdot}1572$

Coefficient of skewness (eqn. 2.29b), $g_1 = +0{\cdot}1572/1{\cdot}328^3 = +0{\cdot}0671$

4th moment (eqn. 2.32c), $m_4 = 17{\cdot}0617 - 4(3{\cdot}4568) + 6(1{\cdot}3142) - 3(0{\cdot}3011) = 10{\cdot}2218$

Corrected 4th moment (eqn. 2.38b), $m_4 = 10{\cdot}2218 - 0{\cdot}0125 - 1{\cdot}7631/2 = 9{\cdot}3278$

Coefficient of kurtosis (eqn. 2.30b), $g_2 = 9{\cdot}3278/1{\cdot}328^4 - 3 = -0{\cdot}00093$

(from eqn. (2.31 d)),

$$s^2 = m_2 = n_2 - n_1^2 \tag{2.32a}$$

and, by derivations similar to that of eqns. (2.31 a) and (2.31 d) it may be shown that

$$m_3 = n_3 - 3n_1n_2 + 2n_1^3 \tag{2.32b}$$

and

$$m_4 = n_4 - 4n_1n_3 + 6n_1^2n_2 - 3n_1^4 \tag{2.32c}$$

where $n_3 = \Sigma f_i x_i^3/n$ and $n_4 = \Sigma f_i x_i^4/n$ from eqn. (2.25 b). Each of the moments in class interval units can be converted to moments in the actual units of the variable scale or powers thereof by multiplying by the appropriate power of the class interval width, h; i.e. $m_2 = h^2 m_2$ etc. However, when the third and fourth central moments are only required for incorporation into dimensionless ratios for measuring skewness and kurtosis they need not be so corrected. In decoding the first moment about the arbitrary mean, $n_1 = \bar{x}$, the location of A must be taken into account, so

$$\bar{x} = A + hn_1 \tag{2.33}$$

and this also applies, *mutatis mutandis*, to the decoding of the mode and median. Equations can be derived for other measures defined in Table 2.6. These and other calculations are illustrated in Tables 2.9 and 2.10. The average deviation is perhaps best found by the direct application of eqn. (2.19 b) but the relationship

$$\Sigma f_i |X_i| = \Sigma f_i |x_i - \bar{x}| = 2\Sigma f_i x_i'' - \bar{x}\Sigma f_i'' \tag{2.34}$$

(where $\Sigma f_i x_i''$ and $\Sigma f_i''$ are the sum and the number of observations greater than $\bar{x}$ respectively) can be applied to grouped data by interpolation in the class interval containing the mean.

The number of moments, or alternatively cumulants, computed depends on the measures required. Thus, if the mean and standard deviation are wanted, calculation can stop at the sum of the values and their squares and the moments or cumulants derived from these totals. The sum of the cubes is found only if a measure of skewness is called for and the sum of the fourth powers for a measure of kurtosis in addition. The same four sums are sufficient for the calculation of the averages given by eqns. (2.9 b) to (2.13 b) but in these cases coded values cause troublesome complications.

TABLE 2.10. THE GEOMETRIC AND OTHER MEANS AND THE GEOMETRIC DEVIATION—SULPHUR SOL SIZES

Class int.	x	f	f/x	fx	fx^2	$f(\log x)$	$\log x$	$f(\log x)^2$
0·5–	1	10	10·00	10	10	0·000	0·000	0·000000
1·5–	2	38	19·00	76	152	11·438	0·301	3·442838
2·5–	3	28	9·33	84	252	13·356	0·477	6·370812
3·5–	4	15	3·75	60	240	9·030	0·602	5·436060
4·5–	5	5	1·00	25	125	3·495	0·699	2·443005
5·5–	6	3	0·50	18	108	2·334	0·778	1·815812
6·5–	7	1	0·14	7	49	0·845	0·845	0·714025
		100	43·72	280	936	40·498		20·222552

$\log m_g = 40{\cdot}498/100 = 0{\cdot}405$ $\qquad -n(\log m_g)^2 = -16{\cdot}400880$

$100)\ 3{\cdot}821672$

$(\log s_g)^2 = 0{\cdot}038217$

Harmonic mean (eqn. 2.6b), $m_h = 100/43{\cdot}72 = 2{\cdot}287$
Geometric mean (eqn. 2.7b), $m_g = \text{anti-log } 0{\cdot}405 = 2{\cdot}541$
Arithmetic mean (eqn. 2.5b), $\bar{x} = 280/100 = 2{\cdot}800$
Quadratic mean (eqn. 2.9b), $m_q = \sqrt{(936/100)} = 3{\cdot}059$
Linear average (eqn. 2.11b), $m_l = 936/280 = 3{\cdot}343$
Geometric deviation (eqn. 2.22b), $s_g = \text{anti-log} \sqrt{0{\cdot}03822} = 1{\cdot}569$

It is desirable to check the sums of powers of the values to whatever degree they are taken and *Charlier's Checks* provide a ready means of doing this. The method is to form values of $x + 1$, raise these to the appropriate power and sum. It is easy to show that

$$\Sigma f_i(x_i + 1) = \Sigma f_i x_i + n \tag{2.35a}$$

$$\Sigma f_i(x_i + 1)^2 = \Sigma f_i x_i^2 + 2\Sigma f_i x_i + n \tag{2.35b}$$

$$\Sigma f_i(x_i + 1)^3 = \Sigma f_i x_i^3 + 3\Sigma f_i x_i^2 + 3\Sigma f_i x_i + n \tag{2.35c}$$

$$\Sigma f_i(x_i + 1)^4 = \Sigma f_i x_i^4 + 4\Sigma f_i x_i^3 + 6\Sigma f_i x_i^2 + 4\Sigma f_i x_i + n \tag{2.35d}$$

(The same relationships apply to uncoded values). Thus, if measures as far as kurtosis or the volume average are required, each of the four necessary sums are checked by the application of eqn. (2.35d). At the expense of a little extra division, these checks can be made in terms of the moments about the origin on the coded scale, i.e. about the arbitrary mean, for example,

$$\Sigma f_i(x_i + 1)^4/n = n_4 + 4n_3 + 6n_2 + 4n_1 + 1$$

At rather greater expense of effort the computations can be checked to this stage by repeating them with a differently located arbitrary mean or by the summation method described below.

One other point must be noted. When the approximation is made that the mid-point of a class interval represents all the values within it a special error is introduced into the computation of even central moments. This is because the true average of the values in a class interval, in which frequency density is changing rapidly, deviates rather less from $\bar{x}$ than does the mid-point. For example, the mean of the twenty-four values in the fourth class interval (1·485–1·494) of Table 2.2 is 1·4908 but they are represented for purposes of calculation by the mid-point 1·4895. These figures differ from $\bar{x}$ by 0·0063 and 0·0076 but it is the latter which appears in the computations as X_i. A similar discrepancy is to be expected on the high side of the distribution and the net effect is that $\Sigma f_i X_i^2$ and $\Sigma f_i X_i^4$ are greater than they ought to be. In the case of X_i and X_i^3 the errors are of opposite sign and tend to cancel, so corrections need not be applied to odd central moments. The corrections for grouping given here, *Sheppard's Corrections*, should be applied when the variable is more or less continuous and the distribution not too skewed. They are not required when the moments are calculated from unclassified data or if the variable is discrete. When the coded values are in steps of unity (like the x of Table 2.9),

$$m_2' = m_2 - 1/12 \tag{2.37}$$

$$m_4' = m_4 + 7/240 - \tfrac{1}{2}m_2 \tag{2.38a}$$

or

$$= m_4 - 1/80 - \tfrac{1}{2}m_2' \tag{2.38b}$$

where the primed quantity denotes the corrected and the unprimed the crude values respectively. The corresponding adjustments to the coded sample cumulants are

$$k_2' = k_2 - 1/12; \quad k_4' = k_4 + 1/120 \qquad (2.39\text{a, b})$$

The main calculations through to the coefficients of skewness and kurtosis are illustrated in Table 2.9 where they are applied to the classified data of Table 2.2. (See Fig. 2.7 for the histogram and some of the estimates). Table 2.9 is set up for the computation of moments since these are perhaps a little more convenient than cumulants and, unless the sample is rather small, the differences are insignificant. Thus, for these data as classified, k_3 is $+0{\cdot}1632$ and the value of g_1 calculated from it is $+0{\cdot}0683$ compared with $m_3 = +0{\cdot}1572$ and $g_1 = +0{\cdot}0671$ in the table.

If the less common measures of central tendency are required it is generally easier to work from the actual class interval mid-points than from coded values since these measures are affected by the scale as well as the distribution of values. They are interrelated and equations can be derived for calculating one from another. For example, from eqns. (2.9 b), (2.11 b) and eqn. (2.32 a) it can be seen that

$$m_q = \sqrt{(\bar{x}^2 + m_2)} \tag{2.40}$$

and

$$m_l = m_q^2/\bar{x} \tag{2.41}$$

In the case of data like those at present under consideration, differences between these various measures are slight ($m_h = 1{\cdot}49678$, $m_g = 1{\cdot}49686$, $\bar{x} = 1{\cdot}49691$, $m_q = 1{\cdot}49696$, $m_l = 1{\cdot}49703$ after grouping) and values for substitution in relationships like eqns. (2.40) and (2.41) must be taken to a large number of significant figures. Obviously also, because of this similarity, they are of little utility in dealing with normally distributed values and we turn to the rather skewed data of Table 2.4(a) the better to illustrate their calculation. The computations are given in Table 2.10.

Cumulative frequencies, which yield quantities called *factorial moments*, can also be used to find to central moments or cumulants. These summation methods, as described for example by Elderton (1938) and by Kendall and Stuart (1958), result in a considerable reduction in the arithmetic, especially if only the mean and standard deviation or variance are required, and may be handy if an adding machine but no calculating machine is available. The simplest of them is illustrated in Table 2.11 in which we return to the carbon dioxide density data. In this, the data are in effect coded in class interval units about an arbitrary mean which is the mid-point of an unpopulated interval immediately preceding the lowest in the table. The first factorial moment is equal to the first moment about this point but between the higher factorial and ordinary moments the relationships are more complex. They are quoted by Kendall and Stuart (1958) and have been combined with eqn. (2.32) to give the central moment formulae (2.43) in Table 2.11. Such formulae can take many different forms (see, for example, the *A.S.T.M. Manual on Quality Control of Materials* (1951)) but those given are designed to make most use of successively computed quantities. Only the basic calculations appear in the table—the remainder follow the same course as those already illustrated in Table 2.9. The slight differences in the results of the two computing procedures are to be attributed to the number of significant figures taken.

Finally, it must be remembered that the results of the calculations on grouped data are subject not only to random sampling error (alluded to in Section 2.1 and dealt with from a more quantitative point of view in the next chapter) but also to computational errors

TABLE 2.11. LARGE SAMPLE COMPUTATIONS—CUMULATIVE METHOD

Class. Int.	f	F	F_2	F_3	F_4
1·455–	1	81			
1·465–	2	80	303		
1·475–	10	78	223	490	
1·485–	24	68	145	267	448
1·495–	21	44	77	122	181
1·505–	15	23	33	45	59
1·515–	6	8	10	12	14
1·525–	2	2	2	2	2
$h = 0{\cdot}01$	$n = 81$	384	793	938	704

Factorial moments, $e_1 = \Sigma F/n = A_1(\text{re } A = 1{\cdot}4495) = 384/81 = 4{\cdot}74074$ (2.42a)

$$e_2 = 2\,\Sigma F_2/n = 2(793)/81 = 19{\cdot}58025 \quad (2.42\text{b})$$
$$e_3 = 6\,\Sigma F_3/n = 6(938)/81 = 69{\cdot}48148 \quad (2.42\text{c})$$
$$e_4 = 24\,\Sigma F_4/n = 24(704)/81 = 208{\cdot}59259 \quad (2.42\text{d})$$

Arithmetic mean (eqn. 2.33), $\bar{x} = 1{\cdot}4495 + 0{\cdot}01(4{\cdot}741) = 1{\cdot}49691$

2nd moment, $m_2 = e_2 + e_1 - e_1^2$ (2.43a)

$$= 19{\cdot}58025 + 4{\cdot}74074 - (4{\cdot}74074)^2 = 1{\cdot}84637$$

Corrected moment (eqn. 2.37), $m_2' = 1{\cdot}84637 - 0{\cdot}08333 = 1{\cdot}76304$

3rd moment, $m_3 = e_3 + 3e_2 + e_1(1 - 3m_2) - e_1^3$ (2.43b)

$$= 69{\cdot}48148 + 58{\cdot}74074 + 4{\cdot}74074\,(-\,4{\cdot}53909) - 106{\cdot}54638$$
$$= +0{\cdot}15719$$

4th moment, $m_4 = e_4 + 6e_3 + 7e_2 + e_1(1 - 4m_3) - e_1^2(6m_2 - e_1^2)$ (2.43c)

$$= 208{\cdot}59259 + 6(69{\cdot}48148) + 7(19{\cdot}58025)$$
$$+ 4{\cdot}74074[1 - 4(0{\cdot}15179)$$
$$- 4{\cdot}74074^2][6(1{\cdot}83637) - 4{\cdot}74074^2]$$
$$= 10{\cdot}21625$$

Notes to Table: F is the cumulative frequency above the lower boundaries, i.e. the downward accumulation of f. F_2 is the accumulation of F in the same sense but stopping at the second lowest interval, and F_3 and F_4 are successive accumulations of values in the preceeding columns stopping short as indicated.

inherent in the practice of grouping. The *average* error from this source is satisfactorily adjusted by Sheppard's corrections but in addition there are errors of a periodic nature due to the accidental location of the class intervals. The grouping method is so commonly recommended and, in fact, is so convenient in the absence of mechanical means of handling large bodies of data that this latter source of error is deserving of some attention. In the present instance, the effects of differences in the location of class intervals of width 0·005, 0·01 and 0·02 (g/cm^3) on the results for selected estimates are shown in Fig. 2.8. Starting with possible scale values (indicated as abscissae) there are between fourteen and fifteen intervals of the smallest, between seven and eight of the intermediate (as used in Tables 2.9 and 2.11) and between four and five of the largest size required to embrace all the values in the array. The central lines on the first four graphs represent the values of the estimates obtained by exact calculation from the original array (Table 2.1) and they are also the averages of the values obtained after grouping. The distances shown vertically on the right of each graph represent the sampling error to be expected in the case of a sample of this size. (They are the standard errors of the estimators and their precise significance is discussed in Section 3.4.) It can be seen that, even for the largest width, the accidental grouping error is on the whole less than the sampling error but that, even if the smallest width is chosen for the calculations, unfortunate location of the intervals can result in a comparatively sizeable discrepancy. On the other hand, the average grouping error (shown in Fig. 2.8 (b) in its effect on the standard deviation) is almost exactly corrected by the application of eqn. (2.37) regardless of the width of the intervals used. In the case of g_2, the corrections to the second and fourth central moments compensate each other and differences between $g_2 = m_4/m_2^2$ and $g_2 = m_4'/m_2'^2$ are slight.

Although it might seem from this comparison that a large number of class intervals (up to, say, fifteen in the present instance) offers a worthwhile increase in computational precision in the case of the mean and standard deviation, this is not true of the mode nor of the median. Kelly (1947) offers an approximate rule for the selection of interval width, based on the variation of frequency density to be expected in normally distributed sample values and the requirements of formulae (2.2c), which is to be found in graphical form in Fig. 2.8 (g). This requires for its application an estimate

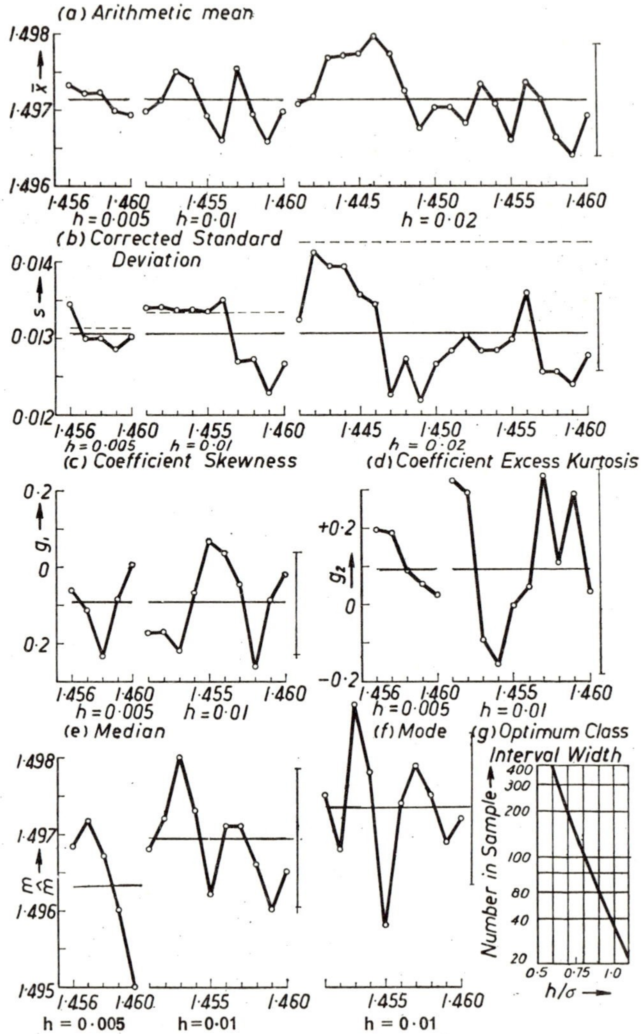

FIG. 2.8. Estimates from Normally Distributed Sample Data.

The figure shows the effect of size and location of class intervals on the calculated values of (a) $\bar{x}$, (b) s, (c) g_1, (d) g_2, (e) m and (f) $\hat{m}$ (four-point). The results using all possible class intervals having $h = 0{\cdot}005$ (from 1·456), $h = 0{\cdot}01$ (from 1·451) and $h = 0{\cdot}02$ (from 1·441) are plotted. (g) gives an approximate relation (based on Kelly (1947)) for the optimum size of class interval in terms of σ for the determination of $\hat{m}$ by eqn. (2.2c). The broken lines in (b) represent the magnitude of Sheppard's corrections which have been applied. The lines on the right-hand side of each graph represent the standard errors of the relevant estimate (see Chapter III) and are a measure of sampling error.

of the standard deviation such as might be obtained from the range. For a sample of size $n = 81$ the factor from the graph is about 0·85 and this, combined with the estimated standard deviation, indicates an optimum width of slightly more than the chosen value, $h = 0{\cdot}01$.

(3) *Graphical Methods*

Graphical methods have already been given for the mode and median which are exactly equivalent to the estimating eqns. (2.2b) and (2.3). An exact method, described by Askovitz (1959), is also available for the arithmetic mean of grouped data but the construction seems overly intricate for general use.†

For approximate estimates, probability paper plots are valuable. Thus, the normally distributed values of our principal example when so plotted (Fig. 2.4) can be fitted by eye with a straight line whose intersection with the $F = 50\%$ line indicates the median and mean, and the standard deviation is found as the difference between the mean and the abscissa at the $F = 16\%$ or the $F = 84\%$ intersection. (These percentages follow from properties of the theoretical Gaussian distribution dealt with in Chapter III.) The graphical results compare well with those calculated in Table 2.9. In the case of a log-normal distribution like that of the data of Table 2.4(a), a line through the points on log-probability paper (Fig. 2.5(a)) indicates the median and geometric mean at the $F = 50\%$ intersection. The ratio of the abscissae, $F_{.16}/m$ or $m/F_{.84}$ (always taken greater than one), is the geometric deviation.

Notation of Chapter II

A	"arbitrary mean" or point on the x scale corresponding to the origin on the coded $\varkappa$ scale
x_r, a_r	universe and sample dimensionless central moments of order r
α, β	lower and upper limits of a universe
$\beta_1, \beta_2, b_1, b_2$	Pearson's universe and sample measures of skewness and kurtosis
c	sample mid-range

† The method is related to the useful procedure for fitting least squares lines described in Section 4.3 below.

c_1, c_2	factors indicating the bias of s and σ with respect to σ
d_2	factor relating w to σ
δ, d	universe and sample average deviation
e_r	sample factorial moment of order r on the coded $\varkappa$ scale
f, sometimes f_i	class interval frequency; frequency density
f_1, f_2, f_3, f_4	class interval frequencies in relation to that containing $\hat{m}$
f''	number of values greater than $\bar{x}$
F	cumulative frequency above or below class interval boundaries
F_2, F_3, F_4	successive accumulations of F
$\mathrm{F}_{\cdot 75}, \mathrm{F}_{\cdot 25}, F_{\cdot 75}, F_{\cdot 25}$	universe and sample upper and lower quartiles
$F_{\cdot 16}, F_{\cdot 84}$	values of x marking off 16% and 84% of total sample frequency in either direction, i.e. the 16th and 84th percentiles
$\gamma_1, \gamma_2, g_1, g_2$	universe and sample coefficients of skewness and excess of kurtosis
h	class interval width
i, also j	subscript indicating individual value or deviation
$\varkappa_r, k_r$	Fisher's universe and sample cumulants of order r
k_2', k_4'	coded second and fourth sample cumulants corrected for grouping
$l_{\hat{m}}, l_m$	boundary of class interval containing $\hat{m}$ or m
M, m	universe and sample median
$\hat{\mu}, \hat{m}$	universe and sample mode
$\mu, \bar{x}$	universe and sample arithmetic mean
μ_c, m_c	universe and sample cubic mean
μ_g, m_g	universe and sample geometric mean
μ_h, m_h	universe and sample harmonic mean
μ_l, m_l	universe and sample linear average
μ_q, m_q	universe and sample quadratic mean or root mean square
μ_r, m_r	universe and sample central moments of order r
μ_s, m_s	universe and sample surface average
μ_v, m_v	universe and sample volume or weight average
μ_z, m_z	universe and sample z-average

m'_2, m'_4	coded second and fourth sample moments corrected for grouping
N	size of universe
n	size of sample
ν_r, n_r	universe and sample moments of order r about the origin of the x scale
Π	multiplication operator
Q, q	universe and sample quartile deviation
r	an exponent in a general expression for measures of central tendency or which indicates the order of a moment or cumulant (generally 1, 2, 3 or 4)
Σ	summation operator†
σ	universe standard deviation
$s, \varsigma, \mathrm{s}, s'$	estimators of σ
σ_g, s_g	universe and sample geometric deviation
σ^2	universe variance
s^2	unbiased sample estimator of σ^2
ς^2	"sample" variance
V, v	universe and sample coefficient of variation
W, w	universe and sample range
x, sometimes x_i	an individual variable value
x_1, x_n	lowest and highest values of x in a sample array
x''	a value of x greater than $\bar{x}$
$\bar{x}$	sample arithmetic mean
X, sometimes X_i	the deviation, $x_i - \bar{x}$
$x. X, \bar{x}, s, n, m$ etc.	values, deviations, estimators, moments etc. on the coded scale

† The strict mathematical notation for the summation of n values of x, x_i $(i = 1, 2 \ldots n)$ is $\sum\limits_{i=1}^{i=n}$ which may be found reduced to $\sum\limits_{i=1}^{n}$, $\sum\limits_{1}^{n}$, $\sum\limits_{i}$, Σ_i or simply Σ. In this book, if the symbol is not used alone, it is qualified only by a superscript to indicate what is being summed and to what extent.

CHAPTER III

ERRORS, PROBABILITY AND TESTS OF SIGNIFICANCE

PROGRESS in the experimental sciences is made by consideration of numbers assigned to particular properties by the process of measurement (or, more primitively, by ordering or counting) and, further, by the comparison of measurements of two or more properties—for example, electric current and voltage, pressure and temperature, density and composition. The number actually assigned in any one measurement is partly the result of the magnitude of the measured property, partly of the units used, and partly of the technique of measurement. Relationships between units and between different properties are limited by their dimensional character, a matter more closely considered in Chapter VII, but the technique or method of measurement imposes on the number assigned an uncertainty or variability whose effect on interpretation is the subject of the present chapter.

It is a common experience that repeated observations of the same property will seldom give completely concordant sets of figures and the scatter in such sets is attributed to errors of measurement. If we regard such a set as a sample from a single universe of possible values and we compute a sample estimate, say the arithmetic mean, the question may be asked: "Within what limits may the true (universe) mean be expected to lie?" or "Is it reasonable to suppose that the true mean, estimated by the sample mean, has some particular value?" The interpretation of a sample estimate along these lines depends on the notion of *probability* and a knowledge of the behaviour of the estimator, i.e. its *sampling distribution*, and these topics, along with the nature of experimental error itself, have first to be considered.

3.1. The Nature of Error in Experimental Science

An error may be defined, perhaps rather loosely, as the difference between an actual measurement and the true value of the magnitude measured. In practice, of course, we do not know this true value. Even where a predicted or "theoretical" value is available, as for example the molar heat of a monatomic gas, we can never be sure, other than by taking more and more accurate measurements, that the theory is complete and that the theoretical and true values are identical. We can, at best, make a provisional estimate of the experimental error of an observation from the difference between its value and that of the weighted mean of an ensemble of all measurements, believed to be of the highest level of accuracy. Ordinarily, we cannot even go so far and must be content with a comparison of similar observations made under the same conditions, i.e. we count the error in the value of a measurement as the difference between it and that of the arithmetic mean (or other measure of central tendency) of the whole set. In this case we can only identify the true value, for the time being, with the mean of all possible experimental values which could be made under the same conditions.

The traditional treatment of the sources of experimental error involves a classification into those which affect the sign of the difference between the experimental and true values and those which can operate in either direction. The former may be said to govern the *accuracy* and the latter the *precision* of an observation. More elaborately, we may distinguish between (a) constant errors, (b) drift, (c) accidental or random errors, and (d) those due to rounding-off. Since these behave rather differently from a mathematical point of view, they will be considered separately.

(a) *Constant Errors*

An error which results from the use of a faulty standard will occur in the whole of a series of measurements. Such errors, therefore, cannot be eliminated or reduced in magnitude by any form of mathematical operation, e.g. averaging, on the data alone. The accuracy of the average is no better, or no worse, than the accuracy of the individual values. Indeed, the existence of a constant error in a set of observations can only be revealed by reference to a different set made using a different standard. Once having found a more reliable standard, however, it is a relatively simple matter to correct for a constant error.

(b) *Drift*

A drift may be described as an error which arises from a single or from a small number of assignable causes. Measurement of the velocity constant of a chemical reaction is subject to drift if the temperature of the reaction mixture is changing during the period of investigation. Similarly, the results of the range testing of artillery material are subject to drift if the wind is changing during the course of firing. The effect of such errors can be corrected by identification and proper allowance (e.g. regression analysis, see Chapter IV) or, alternatively, by compensation for their causes in better experimental technique but on the whole they are less easily eliminated than constant errors.

The term "systematic error" is sometimes used synonymously with "drift" and sometimes more broadly (as in Section 2.1) to cover "constant error" as well. In our distinction here, we imply constant direction and magnitude in the latter and constant direction but variable magnitude in the former. By "systematic" we will continue to imply constant direction in the errors, however caused.

(c) *Accidental or Random Errors*

These are errors which arise from a large number of causes in such a way that they have neither constant direction nor magnitude and it is impossible to eliminate or compensate for all of them. The effect then is a random one which can only be treated on the basis of statistical theory.

It is reasonable to suppose that, given the operation of a large number of disturbances none of which are sufficiently intrusive to be individually identifiable and thus capable of compensation, the frequency of occurrence of such errors will follow a distribution law according to which (i) positive and negative errors occur with equal frequency, (ii) numerically large errors will occur less frequently than small errors, and (iii) those beyond a certain magnitude will not be observed or, at least, on very rare occasions indeed. More precisely, the supposition is that the distribution of random errors follows the normal type exemplified by the data of Table 2.2 and Fig. 2.2. The theoretical Gaussian distribution, which fits this specification well, is discussed in the next section.

(d) *Round-off Errors*

Round-off errors are incurred in the restriction of an experimental value to a limited number of significant figures. In order to avoid systematic error, the rule adopted in rounding off a number to k figures is to eliminate the $(k + 1)$th and following ciphers and (a) leave kth cipher unchanged if the $(k + 1)$th is 0, 1, 2, 3, or 4, or 5 (alone or followed by zeros) when preceeded by an even number, or (b) increase kth cipher by unity if the $(k + 1)$th is 6, 7, 8, or 9, or 5 (alone or followed by zeros) when preceded by an odd number. In this way, relying on an equal distribution of even and odd kth ciphers we round up or down from a terminal 5 on the average an equal number of times. If, of course, the 5 to be eliminated is followed by other ciphers we round up. Thus 1·2345 goes to 1·234, 1·234505 goes to 1·235, and 1·2355 goes to 1·236 when rounded to four significant figures.

This process is not necessarily applied only to sets of written results—it may also be carried out mentally in reading the position of a pointer on scale. In either case the frequency distribution is rectangular but it generally constitutes a minor addition to the normal distribution of the random errors.

3.2. Probability and Random Errors

Before attempting to elaborate further the distribution of random errors it is necessary to consider, at least briefly, the question of probability. The general question is, to say the least, a vexed one. Kendall and Buckland (1957), in their *Dictionary*, describe probability as "a basic concept which may be taken either as undefinable, expressing in some way a 'degree of belief', or as the limiting frequency in an infinite random series". Practical convenience dictates the second approach and useful, brief accounts of both the principle and its "calculus" or rules of operation are to be found in many books on statistical methods, for example those by Hald (1952), Volk (1958), Brownlee (1960) and Parratt (1961).

Kendall and Buckland go on to say that "both approaches have their difficulties and the most convenient axiomatization of probability theory is a matter of personal taste". Broadly, these difficulties stem from the inductive nature of experimental science. The addition of knowledge by inductive methods starts from an assembly of data. An imaginative and constructive hypothesis in explanation of these data is proposed and a further experiment designed, or programme of observations drawn up, to check that hypothesis. Tests may show that the hypothesis is a reasonable one and it is accepted until a more critical test is devised or more data show that its reasonableness is below an acceptable level. (An inductive "proof" along these lines may, unlike a deductive proof which proceeds from definite premises to a definite answer, include the formal possibility of an exception.) The problem lies in the implications of the word "reasonable" (or its synonyms) as used in this connection of working back from the observed data to the "inverse" probability of the explanatory hypothesis. We cannot deal with this issue but an extensive and critical commentary is given by Hogben (1957).

For our present purposes we identify "probability" (or "chance") with "theoretical relative frequency", i.e. we say that the probability or chance that a particular kind of item will be drawn from a universe by a random sampling procedure is equal to the proportion of such items in the universe being sampled. This can be written

$$P_i = f_i/N \tag{3.1}$$

for the probability of drawning an item of type i whose frequency is f_i in a universe of size N. When the probability is known from the nature of the universe, it is said to be *a priori*, but an estimate of it made from a count of such items in a sample of size n,

$$p_i = f_i/n, \tag{3.2}$$

is the observed relative frequency or *a posteriori* or "posterior" probability. A particular estimate, p_i, will not in general be equal to P_i but will approach it as n increases by reason of larger samples or repeated sampling. Thus, P_i is sometimes called the *expectation* of p_i.

This working definition, being quantitative, permits the operation of certain algebraic rules and has certain limiting characteristics.

If the universe contains only items of the specified kind, their relative frequency is unity and this value for the probability of drawing such an item in sampling from the universe corresponds to complete certainty. If the universe contains no such item, the relative frequency and the chance of drawing one are alike zero. $P_i = 0$, therefore, corresponds to impossibility and $P_i = 1$ to certainty. For any other type of event, i.e. drawing a sample of unit size, the probability lies between 0 and 1. For example, the probability of a lottery marble numbered between 1 and 10,000 being drawn for a prize is 0·1 if the barrel contains 100,000 consecutively numbered marbles. The probability of a fairly tossed coin coming down "heads" is 0·5 since the universe can be regarded as infinite but containing "heads" and "tails" in equal proportions, as represented in Fig. 3.1(a).

The probability of a composite event arising from two steps can be derived from the individual probabilities of those events. Sup-

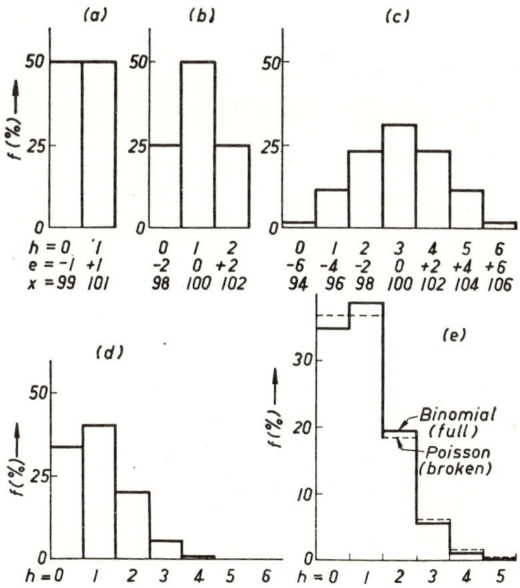

FIG. 3.1. Some Binomial Distributions.

(a), (b) and (c) are distributions of the number, h, of positive errors out of n both positive and negative, of the algebraic sum of positive and negative errors or overall error, e, and of the observed value, x, of a magnitude (100) subject to e. In (a), (b) and (c) $P = 0·5$ and $n = 1, 2$ and 6. In (d) $P = 0·167$ and $n = 6$. In (e) $P = 0·1$ and $n = 10$. The Poisson approximation is given for comparison with the last binomial distribution.

pose the universe contains items of type i and type j, some items having both characteristics. If we denote the drawing of an item having the first characteristic as event I and the second as J, then the probability of the two events occurring simultaneously is

$$P(IJ) = P(I) \,.\, P(J) \tag{3.3}$$

and the probability of either *or* both events occurring is

$$P(I + J) = P(I) + P(J) - P(IJ) \tag{3.4}$$

If the events are mutually exclusive $P(IJ)$ is zero. For example, it is impossible for a single toss of a coin to give both a head and a tail but in this case $P(I)$ (of a head) and $P(J)$ (of a tail) are both 0·5 so the chance of either a head or a tail turning up is $0{\cdot}5 + 0{\cdot}5 = 1$, a certainty. If the event we are considering is a double toss or the drawing of a random sample of size two from the universe of $f_i = f_j = \frac{1}{2}N$, then the outcome could be two heads (from $I + I$), one head (from $I + J$ or from $J + I$), or no heads (from $J + J$). These results can be considered in the form of (a) the values of a discrete variable (number of heads, $h = 2, 1, 0$) or (b) as estimates of the proportion of heads in the sampled universe ($P_i = 0{\cdot}5$; $p_i = 1, 0{\cdot}5, 0$). In either case, we are now considering a new distribution derived from that of the original universe, and in (b) where the distribution is of sample estimates of a universe parameter, we have a *sampling distribution.* The relative frequencies of particular values of h or of p_i in the new distributions, i.e. the probabilities of the possible compound events occurring in two tosses of a coin, can be found from eqns. (3.3) and (3.4):

$$P(h = 2) = P(p_i = 1) = P(I) \,.\, P(I) = 0{\cdot}5 \times 0{\cdot}5 = 0{\cdot}25$$

$$P(h = 0) = P(p_i = 0) = P(J) \,.\, P(J) = 0{\cdot}5 \times 0{\cdot}5 = 0{\cdot}25$$

This same value is also that for any specified order of heads and tails, i.e. $P(IJ)$ and $P(JI)$ are now both 0·25, so

$$P(h = 1) = P(p_i = 0{\cdot}5) = P(IJ) + P(JI) = 0{\cdot}25 + 0{\cdot}25 = 0{\cdot}5.$$

These examples suffice for our present purposes but many more are to be found in the references cited above.

To return to the distribution of random errors. In the absence of systematic error and when rounding error is negligible, the value of an observation can be considered to be of two parts. The first is a true measure of the magnitude of the property and is the only part we are entitled to compare with other magnitudes with a view to finding valid relationships between the properties. The second part is the random error, produced by disturbances in measurement, which places a limitation on the reliability of the whole.

Let x_i be an observed value, μ the true value and e_i the error in x_i:

$$x_i = \mu + e_i \quad \text{or} \quad x_i - \mu = e_i. \tag{3.5}$$

Now suppose $\mu = 100$ and e_i is the result of the operation of sources of disturbance which produce elementary errors of either $+1$ or -1 with equal frequency so that the probability of a positive error from a particular source, which we now denote simply as P, is 0·5. Each source of disturbance, therefore, gives rise to a rectangular distribution whose shape is that of Fig. 3.1 (a). For convenience, we denote by h the number of positive errors out of a total, n, of error sources and therefore of errors attached to each observation.

If the number of sources of disturbance is only one, and if the total number of observations made is 1000 ($= \Sigma f_i = N$), then we could expect to observe 500 of 99 and 500 of 101. If two sources of disturbance operate, the situation is similar to the double tossing of a coin and the results which might be expected in terms of the model are 250 observations of 98 (both errors negative, $h = 0$), 500 of 100 (errors cancel, $h = 1$) and 250 of 102 (both errors positive, $h = 2$). These expectations, together with those for $n = 3$ and for $n = 6$, are given in Table 3.1. It will be noted, in connection with this table, that the expected proportions or probabilities of particular values for observations are also the proportions of the corresponding error values, numbers of positive errors and estimates of the proportion of positive errors in the elementary distributions, i.e. $P(x_i) = P(e_i) = P(h) = P(p \text{ of } P = 0{\cdot}5)$ in the derived or sampling distributions. We denote all of these by P_i and the relative sizes of its values are indicated by the histogram rectangles of Fig. 3.1.

The values in the second last column of Table 3.1 are the terms of a binomial expansion and distributions such as those of x, e, h or p are called *binomial* distributions. In general, for n sources of disturbance and a total of N observations and where P is the probability of a positive and $Q = 1 - P$ the probability of a negative

TABLE 3.1. EXPECTED DISTRIBUTIONS OF 1000 OBSERVATIONS AFFECTED BY ELEMENTARY ERRORS

n	Complexion of errors	h	$p = h/n$	Net error, e_i	Corresponding observation, x_i	Proportion of x_i, P_i	Frequency of x_i, f_i
1	−1	0	0	−1	99	0·5	500
	+1	1	1	+1	101	0·5	500
							1000
2	−1, −1	0	0	−2	98	0·25	250
	−1, +1 +1, −1	1	0·5	0	100	0·5	500
	+1, +1	2	1	+2	102	0·25	250
							1000
3	−1, −1, −1	0	0	−3	97	0·125	125
	−1, −1, +1 −1, +1, −1 +1, −1, −1	1	0·333	−1	99	0·375	375
	+1, +1, −1 +1, −1, +1 −1, +1, +1	2	0·333	+1	101	0·375	375
	+1, +1, +1	3	1	+3	103	0·125	125
							1000

6	All −1	0	0	−6	94	0·016	16
	1(+1), 5(−1) 6 such complexions	1	0·167	−4	96	0·094	94
	2(+1), 4(−1) 15 such complexions	2	0·333	−2	98	0·234	234
	3(+1), 3(−1) 20 such complexions	3	0·5	0	100	0·312	312
	4(+1), 2(−1) 15 such complexions	4	0·667	+2	102	0·234	234
	5(+1), 1(−1) 6 such complexions	5	0·833	+4	104	0·094	94
	All +1	6	1	+6	106	0·016	16
							1000

error from any single source,

$$\Sigma f_i = N = N(Q + P)^n = N\{Q^n + nPQ^{n-1} + n(n-1)\,P^2Q^{n-2}/2! + \ldots + nP^{n-1}Q + P^n\} \quad (3.7\text{a})$$

$$= N\sum^{n}\{n!/i!(n-i)!\}\,P^iQ^{n-i} \quad (3.7\text{b})$$

The terms within braces in (3.7a) all add to unity and are the P_i of Table 3.1, and these, multiplied by N, give the corresponding frequencies, f_i. In the case we are considering $P = Q = 0{\cdot}5$, so the general term given in eqn. (3.7b) becomes

$$P_i = f_i/N = \{n!/i!(n-i)!\}/2^n \quad (3.7\text{c})$$

where the values of the terms in braces are those in Pascal's "triangle" (Table 3.2) and 2^n is a common or "normalizing" factor which reduces their sum to unity.

TABLE 3.2. PASCAL'S TRIANGLE OF BINOMIAL COEFFICIENTS

n	$n!/n!(n-i)!$ for $i = 0, 1, 2 \ldots n$	2^n
0	1	1
1	1 1	2
2	1 2 1	4
3	1 3 3 1	8
4	1 4 6 4 1	16
5	1 5 10 10 5 1	32
6	1 6 15 20 15 6 1	64
7	1 7 21 35 35 21 7 1	128
8	1 8 28 56 70 56 28 8 1	256
9	1 9 36 84 126 126 84 36 9 1	512
10	1 10 45 120 210 252 210 120 45 10 1	1024
11	1 11 55 165 330 462 462 330 165 55 11 1	2048
12	1 12 66 220 495 792 924 792 495 220 66 12 1	4096
Gaussian	2 15 65 211 484 798 946 798 484 211 65 15 2	4096
Experiment	0 7 60 198 430 731 948 847 536 257 71 11 0	4096

It can be shown that the mean and standard deviation of a binomial distribution are

$$\mu = nP \quad \text{and} \quad \sigma = \sqrt{(nPQ)} \quad (3.8\text{a, b})$$

formulae which can be easily verified by calculation from the values in Table 3.1. For example, for $n = 2$ and $P = 0{\cdot}5$, the mean of the variable h is 1 and the standard deviation (from eqn. (2.17a)) is

$$\sqrt{[(250 \times 1^2 + 500 \times 0^2 + 250 \times 1^2)/1000]} = \sqrt{0{\cdot}5}$$

which is the same as $\sqrt{(2 \times 0{\cdot}5 \times 0{\cdot}5)}$ given by eqn. (3.8b). The measures of skewness and of excess of kurtosis are

$$\gamma_1 = (Q - P)/\sigma \quad \text{and} \quad \gamma_2 = (1 - 6PQ)/\sigma^2 \qquad (3.9\text{a, b})$$

Note that if the sample proportion, p_i, is considered to be the variable, the mean is P (which, of course, it is expected to be) and the standard deviation is $\sqrt{(PQ/n)}$.

It can be seen from Figs. 3.1(a), (b) and (c) and also from Table 3.2 that if n were to be increased (and the size of the elementary error correspondingly reduced) a smooth curve would eventually result by reason of the histogram rectangles increasing in number and narrowing in width relative to the range of the variable as plotted. By making these assumptions and by applying Stirling's approximation for the factorials of large numbers, $n! = n^n \exp(-n)\sqrt{(2\pi n)}$, to the general term (eqn. 3.7c) the following equation is obtained

$$f_i = (N/\sqrt{(2\pi nPQ)}) \exp[-(h - nP)^2/2nPQ] \qquad (3.10)$$

The derivation of eqn. (3.10), a form of the theoretical normal or Gaussian distribution, can be found in many books (see Brownlee (1960) and Parratt (1961) for example). It is valid for any value of P provided n is sufficiently large and the size of the elementary error sufficiently small but, as an approximation when n is lower, it is the better the closer are P and Q to $0{\cdot}5$. That it is fair even at quite low values of n can be seen by the results of its application in the case of $n = 12$, $N = 4096$ and $P = Q = 0{\cdot}5$ which are given in the second last line of Table 3.2 for comparison with the true binomial frequencies.

If we substitute for nP and nPQ in terms of eqns. (3.8a) and (3.8b) we obtain the relationship (eqn. 3.10) in a form useful for any appropriately distributed variable. For example, for the variable x or for the error in x, e, we have (now dropping the subscripts)

$$f = \mathrm{d}F/\mathrm{d}x = [N/\sigma\sqrt{(2\pi)}] \exp[-(x - \mu)^2/2\sigma^2] \qquad (3.11)$$

for the frequency density, and

$$F = \int_{\alpha}^{x} f \, . \, dx = [N/\sigma\sqrt{(2\pi)}] \int_{\alpha}^{x} \exp\,[-(x-\mu)^2/2\sigma^2]\, dx \qquad (3.12)$$

for the cumulative frequency or area under the frequency density curve from a chosen point α on the scale up to a particular value of x.

These equations can be simplified, first by taking N as unity so that f and F become relative frequencies or probabilities, and secondly by coding the variable to

$$u = (x - \mu)/\sigma \qquad (3.13)$$

This transformed variable is called the *standardized deviate* and its introduction simply involves shifting the mean from μ_x to 0 and expressing the values of the variable in standard deviation units instead of the original units. Equations (3.11) and (3.12) now become

$$f = [1/\sqrt{(2\pi)}] \exp\,(-\tfrac{1}{2}u^2) \qquad (3.14)$$

and

$$F = [1/\sqrt{(2\pi)}] \int_{\alpha}^{x} \exp\,(-\tfrac{1}{2}u^2)\, du \qquad (3.15)$$

which are called the *standard* normal or Gaussian frequency and distribution functions. The standardizing operation reduces all the possible distributions to which eqns. (3.11) and (3.12) might apply to one single form having a mean of zero and a standard deviation of unity. It is analogous to the single set of curves obtained by the application of the physico-chemical principle of "corresponding states" to the many individual sets of pressure–volume isotherms for non-ideal gases.

The shapes of the curves (3.14) and (3.15) are illustrated in Fig. 3.2 and a selection of values of f and F (the latter between the limits of 0 and u) are given in Table I of the Appendix. The choice of the limits of integration in eqns. (3.12) and (3.15) is a matter of convenience. In Table I the lower bound, α, is taken as 0 and the areas given in that table correspond to the shaded region in Fig. 3.2(a). The cumulative frequencies plotted in Fig. 3.2(b), on the other hand, are from $-\infty$ up to particular values of u and are the areas in Table I plus 0·5 since this is the area under the curve

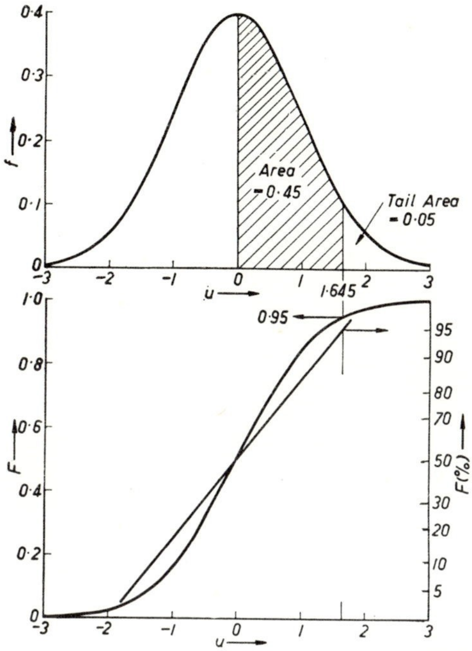

FIG. 3.2. The Gaussian Frequency and Distribution Functions.

The upper graph represents the standardized Gaussian distribution (the "normal curve of errors"), the area under which is unity. The lower curves are those of cumulative frequency, on arithmetic and "probability paper" scales, below the u scale value.

of (3.14) from $-\infty$ to the origin. The application of eqn. (3.14) is direct, for example $f = 1/\sqrt{(2\pi)} = 0{\cdot}3989$ when $u = 0$ and $0{\cdot}3989 \exp(-\frac{1}{2}) = 0{\cdot}2420$ when $u = 1$ and so on, but the integral of eqn. (3.15) is evaluated from the origin to u by series,

$$F = 0{\cdot}3989 \int_0^u \exp(-\tfrac{1}{2}u^2)\,du$$

$$= 0{\cdot}3989(u - u^3/6 + u^5/40 - u^7/336 \ldots) \qquad (3.16)$$

For $u = 1$, this gives

$$F = 0{\cdot}3989(1 - 0{\cdot}1667 + 0{\cdot}0250 - 0{\cdot}0030) = 0{\cdot}3412$$

which is close enough to the correct tabulated value, 0·3413. More approximate values of F as a percentage can be obtained from plots on probability paper which is designed to rectify eqns. (3.12) and (3.15). Tables of the ordinates and integrals of the Gaussian distribution are seldom extended to negative values of u since the function is symmetrical.

The Gaussian distribution fits random error distributions well in general, although they must obviously depart from it in the extreme tails because the theoretical curve is unbounded while the real distributions are limited by possible values of the variable. From our definition of probability, that of finding an error in a random observation within or without particular limits of u is equal to the relative frequency of such values which, in turn, is equal to the area under the relevant section of the curve of eqn. (3.14). Thus, from Appendix Table I, the area under the curve from the origin to $u = +1{\cdot}645$ is 0·4500, so the probability, P, is 0·05 (a "one-in-twenty" chance) of finding a single random error outside $+1{\cdot}645$ *or* $-1{\cdot}645$ standard deviation units from the mean (i.e. in either tail of the distribution), $P = 0{\cdot}1$ or there is one chance in ten of finding an error beyond $\pm 1{\cdot}645$ units (i.e. in both tails), and $P = 0{\cdot}9$ of finding one within these limits. These values are illustrated in Fig. 3.2. Correspondingly, $P = 0{\cdot}95$ of finding a randomly drawn value between $\pm 1{\cdot}96$ units and $P = 0{\cdot}68$ for one between ± 1 unit (twice the value calculated from eqn. (3.16)). This last is the justification for the graphical method of finding the standard deviation illustrated in Fig. 2.4.

3.3. Probability and Physico-chemical Processes

The plausible binomial model, which leads to the Gaussian function in the case of the random error distribution, can be applied to other processes which depend on the element of chance. Simple examples are the results of the tossing of coins and the rolling of dice. If a number of samples of $n = 6$ coin tosses be drawn from an infinite universe having equal proportions of heads and tails, the distribution of the variable h (= number of heads out of six) should resemble that previously derived for errors and illustrated in Fig. 3.1(c). In the case of fair dice, the proportion of times one particular face should turn up in an infinite number of rolls is $1/6 = 0{\cdot}1667$. If the "six" is the face of interest, $P = 0{\cdot}1667$ and the chance of another face coming up in a single roll is $Q = 0{\cdot}8333$.

If a large number of samples of $n = 6$ rolls are drawn and characterized by the number of "six"s per sample, the variable can take the values $h = 0, 1, 2, \ldots, 6$ and its distribution should have as mean, $\mu = nP = 1$, and as standard deviation, $\sigma = \sqrt{(nPQ)} = 0{\cdot}913$. The expected form of the distribution is illustrated in Fig. 3.1 (d). Experiments have been conducted to verify the applicability of the binomial model (the results of one such quoted by Kendall and Stuart (1958), the numbers of "four"s, "five"s and "six"s in $n = 12$ rolls, are given in the last line of Table 3.2) and many real distributions (accident rates, number of defective articles in a sample) follow the same pattern.

The comparison of Figs. 3.1 (c) and 3.1 (d) reveals the effect of a reduction in P for given n on the form of the binomial distribution. If P is very low and n large, i.e. if items of interest are infrequent in the universe ("rare events") but samples are large enough to contain generally a few of them, the binomial frequency function (3.7a) can be approximated by the *Poisson* form

$$\Sigma f_i = N = N(Q+P)^n$$
$$= N[e^{-\mu} + \mu e^{-\mu} + \mu^2 e^{-\mu}/2! + \mu^3 e^{-\mu}/3! \ldots] \quad (3.17)$$

in which N represents a number of samples of size n and the relative frequencies of samples containing $h_i = 0, 1, 2$ etc. items of interest are given by the successive terms within the braces. The actual frequencies of such samples are given by these terms multiplied by N. Equation (3.17) shows that when the mean number of items of interest per sample, μ, is less than 1 the distribution is J-shaped, when $\mu = 1$ the first two terms are the same, and when μ is greater than 1 there is a mode at its value. The mean of the distribution is the same as that of the binomial from which it derives (eqn. 3.8a), but since Q is close to unity the standard deviation given by eqn. (3.8b) becomes

$$\sigma = \sqrt{(nP)} = \sqrt{\mu} \quad (3.18)$$

The Poisson is a skewed and leptokurtic distribution having

$$\gamma_1 = 1/\sqrt{\mu} \quad \text{and} \quad \gamma_2 = 1/\mu \quad (3.19\text{a, b})$$

These are the same as the binomial measures given eqn. (3.18) and $Q \to 1$.

In practice, because it is more easily handled, eqn. (3.17) is used in place of eqn. (3.7a) even when P is as high as 0·1 and n as low as 10. As an example of the *fitting* of a binomial distribution, and of a Poisson as an approximation to it under these limiting conditions, let us consider $N = 100$ samples (say, one per day for twenty weeks' production) each containing $n = 10$ articles which have been classified according to the number, h, of defective articles. The data are given in Table 3.3, together with the calculations, and the computed frequencies are shown for comparison in Fig. 3.1(e). It should be noted that in a case like this the true proportion of defectives, P, is unknown but an estimate can be made from the data themselves. Since 100 defectives were found out of the 1000 articles examined, we find $p = 0{\cdot}1$, so $q = 0{\cdot}9$, and the mean number of defectives per sample, $\bar{h} = np = 1$. These quantities, p, q and $\bar{h}$, are used as estimates of P, Q, and μ when substituting in eqns. (3.7a) and (3.17) as shown in the table. When n is high the direct calculation of binomial frequencies is tedious but the problem is simplified by cumulative frequency tables (see Owen (1962) for example), special probability paper (Mosteller and Tukey (1949)) and by formulae which relate successive terms,

$$p_{i+1} = [(n - i)/(i + 1)]\,(P/Q)\,p_i \qquad (3.20)$$

A similar relationship for the Poisson is

$$p_{i+1} = [\mu/(i + 1)]\,p_i \qquad (3.21)$$

As an example of their application; the first ($i = 0$) term for the binomial in Table 3.3 is 0·3487, the second ($i = 1$) is 0·3487(10/1) (0·1/0·9) = 0·3874, the third ($i = 2$) is 0·3874(9/2)(0·1/0·9) = 0·1937 and so on.

It is not difficult to see that distributions of the binomial or derived forms might arise in a number of physical situations in which the element of chance plays a part. Typical examples are Brownian movement and dependent phenomena in which increments in path length are due to molecular collisions occuring in a purely random manner. The net result of these is an irregular course termed a random flight in which the distance Δx passed over in a time interval Δt at a time t bears no relationship either in magnitude or direction to the distance $\Delta x'$ covered in an identical interval at some other time t′. The actual distance covered in any finite time, say $n\Delta$t, is

h_i	f_i	$f_i h_i$	p_i (binomial)		f_i (calc.)	p_i (Poisson)		f_i (calc.)
0	38	0	$(0{\cdot}9)^{10}$	$= 0{\cdot}3487$	34·9	e^{-1}	$= 0{\cdot}3679$	36·8
1	36	36	$10.(0{\cdot}1)(0{\cdot}9)^9$	$= 0{\cdot}3874$	38·7	$1\,e^{-1}$	$= 0{\cdot}3679$	36·8
2	17	34	$10.9(0{\cdot}1)^2(0{\cdot}9)^8/2!$	$= 0{\cdot}1937$	19·4	$1^2e^{-1}/2!$	$= 0{\cdot}1840$	18·4
3	7	21	$10.9.8(0{\cdot}1)^3(0{\cdot}9)^7/3!$	$= 0{\cdot}0574$	5·7	$1^3e^{-1}/3!$	$= 0{\cdot}0613$	6·1
4	1	4	$10.9.8.7(0{\cdot}1)^4(0{\cdot}9)^6/4!$	$= 0{\cdot}0112$	1·1	$1^4e^{-1}/4!$	$= 0{\cdot}0153$	1·5
5	1	5	$10.9.8.7.6(0{\cdot}1)^5(0{\cdot}9)^5/5!$	$= 0{\cdot}0015$	0·2	$1^5e^{-1}/5!$	$= 0{\cdot}0031$	0·3
6–10	0	0	(by difference)	0·0001	0·0		0·0005	0·1
	100	100		1·0000	100·0		1·0000	100·0

$\bar{h} = \Sigma f_i h_i / \Sigma f_i = 1$

TABLE 3.4. CHARACTERISTICS OF SAMPLING DISTRIBUTIONS

Distribution (sample size n)	Mean	Standard error	Conditions
(a) Means, $\bar{x}$	$\mu_{\bar{x}} = \mu$	$\sigma_{\bar{x}} = \sigma/\sqrt{n}$	(3.22a, b)
(b) Medians, m	$\mu_m = M$	$\sigma_m = c\sigma_{\bar{x}}$	(3.23a, b)
(c) Standard deviations, s	$\mu_s = \sigma$	$\sigma_s = \sigma/\sqrt{(2n)}$	$n > 30$ (3.24a, b)
(d) Variances, s^2	$\mu_{s2} = \sigma^2$	$\sigma_{s2} = \sigma^2\sqrt{(2/n)}$	$n > 100$ (3.25a, b)
(e) Measure of skewness, g_1	$\mu_{g_1} = \gamma_1$	$\sigma_{g_1} \doteqdot \sqrt{(6/n)}$	$n > 100$ (3.26a, b)
(f) Measure of kurtosis, g_2	$\mu_{g_2} = \gamma_2$	$\sigma_{g_2} \doteqdot \sqrt{(24/n)} = 2\sigma_{g_1}$	Large n (3.27a, b)
Differences, d, between means of samples from universes of mean μ_1 and μ_2, so $\mu_1 - \mu_2 = \delta$:			
(g) $n_1 = n_2 = n,\ \sigma_1 = \sigma_2 = \sigma$	$\mu_d = \delta$	$\sigma_d = \sigma\sqrt{(2/n)}$	(3.28a, b)
(h) $n_1 \neq n_2,\ \sigma_1 = \sigma_2 = \sigma$	$\mu_d = \delta$	$\sigma_d = \sigma\sqrt{(1/n_1 + 1/n_2)}$	(3.29a, b)
(i) $n_1 \neq n_2,\ \sigma_1 \neq \sigma_2$	$\mu_d = \delta$	$\sigma_d = \sqrt{(\sigma_1^2/n_1 + \sigma_2^2/n_2)}$	(3.30a, b)
(j) Proportions (binomial)	$\mu_p = P$	$\sigma_p = \sqrt{(PQ/n)}$	Large n (3.31a, b)
(k) Proportions (Poisson)	$\mu_p = P$	$\sigma_p = \sqrt{(P/n)}$	Large n (3.32a, b)

the vector sum of $\Delta x + \Delta x' + \ldots$ (other terms of the same nature), and it can be shown that under these conditions the relationship between distance and time is of the form $x^2 \propto t$. Processes such as diffusion, conduction and viscosity which depend on random molecular motion are all characterized by a variable which has the dimensions of (length)^2/time. The distribution of terminations of one-directional random flights and the related distribution of concentration (i.e. concentration gradient) in diffusion from a given starting point are of the Gaussian form. More complex modifications of this, like Maxwell's molecular speed distribution, come about when account is taken of molecular movement in two or three dimensions.

Other processes which involve spatial movement may be of a chance dependent or stochastic nature. A typical example is the building up of polymer chains in which the arrival of monomer molecules at the point of reaction occurs in a random manner and the resulting distribution of polymer chain lengths is correspondingly random but not necessarily Gaussian. Radioactive decay, in which the individual atomic decompositions are purely random and relatively rare, can be of the Poisson form. An example is the distribution of Table 2.4(c). This and other cases are discussed by Bond (1935) and Parratt (1961). The operation of models to simulate random processes has been made possible by modern computing techniques. Fleundy and Smith (1962) provide a recent account.

3.4. Sampling Distributions

The prime importance of the Gaussian equation lies in the fact that it describes fairly accurately the distributions of values of the more important estimators obtained from random samples and, furthermore, these distributions have means and standard deviations (*standard errors*) which are related to those of the universe from which the samples have come. Suppose samples of size n are drawn at random from a universe of mean, μ, and standard deviation, σ; then Table 3.4 and Fig. 3.3 present some of the characteristics of some of the more useful sampling distributions.

Strictly speaking, the distributions of Table 3.4 are only normal (i.e. well fitted by the Gaussian function, (3.11)) and characterized by the means and standard errors given there if the sampled universe is itself normal or, in the case of (j), if P is close to 0·5. However, the Gaussian approximation is valid for the distribution of sample means under almost any conditions and is fair in the other cases provided the sample size is large enough. Proofs of the relationships in the table are to be found in most statistical texts but for some of the distributions a justification is given later. Here,

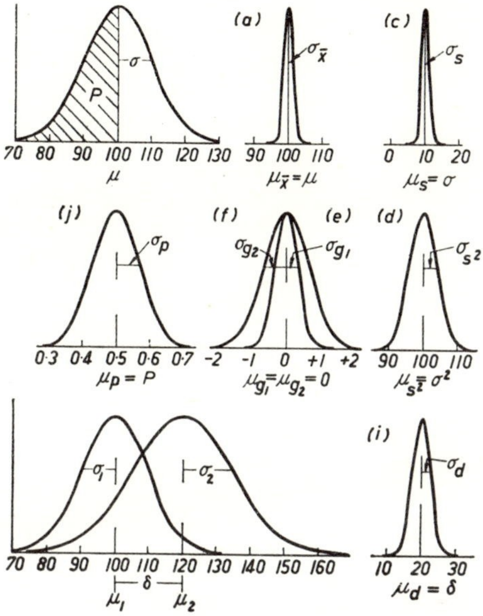

FIG. 3.3. Some Theoretical Sampling Distributions.

The upper and lower graphs on the left hand side represent one and a pair of Gaussian universes. The lettered curves are those of sampling distributions selected from Table 3.4.

we offer a simple demonstration. This consists of the results for samples of size five drawn at random from a universe consisting of numbers from 1 to 100. The universe can be conceived as a lottery barrel containing 100 numbered marbles from which lots of five are drawn and replaced but in fact the numbers come from a special random set (Fisher and Yates (1948)). The results of the experiment are given in Table 3.5 and illustrated in Fig. 3.4.

The discrepancies between the observed average values of the quantities in Table 3.5 and the expected values are due to the operation of random error in the small collection of samples investigated, to grouping in the calculations, and to the fact that

TABLE 3.5. RANDOM SAMPLES FROM A RECTANGULAR UNIVERSE, $\mu = 50{\cdot}5$, $\sigma = 28{\cdot}9$

No. (1)	Sample (2), *m* italicized	$\bar{x}_5$ (3)	s^2 (4)	s (5)	$s_{\bar{x}}$ (6)	$\bar{x} - \mu$ (7)	u (8)	t (9)	F (11)	$\bar{x}_{10}$ (12)	d (13)	p (14)	$\bar{x}_{25}$ (15)
1	17, 53, 77, *58*, 71	55·2	553	23·5	10·3	4·7	0·36	0·46					
2	90, *26*, 59, 21, 19	43·0	951	30·8	13·9	−7·5	−0·58	−0·54	0·58	49·1	12·2	0·4	
3	41, 23, *52*, 55, 99	54·0	790	28·1	12·6	3·5	0·27	0·28					
4	*60*, 20, 50, 81, 69	56·0	535	23·1	10·3	5·5	0·43	0·53	1·48	55·0	−2·0	0·4	
5	91, 25, *38*, 5, 90	49·8	1519	38·9	17·4	−0·7	−0·05	−0·04					51·6
6	34, *50*, 57, 74, 37	50·4	262	16·2	7·2	−0·1	−0·01	−0·01	5·80	50·1	−0·6	0·6	
7	85, 22, 4, *39*, 43	38·6	911	30·2	13·6	−11·9	−0·92	−0·86					
8	9, 79, 13, 77, *48*	45·2	1127	33·6	15·0	−5·3	−0·41	−0·35	0·81	41·9	−6·6	0·7	
9	88, *75*, 80, 18, 14	55·0	1291	35·9	16·1	4·5	0·35	0·28					
10	*90*, 96, 23, 70, 100	75·8	1507	38·8	17·4	25·3	1·96	1·45	0·85	65·4	−20·8	0·3	53·0
11	71, 41, *61*, 50, 72	59·0	180	13·4	6·0	8·5	0·66	1·42					
12	23, 52, *23*, 33, 12	28·6	226	15·0	6·7	−21·9	−1·70	−3·27	0·80	43·8	31·4	0·6	
13	31, 4, *49*, 69, 96	49·8	1616	40·3	18·0	−0·7	−0·05	−0·04					
14	31, 99, 73, *68*, 68	67·8	587	24·2	10·8	17·3	1·34	1·59	3·64	58·8	−18·0	0·4	
15	94, 58, 28, *41*, 36	51·4	688	26·4	11·8	0·9	0·07	0·08					51·3
16	98, 80, 33, 100, *91*	80·4	763	27·6	12·3	29·9	2·31	2·43	0·90	65·9	−29·0	0·4	
17	73, 81, 53, 94, *79*	76·0	224	15·0	6·7	25·5	1·97	3·81					
18	*73*, 82, 97, 22, 21	59·0	1243	35·3	15·8	8·5	0·66	0·54	0·18	76·5	17·0	0·2	
19	22, 95, 75, 42, *49*	56·6	820	28·6	12·8	6·1	0·47	0·48					
20	*39*, 100, 3, 6, 90	47·6	2084	45·5	20·4	−2·9	−0·22	−0·14	0·39	52·1	9·0	0·6	63·9
21	12, *41*, 94, 96, 26	53·8	1520	39·0	17·4	3·3	0·26	0·19					
22	96, 93, 2, 18, *39*	49·6	1853	43·1	19·3	−0·9	−0·07	−0·05	0·82	51·7	4·4	0·6	
23	10, *47*, 48, 45, 88	47·6	763	17·4	7·8	−2·9	−0·22	−0·27					
24	*35*, 81, 33, 3, 76	45·6	1084	32·9	14·7	−4·9	−0·37	−0·33	0·70	46·6	2·0	0·7	
25	45, *37*, 59, 3, 9	30·6	453	21·3	9·1	−19·9	−1·54	−2·19					45·4
26	9, *77*, 93, 19, 82	56·0	1516	39·0	17·4	5·5	0·43	0·32	0·29	43·3	−25·4	0·6	

27	33, 62, *46*, 86, 28	51·0	559	23·6	10·6	0·5	0·04	0·47					
28	5, 3, 27, *24*, 83	28·4	1049	32·4	14·5	−22·1	−1·71	−1·52	0·53	39·7	22·6	0·7	
29	*39*, 32, 82, 22, 49	45·8	529	23·0	10·4	−4·7	−0·36	−0·45					
30	*55*, 85, 78, 38, 36	58·4	505	22·5	10·1	7·9	0·61	0·78	1·05	52·1	−12·6	0·6	47·9
31	*44*, 95, 27, 36, 99	60·2	1166	34·1	15·3	9·7	0·75	0·64					
32	*7*, 2, 18, 36, 7	14·0	186	13·6	6·1	−36·5	−2·83	−6·98	6·27	37·1	46·2	0·8	
33	13, *41*, 43, 89, 20	41·2	907	30·1	13·5	−9·3	−0·72	−0·69					
34	24, *30*, 12, 48, 60	34·8	367	19·2	8·6	−15·7	−1·22	−1·83	2·47	38·0	6·4	0·8	
35	90, *35*, 57, 29, 12	44·6	903	30·1	13·5	−5·9	−0·46	−0·44					39·0
36	*74*, 94, 80, 4, 4	51·2	1909	43·7	19·6	0·7	0·54	0·04	0·47	47·9	−6·6	0·5	
37	8, *31*, 54, 46, 31	34·0	309	18·8	8·4	−16·5	−1·28	−1·96					
38	72, 89, 44, 5, *60*	54·0	1021	32·0	14·3	3·5	0·27	0·24	0·30	44·0	−20·0	0·6	
39	2, 48, 7, 70, *37*	32·8	812	28·5	12·7	−17·7	−1·37	−1·39					
40	94, *37*, 30, 69, 32	52·4	790	28·1	12·6	1·9	0·15	0·15	1·03	42·6	−19·6	0·7	44·9
41	2, 96, *74*, 30, 83	57·0	1560	36·2	16·2	6·5	0·50	0·40					
42	25, 99, *32*, 70, 23	49·8	1120	33·4	14·9	−0·7	−0·05	−0·05	1·39	53·4	7·2	0·5	
43	97, *17*, 14, 49, 17	38·8	1262	35·5	15·9	−11·7	−0·91	−0·74					
44	18, 99, 10, 72, *34*	46·6	1677	41·0	18·3	−3·9	−0·30	−0·21	0·75	42·7	7·8	0·7	
45	82, *62*, 54, 65, 60	64·6	443	21·0	9·4	14·1	1·09	1·50					51·4
46	*45*, 7, 31, 66, 49	39·6	488	22·1	9·9	−10·9	−0·84	−1·10	0·91	52·1	25·0	0·4	
47	53, 94, 13, 38, *47*	49·0	866	29·4	13·2	−1·5	−0·12	−0·11					
48	35, *80*, 39, 94, 88	67·2	812	28·5	12·8	16·7	1·29	1·31	1·07	58·1	18·2	0·5	
49	16, 4, *61*, 67, 87	47·0	1251	35·4	15·8	−3·5	−0·27	−0·22					
50	90, *89*, 100, 76, 33	77·6	694	26·4	11·8	27·1	2·18	2·30	1·80	62·3	−30·6	0·3	56·1
Mean found:	*49·7*	50·4	912	29·2			−0·02	−0·10	1·41	50·4	−0·14	0·54	50·4
Expected:	*50·5*	50·5	833	27·1[b]			0·0	0·0	1·00	50·5	0·0[c]	0·50	50·5
Std. error found:	*19·2*	13·1	503	8·8			1·00	1·35		9·6	19·8	0·16	7·0
Expected:	*18·9*[a]	12·9	527	9·1			1·00			9·1	18·3[c]	0·16	5·7

(For notes (a), (b) and (c) see text below.)

the expected values are calculated for a normal sampled universe from the formulae in Table 3.4 and not for a rectangular universe like that from which the samples have actually come. In the case of the distribution of medians, however, the necessary correction has been made (note a). The factor c in eqn. (3.23b) for the median as defined for small samples by eqn. (2.4), where these are drawn from a normal universe, takes the values 1, 1·160, 1·092 and 1·198 for n from 2 to 5 (even numbered samples being better than odd because of the difference between eqns. (2.4a) and (2.4b)), but the medians of samples from rectangular universes are rather less reliable and for $n = 5$ the factor is 1·464. (A table of c for the normal,

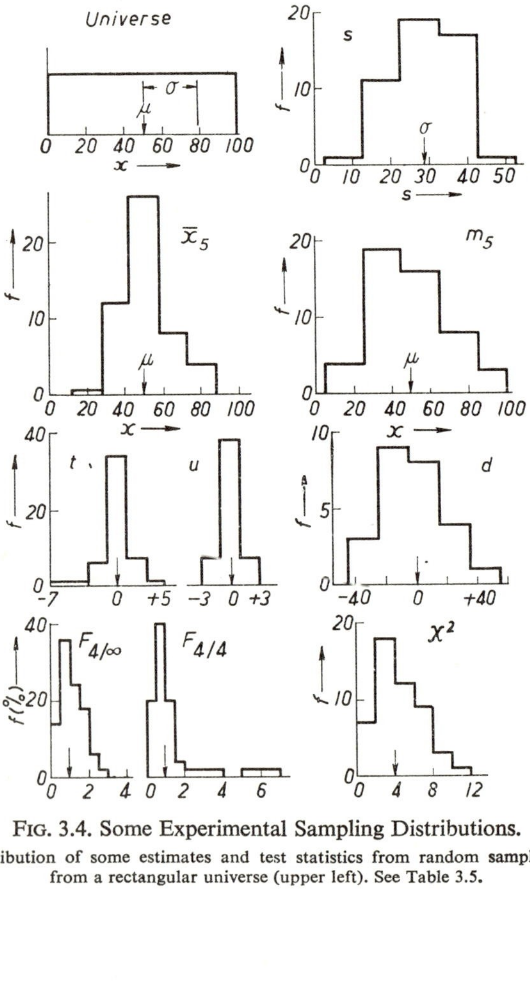

FIG. 3.4. Some Experimental Sampling Distributions.

The distribution of some estimates and test statistics from random samples drawn from a rectangular universe (upper left). See Table 3.5.

rectangular and other distributions is given by Cowden (1957).) In the case of the standard deviation distribution, the expected average is less than σ because of bias (note b, Table 3.5), the correcting factor (the c_1 of eqn. (2.23a) being 1/1·064 for $n = 5$.

On the whole the agreement between the observed and expected means and standard errors is good for this set of random samples and the following conclusions, which are of general application, may be drawn:

(i) The sample estimates do distribute themselves more or less closely, i.e. without obvious bias, round the universe parameter they estimate and their distributions are, in fact, more or less normal.

(ii) Even when, as in this instance, they estimate the same universe value, the median of a sample of given size is less reliable than the mean. The word "reliable", used loosely in Chapter II, can now be given quantitative expression in terms of the standard error. With regard to the comparison of m and $\bar{x}$ we can say (a) the spread of the sampling distribution of m is greater than that of $\bar{x}$ by a factor of $c = 1{\cdot}464$, or (b) $\bar{x}$ is generally closer than m to the universe value (in 40 out of the 50 samples in Table 3.5), or (c) the *efficiency* of m is less than that of $\bar{x}$; here $\sigma_{\bar{x}}^2/\sigma_m^2 = 1/1{\cdot}464^2 = 0{\cdot}47$. Efficiency is measured, as indicated, by the ratio of appropriate variances. It can be *absolute* if the numerator is the lowest possible such variance (as $\sigma_{\bar{x}}^2$ generally is in the case of measures of central tendency), *relative* in a comparison of any two estimators of the same type, or *asymptotic* if the variances are those for n approaching infinity. The asymptotic or limiting efficiency of m for samples from a Gaussian universe is $2/\pi = 0{\cdot}64$.

(iii) The mean of a large sample is more reliable than that of a small sample. It is to be noticed, that, according to the standard error criterion, the advantage of size is a diminishing one since it varies with $\sqrt{n}$. (The mean of a sample of size 25, for instance, is just over twice as reliable as that of a sample of size 5 and not five times better.) The advantage of size is rather greater, however, when the universe is finite (size N) and the sample drawn without replacement. In this case

$$\sigma_{\bar{x}} = \sigma/\sqrt{n}\sqrt{[(N-1)/(N-n)]} \tag{3.33}$$

so that when the universe is completely sampled ($n = N$), the mean is exactly known and its standard error zero. This is, of course, true of any estimator.

(iv) The differences between sample means are distributed about zero as expected.

(v) The mean and standard error of the values of p, in this case estimates of a true value of 0·5 from samples of size ten, are close to the theoretical binomial values although the actual frequencies diverge somewhat from the theoretical.

In addition to the distributions of estimates, certain functions of them are of interest and the general features of their distributions are shown in Table 3.5 and Fig. 3.4. Thus,

(vi) The differences between the sample means and the true mean in true standard error units (the u of eqn. (3.13) with $\bar{x}$ for the variable and $\sigma_{\bar{x}}$ for the standard deviation) are distributed normally with zero mean and unit standard deviation, but the same differences in estimated standard error units, i.e.

$$t = (\bar{x} - \mu)/s_{\bar{x}} \tag{3.34}$$

where $s_{\bar{x}}$ is computed along with $\bar{x}$ from the same sample, are distributed with greater dispersion. For samples of large size $s_{\bar{x}}$ is fairly accurately known and the t distribution approaches that of u as can be seen from Table II of the Appendix.

(vii) The variance ratio, F, for random pairs of samples from the same universe tends to unity as expected (since both variances estimate the same σ^2) but its distribution is very skew. Of the 25 values in Table 3.5, 9 are below 0·8, 9 between 0·8 and 1·2, and 7 between 1·2 and 7.

This is also true of the variance ratio, $F = s^2/\sigma^2$ ($=s^2/833$ here). The values are not given in the table but their distribution is illustrated in Fig. 3.4. It can be seen that the skewness of the F distribution depends on the degrees of freedom, f, appropriate to the variances forming the ratio, that of $F_{4/\infty}$ being much less skewed than that of $F_{4/4}$. The same kind of skewed distribution is obtained for another statistic, "chi-squared", given by

$$\chi^2 = \Sigma(x - \mu)^2/\sigma^2 = \Sigma u^2 \tag{3.35a}$$

or

$$\chi^2 = \Sigma(x - \bar{x})^2/\sigma^2 \tag{3.35b}$$

The quantity defined by eqn. (3.35a) tends to the sample size, n, and that defined by eqn. (3.35b) to the degrees of freedom, $f = n-1$, on the average for large numbers of random samples. This can be seen by substituting for σ^2 according to eqn. (2.16). It can also be seen that the value of χ^2 is the value of $F_{f/\infty}$ multiplied by f. Some quantiles of the χ^2 and F distributions for samples drawn from a Gaussian universe are given in Appendix Tables II and III.

3.5. Tests of Significance and Confidence Intervals

In the last section we have attempted to describe the laws of distribution of estimates between random samples which we can regard as being established, either by algebra or by numerical experiment, from a Gaussian model. With this information we can attempt to answer questions of the kind posed in the introductory paragraphs of this chapter; viz. given an estimate (arithmetic mean, difference between means, standard deviation, proportion etc.) then (a) "Within what limits may the value of the parameter be expected to lie?" (in the sense of "What is the likelihood of limits calculated embracing the parameter?") or (b) "Is it reasonable to suppose that the latter has some particular value?" The first kind of question can be answered by stating *confidence limits* or a *confidence interval* which has a chosen probability of embracing the true value, i.e. a chosen proportion (say 95%) of intervals so calculated will embrace the true value in repeated sampling. The second kind of question is answered by applying a *test of significance* to see if the probability of the difference between the value of an estimator and the hypothetical value of its parameter being due to chance is greater or less than some chosen figure, the *significance level.* This is generally taken as P = 0·05 or 5% and, if the chance of accidental occurrence is less than this, an observed difference is said to be *significant.* In other words, a significant result, indicative of some real difference, is claimed for a test if it is of a kind which occurs less than once in twenty times on the average in ordinary random sampling. An answer to the first kind of question automatically answers the second if we put the same limit on what is to be regarded as the "reasonable" or "likely" operation of chance or to the vagaries of random sampling.

The 5% level, usually chosen as the criterion of significance, is quite arbitrary but, whether it be fixed at this or at 10% or 1%, there is the possibility that an apparently significant result might be a rare accident. There are, however, limits to the range of the variable in which even the rarest of accidental results can lie. To revert to the CO_2 block density data of Chapter II, we found in this instance a mean density of blocks of 1·497 g/cm^3 with a standard deviation of 0·013 based on 81 observations. Applying the ideas of the last section, we can regard this $\bar{x}$ as belonging to a normal sampling distribution whose standard error should be in the vicinity of $s/\sqrt{n} = 0{\cdot}00145$ for samples of this size so long as the same system of causes of variability operates. By further assuming that the Gaussian function fits the sampling distribution passably, we can say that it is unlikely (only about 5% probability) that the true mean estimated by 1·497 g/cm^3 differs from that value by more than about twice 0·00145, say 0·003 g/cm^3, but our certain knowledge of the system tells us that it is quite impossible for μ to be 1 or 2 or even 1·4 or 1·6 g/cm^3.

On the other hand, an apparently non-significant result in a test might be due to accidental error masking a real, though perhaps small and unimportant, difference. In a sense, then, the significance level is a measure of the degree of unimportance of real disturbances which can be safely ignored. Thus, in the industrial application of quality control charts (a kind of running test of significance described below) quite possibly real discrepancies are discounted unless their magnitude is such as to affect the economics of the process and significance levels as low as P = 0·001 are adopted.

In short, the dividing line between the variation of sample values (individuals or estimates as the case may be) due to ordinary non-assignable disturbances on the one hand and that due to secular effects (assignable causes or systematic error) on the other is at best uncertain and any particular significance level is a matter of choice and to some extent dependent on taste and circumstance.†

Before applying a test of significance to an actual problem it is necessary to state this in statistical terms by setting up a null hypothesis (which attributes the observed difference between the estimate and the hypothetical value of its parameter to chance) and alternative hypotheses (which suppose the difference to be due to a real

† It should be noted here that there is a considerable dispute as to the precise logical implications of significance testing and interval estimation. Fisher (1955), writing on statistical methods and scientific induction gives a short account from one point of view and Hogben's (1957) critique may be found illuminating.

effect and of given or unknown sign) and to select the level of significance. Usually in the application of the test, the value of a test statistic (a dimensionless quantity whose formula incorporates the difference being investigated) is found and compared with a critical value depending on the level of significance. Significance is indicated if the value found exceeds the critical value which is the maximum allowable to chance. The commonly used test statistics are those described above: u which can be used when the standard error of the sampling distribution is known, t when only an estimate of it is available, F for comparisons of variability, χ^2 whose distribution is approximated by a function of differences between observed and expected counts, and one or two measures of association which are dealt with in Chapter IV.

The operation of the more generally useful tests of significance and the calculation of confidence limits are illustrated by the following examples. The tests described here are few and conventional. More complete descriptions are to be found in the general references already cited. In addition, there is a very complete treatment with excellent tables by Dixon and Massey (1957) and a briefer description containing an account of the effect of non-ideality of the parent distribution by Davies (1954). In many cases graphical and nomographical methods are available for making quick comparisons. Broadbent (1953) provides some interesting and useful examples.

(1) *Tests involving a Known Standard Error*

Suppose an observer makes six independent measurements of the surface tension, γ, of some water at 25 °C using a tensiometer. His six values average 71·2 dyn/cm with 1·2 dyn/cm standard deviation. He knows from much previous experience that the standard deviation, σ, of routine individual measurements of this magnitude is 0·8 dyn/cm, this covering particular instrumental vagaries and the degree of temperature control he can ordinarily achieve. The literature value for pure water at 25 °C is 71·8 dyn/cm. He might be interested in the following relating to the general level or to the variability of his experimental values.

(a) Could the true value of the surface tension of the water being investigated on this occasion, estimated by $\bar{x} = 71{\cdot}2$ dyn/cm from $n = 6$, be in fact 71·8 dyn/cm as for pure water?

The argument runs as follows:

(i) Null hypothesis, $H_0 : \mu = \mu_0 = 71{\cdot}8$ (i.e. the sample may reasonably be regarded as having been drawn at random from a universe of mean 71·8 so that $\bar{x} = 71{\cdot}2$ estimates this value and differs from it by an amount which can be attributed to chance or accident of random sampling).

Alternatively, $H_a : \mu \neq \mu_0 = 71{\cdot}8$ (i.e. the true μ estimated by $\bar{x}$ differs from 71·8 but, there being no reason to suppose that the error or difference is of particular sign, we apply a *two-sided* test).

(ii) The criterion of reasonable operation of chance is the 5% level of significance (i.e. if the observed difference, $\bar{x} - \mu_0$, is such as can occur more than once in twenty times in random sampling it will be attributed to chance but if less likely it will be attributed to some real effect like contamination of the particular liquid under investigation).

(iii) σ being known, the standard error of the $\bar{x}(n = 6)$ sampling distribution is known ($\sigma/\sqrt{n} = 0{\cdot}8/\sqrt{6} = 0{\cdot}327$) and the statistical question is whether or not the sample mean can be considered to belong to a sampling distribution having this $\sigma_{\bar{x}}$ and a mean of 71·8. The part of the sampling distribution in which $\bar{x}$ should fall (given the 5% significance level) is that within 1·96 standard error units of the mean, i.e.

$\mu_0 \pm 1{\cdot}96\sigma_{\bar{x}} = 71{\cdot}8 \pm 1{\cdot}96(0{\cdot}327) = 71{\cdot}16$ to $72{\cdot}44$ (see Fig. 3.5 (a)), the factor 1·96 coming from Appendix Table I. This is the *acceptance region* and the *critical* or *rejection region* is that outside these limits.

(iv) $\bar{x}$ falls within the acceptance region so the null hypothesis is accepted. There is no clear evidence that the true surface tension of the liquid under investigation is not that of pure water.

An alternative procedure to (iii) and (iv) is just to calculate how far $\bar{x}$ lies, in standard error units, from the hypothetical mean of its sampling distribution, i.e. to find the test statistic (without regard to sign in view of H_a)

$$u = |\bar{x} - \mu_0|/\sigma_{\bar{x}} = |71{\cdot}2 - 71{\cdot}8|/0{\cdot}327 = 1{\cdot}84,$$

and to compare this with the critical value, $u_{.05,\text{2-sided}} = 1{\cdot}96$, which is the value so great in absolute magnitude that it occurs in only 5% of random samples. The observed value, however, is lower and occurs in about 6·6% of cases of random sampling so the null hypothesis is accepted.

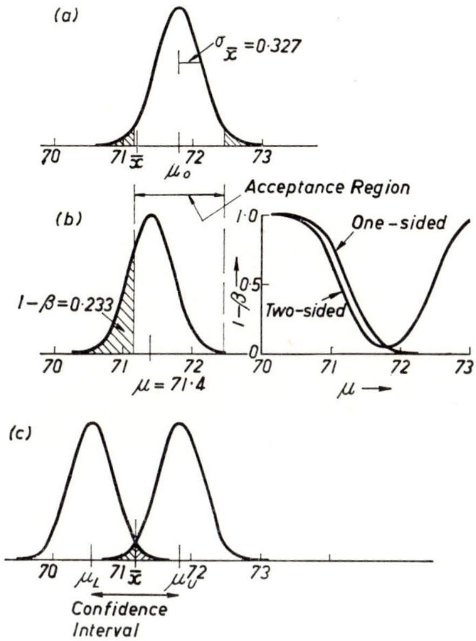

FIG. 3.5. Tests of Significance and the Confidence Interval.

Another way of arriving at the same conclusion is to find the 5% confidence interval for the true mean estimated by $\bar{x}$. This is supposed to be the centre of a sampling distribution whose standard error is 0·327 and of which $\bar{x}$ is a member. $\bar{x}$ is unlikely to lie more than, say, 1·96 $\sigma_{\bar{x}}$ units above μ and the probability, P, of its doing so is only $2\frac{1}{2}$%. In other words μ should not be lower than $\mu_L = 71{\cdot}2 - 1{\cdot}6 \times 0{\cdot}327 = 70{\cdot}56$. By a similar argument μ should not be higher than $\mu_U = 71{\cdot}84$, so that μ should be within the limits 70·56 to 71·84 and there is 95% probability that intervals so calculated embrace the true mean. Figure 3.5(c) illustrates the proposition and to put it in general form

$$100(1 - \text{P})\% \text{ Confidence limits of parameter} = \text{Estimate} \pm u_{\text{P}} \,.\, \sigma_{\text{estimate}} \qquad (3.36\text{a})$$

Here,

$$95\% \text{ Confidence limits of } \mu = \bar{x} \pm 1{\cdot}96\,\sigma_{\bar{x}} = 71{\cdot}2 \pm 1{\cdot}96 \times 0{\cdot}327$$

Since $\mu_0 = 71{\cdot}8$ does come within these limits we cannot reject the hypothesis that this is in fact the value estimated by $\bar{x}$.

(b) The observer has decided beforehand that his water sample might be contaminated and its surface tension might be less than that for pure water. He tests this proposition.

Following the same procedure as before, the argument runs:

(i) $H_0 : \mu = \mu_0 = 71{\cdot}8$; $H_a : \mu < \mu_0$, a one-sided alternative in this case.

(ii) P = 0·05 or 5% as before.

(iii) If μ is less than μ_0 then $\bar{x}$ which estimates μ should also be lower and should fall into the lower tail of the hypothetical sampling distribution of which μ_0 is the mean. This might also happen by accident in random sampling, but less than 5% of the genuine members of a Gaussian sampling distribution lie more than 1·64 standard error units below the mean, i.e. below $71{\cdot}8 - 1{\cdot}64 \times 0{\cdot}327 = 71{\cdot}26$ here. (The acceptance region is therefore that part of the scale above 71·26 and the rejection region that below.) Looking at the distribution in its standardized form (and taking the sign of the difference into account in this case)

$$u = (\bar{x} - \mu_0)/\sigma_{\bar{x}} = (71{\cdot}2 - 71{\cdot}8)/0{\cdot}327 = -1{\cdot}84$$

i.e. $\bar{x}$ lies 1·84 standard error units below the hypothetical mean compared with the critical value, $u_{.05, \text{1-sided}} = -1{\cdot}64$.

(iv) The observer's value might belong to the hypothetical sampling distribution but the chance of its doing so is less than 5% (actually about 3·3%) and it falls in the rejection region. Consequently, he prefers to believe the alternative and concludes that the true surface tension of the water he is investigating is less than that of pure water.

It should be noted that, although the answer to question (b) differs from the answer to question (a), the questions are themselves different and only one of them would be the subject of enquiry in relation to a given set of data. The lowering in magnitude of the critical value in (b) might be thought of as a bonus due to the extra information available compared with (a).

In making these tests the risk of rejecting H_0 when true ("error of the first kind" or Type I or α error) is controlled by the significance level and, when this is 5%, true null hypotheses will be wrongly rejected in 5% of repeated tests. Conversely we may accept H_0 when it is false ("error of the second kind" or Type II or β error) and the probability of so doing depends mainly on the degree of falsity and on sample size. This probability is designated β and the significance level or probability of a Type I error (our P) is generally designated α for comparison. The probability, $1 - \beta$, of correctly rejecting H_0 and of recognizing a truly significant difference is called the *power* of a test.

In the present example, the acceptance region for the two-sided test (question (a)) is 71·16 to 72·44 (given $\mu = \mu_0 = 71{\cdot}8$) but if μ were really, say, 71·4 then the true sampling distribution would have this as mean and 0·327 as standard error and the proportion of it within the acceptance region is (see Fig. 3.5(b))

$$P(\bar{x} = 71{\cdot}16 \text{ to } 72{\cdot}44) = P(u = -0{\cdot}734 \text{ to } +3{\cdot}18)$$
$$= 0{\cdot}268 + 0{\cdot}499 = 0{\cdot}767 = \beta$$

(the values of u being the standardized deviations from 71·4 of the limits of the acceptance region and β being the area derived from Appendix Table I). Thus, the power of the test, $1 - \beta$, is 0·233, i.e. under these conditions we can expect to detect a real *difference* of 0·4 dyn/cm in surface tension in only 23% of repeated tests. In the case of the one-sided test, the acceptance region is 71·26 to $+\infty$, and if means were really distributed about 71·4 with standard error 0·327 the probability of finding one within the region

$$P(\bar{x} = 71{\cdot}26 \text{ to } +\infty) = P(u = -0{\cdot}422 \text{ to } +\infty) = 0{\cdot}664$$

and the power is 0·336 so we can expect to detect a real lowering of 0·4 dyn/cm in 34% of repeated tests. Some other values are:

True mean, μ	70·3	70·8	71·3	71·6	71·8	72·0
$1 - \beta$, 2-sided	0·9957	0·864	0·334	0·087	0·050	0·087
$1 - \beta$, 1-sided	0·9984	0·921	0·451	0·149	0·050	0·012

The power curves are shown in Fig. 3.5(b). It can be seen that the sample size must be increased (to reduce $\sigma_{\bar{x}}$) or the significance level lowered (reducing the acceptance region, but at the expense of incurring more Type I errors) if there is to be a reasonable chance of detecting small errors in H_0. Also, the power of a one-sided test to detect a difference not of interest (a true surface tension higher than 71·8 dyn/cm in this example) is very low. Power curves of this type are most easily drawn on probability paper (see Davies (1957)).

(c) Could the true variability of the current observations, measured by a standard deviation, $s = 1{\cdot}2$ dyn/cm, be in fact 0·8 like the many made previously with the same instrument?

Answering this question involves comparing measures of variability, one being a given value. When samples are small the ratio of variances, F, is most suitable. The critical values are given in Appendix Table III. The question is whether or not the true σ^2 estimated by the sample s^2 is the given value $0{\cdot}8^2$. Setting out the argument as before:

(i) $H_0 : \sigma^2 = \sigma_0^2 = 0{\cdot}8^2$; $H_a : \sigma^2 \neq 0{\cdot}8^2$

(ii) P = 0·5

(iii) $F = s^2/\sigma^2 = 1{\cdot}2^2/0{\cdot}8^2 = 2{\cdot}25$

Critical value, $F_{\cdot 05, \text{2-sided}, 5/\infty} = F_{\cdot 025, \text{1-sided}, 5/\infty} = 2{\cdot}57$

(iv) F is less than the critical value so H_0 is accepted. There is no clear evidence that the variability of the current measurements differs from the usual.

The F distribution is highly skewed (see Fig. 3.4 above) and the tables of its critical values are designed for one-sided testing. In the data on which the answer to question (c) is based there is nothing to indicate that the current measurements are more variable than usual and a two-sided test is called for. For this reason the one-sided $2\frac{1}{2}\%$ value is selected from the table. Ratios of random pairs of variances which differ significantly should be either very high or very low. Ratios of values estimating the same true value should be in the vicinity of unity but the chance 5% of exceptions can be divided into $2\frac{1}{2}\%$ above the tabulated value, 2·57, and $2\frac{1}{2}\%$ below $1/F_{\cdot 025, \infty/5} = 0{\cdot}17$. These limits, 0·17 to 2·57, bound the acceptance region in terms of the ratio and the observed value, 2·25, falls within them. Had there been prior grounds for suspicion that the sample observations were more variable than previously, the alternative hypothesis that $\sigma^2 > \sigma_o^2$ would have been permissible and this would have been accepted since the one-sided 5% value, $F_{\cdot 05, 5/\infty}$ as tabulated, is 2·21.

The test can also be made on the basis of the standard deviation sampling distribution (Fig. 3.3(c)). This is more suitable in the case of large samples but, approximately from eqns. (3.24a) and (3.24b),

$$u = (s - \sigma_0)/\sigma_s = (1{\cdot}2 - 0{\cdot}8)/(0{\cdot}8\sqrt{(2\times 6)}) = +1{\cdot}74$$

Again, this is within the two-sided limits $\pm 1{\cdot}96$ but greater than the one-sided $+1{\cdot}64$, and the conclusion reached in either case is the same as that derived from the corresponding F test.

Other tests in which the standard error of the sampling distribution is known rather than estimated are those on sample values of the measures of skewness and excess of kurtosis, g_1 and g_2. For the eighty-one CO_2 block densities of Table 2.1 we have $+0{\cdot}067$ and $-0{\cdot}0009$ respectively, as calculated in Table 2.9. If we wish to test the proposition that the universe from which this sample has come is Gaussian and the universe values of γ_1 and γ_2, therefore, zero we can make use of the sampling distributions (e) and (f) of Table 3.4. The standard errors depend only on the sample size according to eqns. (3.26b) and (3.27b), being approximately $\sqrt{(6/81)} = 0{\cdot}27$ and $\sqrt{(24/81)} = 0{\cdot}54$. The sample values are quite close to the hypothetical universe values, the absolute differences in standard error units being

$$u = |0{\cdot}067 - 0|/0{\cdot}27 = 0{\cdot}25 \quad \text{for} \quad g_1,$$

and

$$u = |0{\cdot}0009 - 0|/0{\cdot}54 = 0{\cdot}0017 \quad \text{for} \quad g_2.$$

Both are well below the critical value, 1·96, and the null hypotheses are accepted. In other words, there is no reason to suppose the parent distribution is either skewed or has other than normal kurtosis.

In general, any comparisons involving very large samples fall into the category of u tests because, when n is large, the standard deviation is estimated very precisely and approximates σ.

(2) *Tests involving Estimated Standard Errors*

Consider the surface tension data given above ($n = 6$, $\bar{x} = 71{\cdot}2$ dyn/cm, $s = 1{\cdot}2$ dyn/cm) but suppose no prior information is available as to the usual precision of measurement. The following question would be of interest.

(a) Could the true surface tension be 71·8 dyn/cm?

In this case we are still concerned with the sampling distribution of means but allowance must be made for the fact that only an estimate of the standard error is at hand and this from the same sample as the mean itself. Accordingly we use the distribution of t defined by eqn. (3.34). The critical values are tabulated in Appendix

Table II and (because the estimates, s and $s_{\bar{x}}$, are the better as the sample increases in size) they decrease towards the corresponding value of u as n increases. The form of the distribution can be seen in Fig. 3.4. For two-sided tests the tabulated 5% value which tends to 1·96 is used. For one-sided tests at the same significance level the tabulated 10% value, which tends to 1·64, is appropriate.

To make the test:

(i) $H_0 : \mu = \mu_0 = 71{\cdot}8; \quad H_a : \mu \neq \mu_0.$

(ii) P = 0·05.

(iii) The estimated standard error is $s/\sqrt{n} = 1{\cdot}2/\sqrt{6} = 0{\cdot}49.$

$$t = |\bar{x} - \mu_0|/s_{\bar{x}} = |71{\cdot}2 - 71{\cdot}8|/0{\cdot}49 = 1{\cdot}22$$

$$\text{Critical } t = t_{.05,5} = 2{\cdot}57.$$

(iv) t is less than the critical value so H_0 is accepted. The true value estimated by $\bar{x} = 71{\cdot}2$ could easily be 71·8 dyn/cm.

Alternatively: in this test the extreme value of μ_0 which would lead to just acceptable null hypotheses are those which differ from $\bar{x}$ by the maximum amount attributable to chance at the 5% level of significance, i.e. which would make $t = 2{\cdot}57$. This amount is $2{\cdot}57 \times 0{\cdot}49 = 1{\cdot}26$ and the extreme values of μ_0 are $\bar{x} \pm 1{\cdot}26$ or $\mu_L = 69{\cdot}94$ and $\mu_U = 72{\cdot}46$, these constituting the 95% confidence limits for μ. Where the t distribution is applicable, we have in general

$$100\ (1-\text{P})\%\ \text{Confidence limits of parameter}$$

$$= \text{Estimate } \pm t_{\text{P},f} \cdot s_{\text{estimate}} \tag{3.36b}$$

Here,

$$95\%\ \text{Confidence limits of } \mu = \bar{x} \pm t_{.05,f}\, s_{\bar{x}} = 71{\cdot}2 \pm 2{\cdot}57 \times 0{\cdot}49$$

$$= 69{\cdot}9 \text{ to } 72{\cdot}5$$

Since this interval embraces 71·8 there is nothing against the proposition that the values and their mean are estimates of this, the pure water value.

Let us now suppose that our observer has two tensiometers which he wishes to compare and to do this he takes a few readings on each, obtaining the results 71·4, 72·1, 69·6, 70·6, 72·3 and 71·8 with the first instrument and 73·1, 72·6, 71·0, 71·6, 72·3, 72·2 and 71·2 with the second. To summarize the results:

$$n_1 = 6;\ \bar{x}_1 = 71{\cdot}3;\ \Sigma X_1^2 = 5{\cdot}28;\ s_1^2 = 1{\cdot}053;\ s_1 = 1{\cdot}026;\ f_1 = 5$$

$$n_2 = 7;\ \bar{x}_2 = 72{\cdot}0;\ \Sigma X_2^2 = 3{\cdot}50;\ s_2^2 = 0{\cdot}583;\ s_2 = 0{\cdot}764;\ f_2 = 6.$$

All variation in the observations is attributed to measuring error.

Now, (a) are the instruments comparable with respect to random error?

The samples being small, it is best to compare variances in answering this question. H_0 takes the form $\sigma_1^2 = \sigma_2^2$ or $\sigma_1^2 - \sigma_2^2 = 0$ or $\sigma_1^2/\sigma_2^2 = 1$, and H_a is the two-sided alternative $\sigma_1^2 \neq \sigma_2^2$ unless there is prior knowledge that one instrument is less reliable than the other. Here, $F = 1{\cdot}053/0{\cdot}583 = 1{\cdot}81$ and the critical value is $F_{\cdot 05,\text{2-sided},5/6} = F_{\cdot 025,\text{1-sided},5/6} = 5{\cdot}99$ from Appendix Table III. Since the calculated value is less than the critical, H_0 is accepted and the two sets of readings regarded as being of comparable variability.

(b) Are the instruments comparable with respect to systematic error, i.e. can the two sample means be considered to represent the same universe mean?

H_0 takes the form $\mu_1 = \mu_2$ or $\delta = \mu_1 - \mu_2 = 0$ given $\sigma_1 = \sigma_2 = \sigma$. The qualification is by reason of the answer to question (a) which permits us to assume that any significant difference between $\bar{x}_1$ and $\bar{x}_2$ is due to their estimating different true means and not a stray effect of differing variability and, therefore, allows us to combine or pool the sample estimates of σ and σ^2. This is done by adding the sums of squares of deviations and by dividing by the total of the available degrees of freedom, viz.

$$s_p^2 = (\Sigma X_1^2 + \Sigma X_2^2)/(f_1 + f_2) \qquad (3.37\text{a})$$

$$= (5{\cdot}28 + 3{\cdot}50)/(5 + 6) = 0{\cdot}798$$

so the pooled standard deviation, $s_p = \sqrt{0{\cdot}798} = 0{\cdot}893$.† The standard error of the difference between means of samples in this situation is given by eqn. (3.29b), s_p being substituted for the supposed common σ:

$$s_d = s_p\sqrt{(1/n_1 + 1/n_2)} = s_p\sqrt{[(n_1 + n_2)/n_1 n_2]} \qquad (3.37\text{b})$$

$$= 0{\cdot}893\sqrt{(13/42)} = 0{\cdot}497$$

† This is very little different to the crude or weighted average of s_1 and s_2.

A t test compares the observed difference in terms of this estimated standard error which, being derived from s_p, is based on $n_1 + n_2 - 2 = f_1 + f_2$ degrees of freedom.

Since the true difference, δ, is zero on the null hypothesis we have

$$t = |d - 0|/s_d = |71{\cdot}3 - 72{\cdot}0|/0{\cdot}497 = 1{\cdot}41, \text{ cf. } t_{.05,11} = 2{\cdot}20.$$

Clearly, H_0 is acceptable and any differences between the two sets of readings can be attributed to random error. The same conclusion is reached by calculating the 95% confidence limits of δ according to eqn. (3.36b) which shows that δ could be zero with 95% confidence:

$$95\% \text{ Confidence lts. } \delta = d \pm t_{.05,f}\, s_d = 0{\cdot}70 \pm 2{\cdot}20 \times 0{\cdot}497$$
$$= -0{\cdot}4 \text{ to } + 1{\cdot}8.$$

Since, by accepting the null hypothesis, we assume $\mu_1 = \mu_2 = \mu$, a combined estimate of this single true mean can be made. This is done by treating the two samples as one, i.e. by weighting $\bar{x}_1$ and $\bar{x}_2$ according to sample size:

$$\bar{\bar{x}} = (n_1\bar{x}_1 + n_2\bar{x}_2)/(n_1 + n_2) = (\Sigma x_1 + \Sigma x_2)/(n_1 + n_2) \quad (3.38)$$

which in this case is 71·7 dyn/cm. The standard error of $\bar{\bar{x}}$ can be obtained by finding the deviations of all the values from it but a reasonable estimate is given by

$$s_{\bar{\bar{x}}} = s_p/\sqrt{(n_1 + n_2)} \quad (3.39)$$
$$= 0{\cdot}893/\sqrt{13} = 0{\cdot}247$$

here, and the 95% confidence interval from eqn. (3.36b) is $71{\cdot}7 \pm 2{\cdot}20 \times 0{\cdot}247 = 71{\cdot}7 \pm 0{\cdot}5$ dyn/cm.

When small samples whose means are to be compared have widely differing variances the above procedure is inappropriate. Several tests have been devised for this situation. One, due to Welch (1947, 1949), is an approximate t test in which the difference under test, $d - \delta$, is divided by the large sample standard error (eqn. 3.30b) with the sample variance substituted, i.e.

$$s_d = \sqrt{(s_1^2/n_1 + s_2^2/n_2)} = \sqrt{(s_{\bar{x}_1}^2 + s_{\bar{x}_2}^2)} \quad (3.40)$$

and the t value so found referred to the tabulated value for the number of degrees of freedom given by

$$1/f = C^2/f_1 + (1 - C)^2/f_2, \quad \text{where } C = [s_{\bar{x}_1}^2/(s_{\bar{x}_1}^2 + s_{\bar{x}_2}^2)] \quad (3.41\text{a,b})$$

this f being rather less than the $n_1 + n_2 - 2$ used when the sample variances are comparable.

Again suppose that two sets of surface tension readings are to be compared. The first set is the same as before,

$$n_1 = 6; \quad \bar{x}_1 = 71{\cdot}3; \quad s_1^2 = 1{\cdot}053; \quad s_{\bar{x}_1}^2 = 0{\cdot}176; \quad f_1 = 5,$$

but the second set comes from a much better instrument and

$$n_2 = 7; \quad \bar{x}_2 = 72{\cdot}3; \quad s_2^2 = 0{\cdot}170; \quad s_{\bar{x}_2}^2 = 0{\cdot}024; \quad f_2 = 6.$$

so that the sample variances ($F = 1{\cdot}053/0{\cdot}170 = 6{\cdot}20$) differ significantly. To see if there is any systematic difference between the two sets we find from eqn. (3.40)

$$s_d = \sqrt{(0{\cdot}176 + 0{\cdot}024)} = \sqrt{0{\cdot}200} = 0{\cdot}447,$$

whence

$$t \doteqdot |d - \delta|/s_d = |71{\cdot}3 - 72{\cdot}3|/0{\cdot}447 = 2{\cdot}24.$$

This is just significant at the 5% level by comparison with $t_{.05,11} = 2{\cdot}20$ but from eqns. (3.41 b) and (3.41 a) we have

$$C = 0{\cdot}176/0{\cdot}200 = 0{\cdot}88 \quad \text{and} \quad f = 1/(0{\cdot}88^2/5 + 0{\cdot}12^2/6) = 6{\cdot}36,$$

say 6. Now, $t_{.05,6} = 2{\cdot}45$ so the null hypothesis ($\mu_1 = \mu_2 = \mu$) is accepted at the 5% level and a combined estimate of μ can be made if desired.

A t test is also appropriate in dealing with paired values. Suppose the observer again has two tensiometers to compare but this time he decides to make the comparison over a reasonable range of surface tensions and uses various readily available substances for the purpose. He uses the same liquid samples for each pair of determinations and makes them at the same temperature so that differences between the observations are attributable to instrumental error alone. The results are:

Substance	Ether	Benzene	Chlor-benzene	Aniline	Glycerol	Water
1st tensiometer	16·8	28·7	36·9	48·0	67·2	71·3
2nd tensiometer	17·3	28·6	37·4	47·8	67·6	72·0
Difference, $x_2 - x_1 = x$	+0·5	−0·1	+0·5	−0·2	+0·4	+0·7

There is no question here of comparing the sets of original values because they are not random (or, even if the values of one set were, the values in the second, being tied to those of the first, could not be). The differences, however, can be regarded as a random sample characterized by

$$\bar{x} = +0{\cdot}3;\ s = 0{\cdot}245;\ s_{\bar{x}} = 0{\cdot}245/\sqrt{6} = 0{\cdot}1;\ f = 5.$$

If the differences between the readings are not systematic then $\bar{x}$ should estimate a true mean $\mu = 0$ and this is the null hypothesis. The test gives:

$$t = |0{\cdot}3 - 0|/0{\cdot}1 = 3, \text{ cf. } t_{.05,5} = 2{\cdot}57,$$

so the null hypothesis is rejected and it is concluded that the two instruments read differently, the second giving higher results than the first.

(3) *The Comparison of Several Samples*

Tests on pairs of samples like those above are based on the assumptions that the individuals within the samples are random and the samples themselves constitute a random pair. Such tests cannot be applied directly to any two samples *selected* from a larger number. There are several methods available for the comparison of several samples ranging from formal "analysis of variance" procedures to simple graphical tests like the industrial "quality control" charts. The latter are generally adequate to the task but they are most commonly applied when the samples are of equal size.

In the general case of k samples of unequal size, let an individual sample be designated $j\ (j = 1, 2, \ldots k)$, and n_j, f_j, $\bar{x}_j$, s_j^2 etc. represent the number of observations, degrees of freedom, arithmetic mean, variance etc. of the jth sample. The individual values within a sample are designated $i\ (i = 1, 2 \ldots n_j)$ but there is no connection between the ith value of one sample and that of another. The total number of values is $N = \sum^{k} n_j$ and their grand mean is $\bar{\bar{x}}$. Now it can be shown that the total sum of squares of deviations of the individual values from the grand mean (total SS) can be divided into a "between samples" component and a residual or "within samples" component as follows

$$\sum^{k}\sum^{n} (x_{ij} - \bar{\bar{x}})^2 = \sum^{k} n_j(\bar{x}_j - \bar{\bar{x}})^2 + \sum^{k}\sum^{n} (x_{ij} - \bar{x}_j)^2 \quad (3.42)$$

(The double summation, $\sum^{k}\sum^{n}$, indicates the addition of quantities first over the n values in a sample and then over all the k samples. Subsequently this will be indicated without superscripts. Sums of squares of deviations are referred to as SS for brevity.) If all the values were part of a single sample the total number of degrees of freedom would be $N - 1$. This can also be split into $k - 1$ between samples and $N - k = \Sigma f_j$ within samples. It can also be shown that the division of the sums of squares of deviations by the appropriate number of degrees of freedom yields mean squares or variances (MS) which, if the samples have in fact come from a single universe of variance σ^2, are independent estimates of that variance (one from the sample values, the other from eqn. (3.22b) in effect). An F test allows judgement to be passed on this proposition—if the ratio of the "between samples" MS to the "within samples" MS is reasonably low it can be accepted; if the ratio is too high it is rejected. This F test is one-sided because, if the various sample values differ significantly in level, the effect can only be to increase the quantities $(\bar{x}_j - \bar{\bar{x}})$ and raise the "between samples" SS relative to the residual. As in the case of pairs of samples, considerable differences in variability can lead to apparent differences between means and, if there are grounds for doubt, a test should be applied to some measure of dispersion of the samples as a preliminary.

Table 3.6(a) contains some data of Lord Rayleigh for the weights of nitrogen gas, derived from various sources, in a bulb of fixed volume. Of the five sets of observations, the first three were obtained after extraction of oxygen from air by various means and the last two by decomposition of N_2O and NH_4NO_2 over hot iron. The data have already been coded by subtraction of 2·31000 g, i.e. the first value of the first set was 2·31035 g and of the last 2·29849 g and so on. Part (a) of Table 3.6 also contains various sample calculations (some superfluous but included for the sake of the example) and the tests for comparability of variances and of means are given in parts (b) and (c).

The test in (b) is known as Bartlett's Test. It involves the computation of

$$M = \Sigma f_j \,.\, \ln s^2 - \Sigma f_j \ln s_j^2 = 2{\cdot}303[\Sigma f_j \,.\, \log s_p^2 - \Sigma(f_j \log s_j^2)] \tag{3.43}$$

TABLE 3.6

(a) DENSITY OF NITROGEN FROM VARIOUS SOURCES (CODED DATA)

Source, j / Observation, i	1	2	3	4	5	ΣΣ
1	35	17	24	−1131	−1151	
2	26	−14	10	−1060	−1111	
3	24	10	28			
4	12	1				
5	27					
Σx	124	14	62	−3969	−2262	−4253
n	5	4	3	2	2	16
f	4	3	2	1	1	11
$\bar{x}$	24·8	3·5	20·7	−1095·5	−1131·0	
Σx^2	3250	586	1460	2,402,761	2,559,122	4,967,179
$(\Sigma x)^2/n$	3075·2	49	1281·3	2,400,240·5	2,558,322	4,962,968
ΣX^2	174·8	537·0	178·7	2520·5	800·0	4211
s^2	43·7	179·0	89·4	2520·5	800·0	
$\log s^2$	1·6405	2·2529	1·9513	3·4015	2·9031	
$f.\log s^2$	6·5620	6·7587	3·9026	3·4015	2·9031	23·5279

$s_p^2 = 4211/11 = 382{\cdot}8 \quad \log s_p^2 = 2{\cdot}5830 \quad \Sigma f.\log s_p^2 = 11 \times 2{\cdot}5830 = 28{\cdot}4130$
$1/\Sigma f = 1/11 = 0{\cdot}0909 \quad \Sigma(1/f) = 0{\cdot}2500 + 0{\cdot}3333 + 0{\cdot}5000 + 1 + 1 = 3{\cdot}0833$
$(\Sigma\Sigma x)^2/N = 4253^2/16 = 1{,}130{,}501 \quad \bar{\bar{x}} = -4253/16 = -265{\cdot}8$

From eqn. (3.46), $\Sigma\Sigma(x - \bar{\bar{x}})^2 = 4{,}967{,}179 - 1{,}130{,}501 = 3{,}836{,}678$
From eqn. (3.47), $\Sigma n(\bar{x} - \bar{\bar{x}})^2 = 4{,}962{,}968 - 1{,}130{,}501 = 3{,}832{,}467$

$$\Sigma\Sigma(x - \bar{x})^2 = \Sigma\Sigma X^2 = 4211$$

(b) BARTLETT'S TEST ON THE SAMPLE VARIANCES

From eqn. (3.45), $c = 3{\cdot}0833 - 0{\cdot}0909 = 2{\cdot}9924 \doteqdot 3, \quad k = 5$
From eqn. (3.43), $M = 2{\cdot}303(28{\cdot}4130 - 23{\cdot}5279) = 11{\cdot}25$,
cf. $M_{.05,3/5} = 12{\cdot}86$
$(\chi^2 = 11{\cdot}25/(1 + 3/3 \times 4) = 11{\cdot}25/1{\cdot}25 = 9{\cdot}0, \quad \text{cf. } \chi^2_{.05,k-1=4} = 9{\cdot}49)$

(c) THE ANALYSIS OF VARIANCE

Source of variation	SS	f	MS	Ratio
Between samples	3,832,467	4	958,117	2503
Within samples	4211	11	382·8	cf. $F_{05.4/11} = 3{\cdot}36$
Total	3,836,678	15		

where s_p^2 is the pooled or weighted average of all the sample variances and is given by the generalized form of eqn. (3.37 a)

$$s_p^2 = \Sigma f_j s_j^2 / \Sigma f_j = \Sigma\Sigma X_j^2 / \Sigma f_j \tag{3.44}$$

The critical values of M are given in Appendix Table IV for various values of k and

$$c = \Sigma(1/f_j) - 1/\Sigma f_j \tag{3.45}$$

When the n_j are not very small, the quantity $M/[1 + c/3(k-1)]$ is distributed as the statistic χ^2 for $f = k - 1$ degrees of freedom and can be compared with the critical values in Appendix Table II(b). When the samples are larger and more numerous, the denominator in this χ^2 expression is substantially unity and can be ignored.

The result of both the M and the χ^2 calculation are the same for this example—the null hypothesis ($\sigma_1^2 = \sigma_2^2 \ldots = \sigma^2$) is accepted and we are free to proceed with the comparison of the sample means. Part (c) of the table follows the F test described above, the total sum of squares of deviations being generally found from the squares of the actual values,

$$\Sigma\Sigma(x_{ij} - \bar{\bar{x}})^2 = \Sigma\Sigma x_{ij}^2 - (\Sigma\Sigma x_{ij})^2/N \tag{3.46}$$

The "between samples" SS can be found directly from the differences between the sample means and grand mean but more conveniently from

$$\Sigma n_j(\bar{x}_j - \bar{\bar{x}})^2 = \Sigma[(\Sigma x_j)^2/n_j] - (\Sigma\Sigma x_{ij})^2/N \tag{3.47}$$

The residual SS can also be found directly or by difference from the results of eqns. (3.46) and (3.47). Both methods are used in the table.

The result of the F test shows that there are very highly significant differences between the samples. It is obvious enough in this case (although not quite so apparent from the original values) that these differences lie in the main between the atmospheric N_2 weights (samples 1, 2 and 3) and the chemical N_2 weights (samples 4 and 5) but in less startling situations the guidance of special tests is required. The difference between any two sample means can be checked for significance by ordinary t test ($t = |d|/s_d$ as above), the standard error of the difference being still given by eqn. (3.37 b) but the pooled s_p is now the overall "within samples" value from eqn. (3.44).

Special tests for comparing groups of sample means are to be found in the general references already cited.

Samples of equal size can be examined in the same way as the data of Table 3.6(a) and, if so, the eqns. (3.43)–(3.47) can be simplified. Graphical "control chart" techniques are, however, generally adequate. When the problem is one of comparing samples for consistency between themselves and the null hypothesis is that all are derived from the same universe, the sample estimates are compared with their own mean value (as best estimate of the universe parameter) and reasonable limits put to chance variation from it. In American practice these are set at three standard errors on either side of the mean ("3 σ" limits, corresponding to a one-sided $P = 0{\cdot}0013$ level of significance). In British practice, "warning" and "action" limits are used and these correspond to the one-sided 0·025 and 0·001 levels respectively. Extensive descriptions and many examples are to be found in the *A.S.T.M. Manual* (1951), Dudding and Jennett (1955) and in special monographs like that of Cowden (1957).

For our present purposes we require an ordinary two-sided 5% level test on means and on a measure of variability (the range for preference) and the British Standard "warning" limits suffice for this. The mean range, $\bar{w}$, takes the place of s_p in the previous example and from it an estimate of σ can be obtained by eqn. (2.24b) using d_2 in Table V of the Appendix. From this, in turn, $\sigma_{\bar{x}}$ can be estimated and limits set to the allowable variation of the means, but the required factor is incorporated in the tabulated A' which is applied directly to $\bar{w}$. The table is restricted to samples of a size not greater than 12 (since $\bar{w}$ is not very reliable above this) but values for n up to 25 can be calculated from those given in the *A.S.T.M. Manual* (1951) which also includes factors for use with the sample standard deviation, σ.

As an example of the technique, consider the values of for the standard e.m.f., E^0, of an antimony electrode at several points on the pH scale. The data constitute seven samples of size $n = 2$ and are given, together with the necessary calculations, in Table 3.7. x is taken as $E^0 - 255$ (mV) without further coding. The range chart, in Fig. 3.6, shows that there are considerable and possibly non-random discrepancies in the differences between the duplicate determinations—two identical values as at pH 3 are not to be expected and the two at pH 7 are questionably far apart—but they

TABLE 3.7. E^0 VALUES FOR THE ANTIMONY ELECTRODE

pH	E^0 (mV)	$\bar{x}$	w	Control chart limits
2·00	255·38, 255·34	0·36	0·04	$\bar{x}$ chart: $\bar{\bar{x}} \pm A'_{.025}\,\bar{w}$
3·00	255·12, 255·12	0·12	0·00	$= 0{\cdot}52 \pm 1{\cdot}229 \times 0{\cdot}071$
4·00	255·06, 255·12	0·09	0·06	$= 0{\cdot}432$ to $0{\cdot}608$
5·00	255·02, 255·00	0·01	0·02	
6·00	255·59, 255·43	0·51	0·16	w chart: $D'_{.025}\,\bar{w}$ to $D'_{.975}\bar{w}$
7·00	255·93, 255·73	0·83	0·20	$= 0{\cdot}036 \times 0{\cdot}071$ to $2{\cdot}810 \times 0{\cdot}071$
8·00	256·70, 256·72	1·71	0·02	$= 0{\cdot}003$ to $0{\cdot}200$
		7) 3·63	7) 0·50	
		$\bar{\bar{x}} = 0{\cdot}52$	$\bar{w} = 0{\cdot}071_4$	

are not of so gross a nature as to account for the variation between the means. The latter are spread well beyond the allowed limits and the null hypothesis ($\mu_1 = \mu_2 \cdots = \mu$) would be rejected at a very high level of significance. The advantage of the procedure illustrated here lies in its ease of application and in the visual representation of the pattern of variation it provides. It is obvious in the case of the

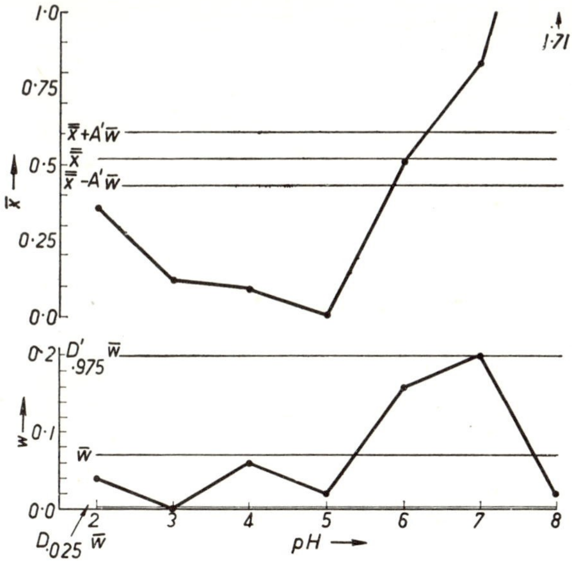

FIG. 3.6. Quality Control Charts.

The upper chart shows the means of the duplicate e.m.f. observations of Table 3.7 (coded) and the lower the differences between the duplicates or ranges.

example that not only do the sample means differ significantly but that the differences are systematic and probably of a curvilinear nature. It must be remembered, in applying control chart (P = 0·05 two-sided) limits to very large numbers of samples, that one in twenty on the average (more if the samples are small and drawn from a platykurtic parent universe) will fall outside the calculated limits because of purely random variation.

(4) *Some Tests on Frequencies and Proportions*

The tests described so far deal with variable data but others are available for the investigation of the effects of chance on attribute data. These depend in the main on the binomial distribution and the approximations to it discussed in Section 3.2 above.

Suppose, as an example, bars (randomly collected) of two grades of steel have been put to a particular test and counts made of the numbers passing and failing:

	Passed	Failed	Total	Proportion failed, p
Grade A	200	50	250	0·200
Grade B	400	150	550	0·273

Questions like the following might be of interest.

(a) Is the quality of this batch of Grade A steel consistent with that of previously produced material for which the proportion of failed test bars is only 0·15?

We denote by P the true proportion, of which p is the sample estimate, and by P_0 a hypothetical value of P. The argument may be presented as for the first u test:

(i) $H_0 : P = P_0 = 0{\cdot}15$; $H_a : P \neq P_0$, i.e. two-sided alternatives.
(ii) P = 0·05, i.e. the usual 5% level.
(iii) Since the distribution of p is approximately normal and characterized by eqns. (3.31 a) and (3.31 b), we have

$$u = |p - P_0|/\sigma_p = |p - P_0|/\sqrt{(P_0 Q_0/n)}$$

$$= |0{\cdot}2 - 0{\cdot}15|/\sqrt{(0{\cdot}15 \times 0{\cdot}85/250)} = 2{\cdot}21, \text{ cf. } u_{\cdot 05} = 1{\cdot}96.$$

(iv) u is greater than the critical value so we reject H_0. There is good reason for supposing this batch of bars to be below the average quality of the grade.

(b) What are the 95% confidence limits of P, given the estimate $p = 0{\cdot}2$ based on a random sample of 250?

The confidence interval can be found by application of eqn. (3.36a), $p \pm 1{\cdot}96\,\sigma_p$, but to answer this question on the data given σ_p must be found using p as an approximation whereas in the test above P_0 was available. Here,

$$95\%\ \text{Confidence lts. } P \doteqdot 0{\cdot}2 \pm 1{\cdot}96\sqrt{(0{\cdot}2 \times 0{\cdot}8/250)} \doteqdot 0{\cdot}15 \text{ to } 0{\cdot}25.$$

This confidence interval just touches the P_0 value of question (a) but when the true P is small (or large) the Gaussian approximation to the sampling distribution of p is poor. Tables which give accurate intervals are available (Hald (1952)) and also charts (Pearson and Hartley (1958)). Hald's table gives $0{\cdot}152 - 0{\cdot}255$ for the interval. When n is small the p distribution is not even approximately continuous and exact tests are made using the binomial formula itself.

(c) Is Grade B steel inferior to Grade A as expected? The available estimates of proportions of bars failing are 0·273 (out of 550) and 0·200 (out of 250).

This question turns on the observed difference between two values of p which should estimate the same true P if the samples are comparable. If this is so, a pooled estimate can be made,

$$\bar{p} = \text{total failed/grand total} = 200/800 = 0{\cdot}250,$$

and in addition the variance of the difference between estimates of P can be taken as the sum of the variances of the estimates themselves, so

$$\sigma_d = \sqrt{(PQ/n_A + PQ/n_B)} \doteqdot \sqrt{[\bar{p}\bar{q}(1/n_A + 1/n_B)]} \qquad (3.48)$$

The argument, then, runs

(i) $H_0: P_A = P_B = P$ or $\delta = (P_B - P_A) = 0$;
$H_a: P_B > P_A$ or $\delta > 0$.

(ii) P = 0·05, but one-sided as required by the question.

(iii) $u = (d - 0)/\sigma_d = (p_B - p_A)/\sqrt{[\bar{p}\bar{q}(1/n_A + 1/n_B)]}$
$= (0{\cdot}273 - 0{\cdot}2)/\sqrt{[0{\cdot}2 \times 0{\cdot}8/(1/250 + 1/550)]}$
$= 0{\cdot}073/0{\cdot}033 = +2{\cdot}21,$

cf. $u_{.05,\text{1-sided}} = +1{\cdot}64$.

(iv) u is greater than the critical value so we reject H_0. The Grade B steel is almost certainly inferior.

Data like these may also be tested by a measure of difference between observed (O) and expected (E) frequencies. The statistic is

$$\chi^2 = \Sigma\,[(O - E)^2/E] \qquad (3.49)$$

It is not quite identical with but is distributed approximately as the χ^2 of eqn. (3.35) and Appendix Table II. In order to apply a test based on this statistic the data are drawn up in the form of a *contingency* table and the expected frequencies consistent with the null hypothesis calculated. If this takes the form that the division of items according to one classification is in no way dependent, i.e. contingent, upon their division according to another, then the number of items in any cell should be

$$E = (\text{row total}) \times (\text{column total})/(\text{grand total}) \qquad (3.50)$$

For example, in the present instance the overall proportion of failed bars is 0·25, so if the grades of steel are of equal quality we can expect 250/4 or 62·5 Grade A bars and 550/4 = 137·5 Grade B bars to fail. These are the numbers given by eqn. (3.50). We can also expect, however, some departure of the actual frequencies from these calculated values because of the accidents of random sampling and the critical values of χ^2 set the limits on this. The number of degrees of freedom appropriate to χ^2 is the number of independent differences of the kind $(O - E)$ which go into its total and, in a contingency table, this is given by the product: (number of rows − 1) × (number of columns − 1). In a 2 × 2 table, therefore, $f - 1$ and this can be seen by an inspection of Table 3.8 in which all the $(O - E)$ are the same. In other words, given one $(O - E)$ and the marginal totals the remaining expected frequencies can be found.

Now, the question is:

(d) Do the grades differ in quality?

The null hypothesis is that they do not, i.e. that the observed and expected frequencies differ by an amount which can be reasonably attributed to chance. In Table 3.8 the value of χ^2 found is greater than the critical value at the 5% level and, in consequence, H_0 is rejected, but had χ^2 been less we would have accepted the proposition that the $(O - E)$ could have been attributed to chance regardless of the actual nature of the differences. Thus, this test, like the F test on the means of several samples, is one-sided in its operation but two-sided in its effect. The probability that the value of χ^2 found here ($4{\cdot}85, f = 1$) is due to chance is 0·027 like that of $u = |2{\cdot}21|$ which we would have obtained in answer to question (c) had that involved a two-sided alternative.

TABLE 3.8. CONTINGENCY TABLE AND χ^2 TEST

	Passed	Failed	Total	$(O - E)^2/E$
Grade A	$O = 200$ $E = 187{\cdot}5$ $O-E = +12{\cdot}5$	$O = 50$ $E = 62{\cdot}5$ $O-E = -12{\cdot}5$	250	$12{\cdot}5^2/187{\cdot}5 = 0{\cdot}834$ $12{\cdot}5^2/62{\cdot}5 = 2{\cdot}500$
Grade B	$O = 400$ $E = 412{\cdot}5$ $O-E = -12{\cdot}5$	$O = 150$ $E = 137{\cdot}5$ $O-E = +12{\cdot}5$	550	$12{\cdot}5^2/412{\cdot}5 = 0{\cdot}379$ $12{\cdot}5^2/137{\cdot}5 = 1{\cdot}137$
Total	600	200	800	Total, $\chi^2 = 4{\cdot}85$, cf. $\chi^2_{\cdot 05,1} = 3{\cdot}84$

The χ^2 test is not restricted to dichotomous classifications and the above procedure could be used if the test-bars were classified into several grades of steel and several designations of quality. In such cases the values of $O - E$ are not the same for each cell.

In the special case of 2×2 tables (and one or two others) a shorter method of finding χ^2 is available,

$$\chi^2 = [(ad - bc)^2\, n/(a + b)(c + d)(a + c)(b + d)] \quad (3.51a)$$

where a, b, c, d are the O for each cell, $(a + b)$ etc. are column and row totals and n is the total frequency. See, also, Kupperman (1959) for a very quick method which requires only tables of n. ln n (Kullback, 1959) and an adding machine. Note also, since χ^2 as found is not strictly continuous especially when cell frequencies are low, a correction for continuity (Yates')—to reduce the magnitude of each $O - E$ by $\frac{1}{2}$ and so reduce χ^2—is made in borderline cases. In the example, each $O - E$ becomes 12 and χ^2 becomes 4·47 which is still significant. The corrected 2×2 formula is

$$\chi^2 = [(|ad - bc| - \tfrac{1}{2}n)^2\, n/(a + b)(c + d)(a + c)(b+d)] \quad (3.51\text{b})$$

There is no need to apply the correction if the uncorrected value is less than significant. The χ^2 test is not appropriate if the cell frequencies are very low. For special tables and an explanation of their application see Pearson and Hartley (1958).

The $\Sigma (O - E)^2/E$ statistic is convenient for comparing class-interval frequencies with those calculated from a particular distribution law and thus tests the "goodness of fit" of such laws. In this application the number of degrees of freedom available is the number of cells or class intervals, k, less the number of constants used in the calculation of, or constraints on, the expected frequencies. When fitting a rectangular distribution the only constraint on the expected frequencies in the class intervals is that they add to the total observed frequency, n, so $f = k - 1$. When fitting a binomial or Poisson distribution the expected frequencies are usually found from n and p or n and $\bar{h}$ (i.e. from sample estimates) so $f = k - 2$, but if the parameters (P, μ) are given then $f = k - 1$. Thus, in the example below in which a binomial distribution is fitted to the data $f = 5 - 2 = 3$, but if the observed frequencies in the last line of Table 3.2 were to be compared with the theoretical frequencies (for $n = 12$) two lines above it $f = k - 1$ would be used. In the case of a comparison of observed with theoretical Gaussian frequencies, $f = k - 3$ for n, $\bar{x}$ and s used in the fitting.

As a simple example, consider the data of Table 3.3 in which we have an observed classification of samples and the corresponding fitted binomial frequencies. The data are reproduced and a χ^2 test applied in Table 3.9. The value of χ^2 found is low and we accept the null hypothesis that the differences between the observed and expected frequencies can be attributed to chance, i.e. that the observed frequencies can be considered to constitute a binomial distribution. Note that the last three class frequencies of Table 3.3 have been combined—this should be done to avoid too low values of the observed frequency.

TABLE 3.9. A χ^2 TEST OF GOODNESS OF FIT

h	O	E	$O-E$	$(O-E)^2$	$(O-E)^2/E$
0	38	34·9	+3·1	9·61	0·275
1	36	38·7	−2·7	7·29	0·188
2	17	19·4	−2·4	5·76	0·297
3	7	5·7	+1·3	1·69	0·296
4+	2	1·3	+0·7	0·49	0·377
	100	100·0	0·0		$\chi^2 = 1·433$, cf. $\chi^2_{·05,3} = 7·81$

3.6. The Expression, Combination and Manipulation of Experimental Results

We deal in this section with various problems which arise in the treatment of experimental observations.

(1) *Expression of Results*

The final and most complete statement of the results of a simple set of measurements is an interval estimate of the true quantity assumed to underly the values. This is obtained in general by the application of eqns. (3.36a) or (3.36b) and given in the form of a mean or other estimate with attached "error", the latter being half the width of the confidence interval. This form of expression is, strictly speaking, only appropriate when the relevant sampling distribution is symmetrical but this is in fact the usual situation.

Some choice is available, and must be exercised, with respect to the "error" quoted. In the examples above, the 95% limits were calculated for purposes of illustration. These are often favoured because they are in accord with the almost universally adopted level in significance testing but narrower limits are frequently to be found in the literature, e.g. the mean ± the "probable" error (in effect, the 50% limits) or ± the standard error (68% limits). The *A.S.T.M. Manual* (1951) recommendation is to use the 90%, 95% or 99% errors and the first might well be preferred as a compromise between the unduly narrow and apparently flattering probable error (which is as likely to be exceeded as not) and the rather conservative values based on higher levels of confidence.

In writing results, say for a set of surface tension observations, the following would be satisfactory:

$$\gamma(H_2O, 0°C) = 75{\cdot}0 \pm 0{\cdot}3 \text{ dyn/cm } (90\% \text{ confidence})$$

or

$$= 74{\cdot}96 \pm 0{\cdot}29 \text{ dyn/cm } (P = 0{\cdot}90,\ n = 10).$$

The first of these versions, with only one doubtful figure retained in the mean, is only proper to a general conclusion. The second contains the minimum information needed for comparison with other series of observations. The A.S.T.M. rule, taking account of the number of observations, may be summarized:

Single values to nearest			No. of observations		
0·1	1	10		2–20	21–200
0·2	2	20	<4	4–40	41–400
0·5	5	50	<10	10–100	101–1000
No. of significant figures retained in mean and error			As in single values	One more	Two more

The standard deviation should be reported to three significant figures and the "$\pm$ error" or confidence limits should contain the same number of significant figures as the mean itself.

(2) *Testing of Suspected Observations or Detection of "Outliers"*

This problem arises frequently in experimental work and many rules or criteria have been proposed. Some call for rejection of an observation at a fixed deviation from the observed mean (e.g. 3 standard or $2\frac{1}{2}$ or 4 average deviations)—others allow for the possible occurrence of larger deviations in larger samples (e.g. Chauvenet's criterion—see Parratt (1961), Bond (1935)). A critical list is given by Blaedel, Meloche and Ramsey (1951). Some of the recommended rules too easily reject fair values while others are too strict, particularly if tied to the rule that no observation, however doubtful statistically, should be rejected unless the statistical test is supported by direct evidence of experimental failure. Strictly, this latter consideration is irrelevant because all observations, whether their values be apparently good, bad or indifferent, should be subjected to the same degree of critical scrutiny. The problem is to pass judgement on an observation which has already been accepted from an experimental point of view but whose deviation from the remainder of the set is so great as to raise the suspicion that an unknown source of error has supervened.

A discrepant observation might belong to a universe which differs from that of the remainder in location (systematic error) or spread (differing random error) or both. A reasonable criterion for rejection is failure of the null hypothesis (that the suspected value belongs with the rest) at the usual level of significance. For small samples ($n \leqq 25$) the best criterion is Dixon's (1953) r test. In this, the difference between the suspected value and its immediate or a close neighbour is divided by the overall range or a *modified range*. The precise specification of r for best results depends on the sample size, thus

$$r_{10} = (x_2 - x_1)/(x_n - x_1) \qquad (n = 3 \text{ to } 7) \qquad (3.52\text{a})$$

$$r_{11} = (x_2 - x_1)/(x_{n-1} - x_1) \qquad (n = 8 \text{ to } 10) \qquad (3.52\text{b})$$

$$r_{21} = (x_3 - x_1)/(x_{n-1} - x_1) \qquad (n = 11 \text{ to } 13) \qquad (3.52\text{c})$$

$$r_{22} = (x_3 - x_1)/(x_{n-2} - x_1) \qquad (n = 14 \text{ to } 25) \qquad (3.52\text{d})$$

In these equations x_1 represents the lowest and suspected value and x_n the highest—the necessary changes if x_n is suspected are obvious. The one-sided critical values of $r_{..}$ are given in Appendix Table VI.

Dixon's test uses only the internal evidence provided by the sample itself but additional evidence as to the extent of random error in the system under investigation should, if available, be taken into account. For example, the measurements in the sample containing the suspected value might have been obtained from an instrument so much used in the past that the true measure of random error, σ, is virtually known. On the other hand, only a few sets of similar measurements might have been made previously but they still provide a better estimate of σ than can be obtained from the sample itself. A convenient statistic for the purpose is the *studentized* extreme deviate from the sample mean which we denote by q,

$$q = |\bar{x} - x_1|/s_p \quad \text{or} \quad q = |x_n - \bar{x}|/s_p \qquad (3.53)$$

where $\bar{x}$ is the mean of the sample containing the suspected value, x_1 or x_n, and s_p is a pooled estimate of σ from other samples. Critical values are given in Appendix Table VII for sample sizes from 3 to 9 and degrees of freedom appropriate to the independently obtained s_p, from 10 to ∞.

As an example, consider some surface tension measurements of the kind previously described. Suppose they are 71·4, 72·1, 69·6, 70·6, 72·3, 71·8 and 67·0 dyn/cm, having as mean 70·7 dyn/cm. If no other information is available but that the *lowest* value be suspect, we calculate

$$r_{10} = (69{\cdot}6 - 67{\cdot}0)/(72{\cdot}3 - 67{\cdot}0) = 2{\cdot}6/5{\cdot}3 = 0{\cdot}49.$$

For $n = 7$ we find the 5% critical value in Table VI to be 0·507 and, since the calculated r_{10} is less than this, the value is retained. If, on the other hand, we know from past experience that the standard deviation of routine measurement is 0·8 and we take this as σ, then we compute

$$q = (70{\cdot}7 - 67{\cdot}0)/0{\cdot}8 = 4{\cdot}63.$$

This exceeds the 5% value, 2·27, in Table VII for $n = 7$ and $f = \infty$ (appropriate to a known σ) and the suspected observation is rejected. Lastly, suppose these measurements had been preceded by five series containing 29 concordant observations in all so that a pooled estimate, $s_p = 1{\cdot}01$ say, obtained from them by eqn. (3.44) is based on $f = 29 - 5 = 24$ degrees of freedom. We now have

$$q = (70{\cdot}7 - 67{\cdot}0)/1{\cdot}01 = 3{\cdot}66$$

which we compare with the critical value, $q_{\cdot 05,7,24} = 2{\cdot}44$, and again reject the suspected value.†

A test, similar to that based on q, can be carried out by constructing a "quality control" range chart with the aid of the factors in Table V of the Appendix. A bad observation will throw the range of the sample which contains it outside the calculated limits for the likely operation of random error. This test is most conveniently applied if all the samples are of equal size but it can be adapted to unequal samples as well. Proschan (1957) describes the procedure.

If a statistical test rejects an observation and this draws attention to a valid experimental reason for omitting it (a reason which should of course, have been evident beforehand), a confidence interval for the true mean calculated from the reduced data has its usual significance. If no experimental support is to be found for rejection and a value omitted on statistical grounds only, then there is the possibility of mistake and the degree of confidence to be placed on intervals calculated in the usual way will not be known exactly. In view of this, it is sometimes recommended that a statistically rejected value be ignored in computing the sample mean but retained in estimating its precision. This practice seems unduly conservative

† It should be noted that if (as is usual) a value comes under suspicion after the measurements have been obtained and simply from inspection of them, the level of significance is double that given in the tables, e.g. 10% in the examples given.

if the values as a whole are more or less normally distributed but it would afford some protection if normality could not be assumed. The question of non-normality is discussed by van Strik (1960).

Broadly speaking, the two small sample criteria described are the best (in their respective fields) for detecting outliers whether the latter are due to change of mean or variability. The systematic type of error, however, is always the easier to detect. Dixon (1950) presents some interesting comparisons of the performance of these and other criteria for one, two and several outliers. All perform badly when one or more erroneous values occur at each end of the sample array.

This topic is obviously of great moment when observations are few and costly to obtain but the problem diminishes in importance with increasing sample size because one or even a few bad observations can have little effect on the estimates found from a large sample. Chauvenet's criterion is useful in this situation, the principle of rejection being to exclude values whose deviations from the mean are of a magnitude having a probability of occurrence less than $1/2n$. This means in effect a sliding level of significance as to magnitude and the maximum allowable discrepancy in standard deviation units, $u = (x - \mu)/\sigma$ (μ and σ being virtually known when n is large), is taken as 2·4 for $n = 30$, 2·6 for 50, 2·8 for 100, 3·0 for 200 and 3·3 for 500. The criterion can be applied repeatedly.

(3) *The Weighted Mean and its Confidence Limits*

If it can be shown that two or more sample sets of results differ in their means and standard deviations by amounts small enough to be attributed to chance, then the means can be combined to get a better estimate of the universe mean. Since the individual sample means may vary in precision, the desired result is obtained by weighting them according to the confidence to be placed in them, i.e. as $1/s_{\bar{x}}^2$, which is equivalent to weighting according to the sample size ($ns_{\bar{x}}^2$ being approximately σ^2, the supposed common variance). This has already been done in a previous example by application of eqn. (3.38).

As a further example, consider the following. Three results have been obtained for the surface tension, γ, of an organic liquid—30·0 ± 0·2, 30·2 ± 0·5 and 29·9 ± 0·4 dyn/cm, all 90% limits. The calculations, including coding and weighting, are shown in Table 3.10

in which $e_{.9}$ represents the 90% error or half the width of the 90% confidence interval.

TABLE 3.10. THE WEIGHTED MEAN AND ITS CONFIDENCE LIMITS

$\bar{x}$	$e_{.9}$	$w = 1/e_{.9}^2$	$w\bar{x}$	$w\bar{x}^2$
0·0	0·2	25	0·000	0·0000
0·2	0·5	4	0·800	0·1600
−0·1	0·4	6·25	−0·625	0·0625
		$35{\cdot}25 = \Sigma w)$	$0{\cdot}175 = \Sigma w\bar{x}$	$0{\cdot}2225 = \Sigma w\bar{x}^2$
			$0{\cdot}00496$	$-0{\cdot}0009 = -(\Sigma w\bar{x})^2/\Sigma w$

$x = \gamma - 30$ $\qquad (k-1)\,\Sigma w = 70{\cdot}5)\ 0{\cdot}2216 = \Sigma w(\bar{x} - \bar{\bar{x}})^2 = \Sigma w\bar{\bar{\bar{x}}}^2$

$\bar{\bar{x}} = 30 + 0{\cdot}00496 = 30{\cdot}005$ $\qquad 0{\cdot}00314 = \Sigma w\bar{X}^2/(k-1)\,\Sigma w$

$e_{.9} = 1/\sqrt{35{\cdot}25} = 0{\cdot}169$ $\qquad \mathrm{e} = \sqrt{0{\cdot}00314} = 0{\cdot}056$

$\gamma = 30{\cdot}00 \pm 0{\cdot}17$ dyn/cm $\qquad \mathrm{e}/e = 0{\cdot}056/0{\cdot}169 = 0{\cdot}33$

(90% confidence)

Random error is sufficient to account for the discrepancies between the means of the samples in Table 3.10 and the error in the weighted mean is based solely on the errors of the individual means, viz.

$$(e = 1/\sqrt{(1/e_1^2 + 1/e_2^2 + \cdots 1/e_k^2)} = 1/\sqrt{(\Sigma w_j)} \quad (j = 1, 2 \ldots k) \tag{3.54}$$

where w_j is the weight attached to the jth mean and this is taken as being inversely proportional to the square of its standard (or related) error, e_j. In the older literature of physics dealing with errors, the e of eqn. (3.54) is called a measure of "internal" consistency, whereas

$$\mathrm{e} = \sqrt{[\Sigma(w_j(\bar{x}_j - \bar{\bar{x}})^2)/(k-1)\,\Sigma w_j]} \tag{3.55}$$

is a measure of "external" consistency. e and e correspond to the "within" and "between samples" measures of variability in the example of Table 3.6. Their ratio

$$\mathrm{e}/e = \sqrt{[\Sigma(w_j(\bar{x}_j - \bar{\bar{x}})^2)/(k-1)]} \tag{3.56}$$

is distributed approximately normally with a mean of unity and a standard error of $1/\sqrt{[2(k-1)]}$. The ratio for the data of the example given in Table 3.10 is only 0·33, a value well below expectation so no formal test is required. An account of this test, due to Birge (1932), is to be found in Topping (1955) and Worthing and

(1943). It is convenient when only standard errors or confidence intervals are available for weighting but the test described in Table 3.6 is to be preferred if the original data including numbers of observations are at hand.

Even if the variability or degree of random error differs significantly between samples weighting according to the above principle is still appropriate, but when there are significant differences between the sample means due to systematic errors (revealed by Birge's or the F test if not clear from inspection) the problem is difficult. It involves an attempt to estimate some single true value, μ, from estimates of subsidiary true values, μ_1, μ_2 etc., when little can be said about the relationship of the latter to the former. Suppose the following results of different observers for Avogadro's Number to be the only data available; $60{\cdot}44 \pm 0{\cdot}36$ (Westgren by gold sol sedimentation), $59{\cdot}0 \pm 4{\cdot}9$ (Nordlund from the Brownian movement of a single particle), $71{\cdot}1 \pm 4{\cdot}6$ (Perrin by gamboge and mastic particle sedimentation), all $/10^{22}$. (The data are derived from Burton (1938).) To combine values such as these one might (a) assume all to be of equal weight and treat them as single observations, (b) assign arbitrary weights on the basis of experience (in which case the errors should be weighted in the same way, using eqn. (3.55)), or (c) weight according to precision, as in the example above, on the dubious assumption that the most precise value is likely to be the least subject to systematic error by virtue of the number and/or consistency of the observations on which it is based. (a) is the safe procedure in the absence of further information although it assumes, when it comes to calculating a confidence interval for the overall μ, that the subsidiary experimental means or their estimates are a random collection out of a much larger possible number. This might be a fair assumption with respect to observers but the number of different ways in which physico-chemical properties can be experimentally determined is in general limited. To avoid the difficulty by adopting the most precise result is no better than proceeding according to (c). Worthing and Geffner (1943) should be consulted for a more extensive discussion of the problem.

(4) *Propagation of Error*

When a quantity is not measured directly, but must be calculated from an equation involving other quantities, its precision has also to be found indirectly from the reliabilities of the estimates of the independent quantities.

If y is a function of x only then, clearly, a small change in x, e_x, involves a small change in y, e_y, given by

$$e_y = (\mathrm{d}y/\mathrm{d}x)\, e_x$$

If y is a function of several different variables, say

$$y = \varphi(x_1, x_2 \ldots) \tag{3.57}$$

then, for small changes e_1, e_2 etc. in the independent variables, we have

$$e_y = (\partial y/\partial x_1)\, e_1 + (\partial y/\partial x_2)\, e_2 \ldots \tag{3.58}$$

This equation can be used to find the specific or maximum change in y due to specific or maximum changes in x_1, x_2 etc. but if e_1 and e_2 are statistically determined errors (standard errors or half confidence intervals) attached to the available values of the independent variables (these generally being sample means $\bar{x}_1$, $\bar{x}_2$ etc.) then eqn. (3.58) will over-estimate the corresponding measure of the reliability of the calculated y value. In this case, assuming the errors of measurement of x_1 and x_2 to be independent, it can be shown that

$$e_{\hat{y}} = \sqrt{[(\partial y/\partial \bar{x}_1)^2\, e_1^2 + (\partial y/\partial \bar{x}_2)^2\, e_2^2 + \ldots]} \tag{3.59}$$

where $e_{\hat{y}}$ is now the measure of reliability of the best estimate of a true value μ_y obtainable from the available estimates of the true values of the independent variables, i.e. from eqn. (3.57) in the form

$$\hat{y} = \varphi(\bar{x}_1, \bar{x}_2 \ldots)$$

The familar forms to which eqns. (3.58) and (3.59) reduce in special cases are given in Table 3.11. In this table x represents a single value or sample mean and e represents a specific change, maximum error or statistical error as the case may be. It sometimes happens that the errors in the values to be combined are not equivalent, e.g. e_1 might be the standard error of a sample mean $x_1 = \bar{x}_1$ while e_2 is simply a limitation on measurement and is, in effect, half the width of a rectangular distribution. Thus, $25 \pm 0{\cdot}5\,°\mathrm{C}$ implies that the value 25° has been read on a thermometer graduated to 1 °C and it is "almost certain" (as far as random errors are concerned) that the actual temperature lies between 24·5° and 25·5°. It is recommended

TABLE 3.11. PROPAGATION OF ERROR

Relationship	Specific change	Maximum error	Statistical error	
$\hat{y} =$	$e_{\hat{y}} =$	$e_{\hat{y}} =$	$e_{\hat{y}} =$	
(1) $x_1 \pm x_2 \pm \cdots$	$\pm e_1 \pm e_2 \pm \cdots$	$e_1 + e_2 + \cdots$	$\sqrt{(e_1^2 + e_2^2 + \cdots)}$	(3·60a)
(2) ax	$\pm ae$	ae	$a\sqrt{e^2} = ae$	(3·60b)
(3) b/x	$\mp be/x^2$	be/x^2	be/x^2	(3.60c)
(4) x^c	$\pm cex^{c-1}$	cex^{c-1}	cex^{c-1}	(3·60d)
(5) x_1x_2	$\pm e_1x_2 \pm e_2x_1$	$e_1x_2 + e_2x_1$	$\sqrt{(e_1^2x_2^2 + e_2^2x_1^2)}$	(3.60e)
(6) x_1/x_2	$\pm e_1/x_2 \mp e_2x_1/x_2^2$	$e_1/x_2 + e_2x_1/x_2^2$	$\sqrt{(e_1^2/x_2^2 + e_2^2x_1^2/x_2^4)}$	(3.60f)
(7) $\exp x$	$\pm e \,.\, \exp x$	$e \,.\, \exp x$	$e \,.\, \exp x$	(3.60g)
(8) $\log x$	$\pm 0{\cdot}4343\, e/x$	$0{\cdot}4343\, e/x$	$0{\cdot}4343\, e/x$	(3.60h)
(9) $\sin x$	$\pm e \,.\, \cos x$	$e \,.\, \cos x$	$e \,.\, \cos x$	(3.60i)
	(e taken + or −)	(e taken +)	(e taken ±)	

(Baird (1962) discusses the matter) that the "almost certain" error, $\pm 0{\cdot}5\,°C$, be regarded as equivalent to two-thirds of a standard error. In other words, $e_2 = 0{\cdot}33\,°C$ for combination with other standard errors.

It will be noticed, from entry (8) in the Table 3.11, that the error in a logarithm is the relative error of the original value so that, if x is a sample mean and e its is a standard error, the standard deviation of the logarithms or the standard error of their mean is proportional to a relative error or a coefficient of variation (from eqn. (2.18 b)) in x. Most of the formulae in the table can be expressed more simply in terms of relative errors. Thus, division by $\hat{y}$ shows that in both case (2) and case (3) the relative error in $\hat{y}$, $e_{\hat{y}}/\hat{y}$, is equal to the relative error in x, e/x. Similarly, from (5) and (6), it can be seen that for any combination of products and quotients the maximum relative error in $\hat{y}$ is the sum of the relative errors in the values of the independent variables,

$$e_{\hat{y}}/\hat{y} = e_1/x_1 + e_2/x_2 + \cdots$$

and also that the coefficient of variation in $\hat{y}$ is given by

$$v_{\hat{y}} = \sqrt{(v_1^2 + v_2^2 + \cdots)}$$

Lastly, from (4), the relative error in x^c is c times the relative error in x and, to generalize again, if $\hat{y} = x_1^a x_2^b \ldots$, then

$$v_{\hat{y}} = \sqrt{(av_1^2 + bv_2^2 + \cdots)}$$

As an example, consider the viscosity of a liquid, η (centipoise), as related to the time of efflux, t (seconds), from a Saybolt cup and its density, ϱ(g/ml), by the formula

$$\eta/\varrho = 0{\cdot}22\text{t} - 180/\text{t}$$

Suppose (a) t = 200 sec and $\varrho = 0{\cdot}950$ g/ml and it is required to find the effect of a 1 sec decrease in time and a 0·005 g/ml increase in density; (b) t = 200 sec, accurate to 1 sec, and $\varrho = 0{\cdot}950$ g/ml, accurate to 0·005 g/ml, and it is required to find the maximum error in η; and (c) t = 200 ± 1 sec (95% confidence) and $\varrho = 0{\cdot}950 \pm 0{\cdot}005$ g/ml (95% confidence) from several measurements of each and it is required to find the 95% limits for the true value of η. Then

$$\eta = 0{\cdot}22\,\text{t}\varrho - 180\,\varrho/\text{t} = 40{\cdot}94\text{ cP.}$$

(a) By eqn. (3.58),

$$e_\eta = (\partial\eta/\partial\varrho)\,e_\varrho + (\partial\eta/\partial t)\,e_t$$
$$= (0{\cdot}22t - 180/t)\,e_\varrho + (0{\cdot}22\varrho + 180\varrho/t^2)\,e_t$$
$$= (0{\cdot}22 \times 200 - 180/200)\,0{\cdot}005 + 0{\cdot}95(0{\cdot}22 + 180/200^2)\,(-1)$$
$$= 43{\cdot}1 \times 0{\cdot}005 - 0{\cdot}213 \times 1 = +0{\cdot}003 \text{ cP}.$$

(b) By eqn. (3.58),

$$e_\eta = 43{\cdot}1 \times 0{\cdot}005 + 0{\cdot}213 \times 1 = \pm 0{\cdot}43 \text{ cP (maximum error)}.$$

(c) By eqn. (3.59),

$$e_\eta = \sqrt{[(\partial\eta/\partial\varrho)^2 e_\varrho^2 + (\partial\eta/\partial t)^2\, e_t^2)]}$$
$$= \sqrt{(43{\cdot}1^2 \times 0{\cdot}005^2 + 0{\cdot}213^2 \times 1^2)}$$
$$= \pm 0{\cdot}31 \text{ cP} \quad (95\% \text{ error})$$
$$\therefore\ \eta = 40{\cdot}94 \pm 0{\cdot}31 \quad (95\% \text{ confidence})$$

Some of the results of eqn. (3.59) have already been used. Thus eqn. (3.60a) is the basis of eqn. (3.30b) and similar equations for standard errors of differences in Table 3.4. To take another example; let $\hat{y} = (x_1 + x_2 + x_3 \ldots)/n = x_1/n + x_2/n + x_3/n + \ldots$ Now, from eqn. (3.59) or the generalization of eqn. (3.60a), we have

$$e_{\hat{y}}^2 = e_1^2/n^2 + e_2^2/n^2 + e_3^2/n^2 + \ldots$$

If x_1, x_2 etc. are independent values from a Gaussian distribution their errors, in the sense in which the term is used here, are all representable by σ, so, correspondingly for y,

$$\sigma_{\hat{y}}^2 = \Sigma(\sigma^2/n^2) = n\sigma^2/n^2 = \sigma^2/n$$

and, since $\hat{y}$ is in fact $\bar{x}$, this gives eqn. (3.22b), $\sigma_{\bar{x}} = \sigma/\sqrt{n}$.

Although eqn. (3.59) shows how an error is propagated it gives no indication as to the form of the derived error distribution. In the case of simple linear functions like (1) and (2) in Table 3.11 there is no difficulty—if e_1 and e_2 represent the standard deviations of Gaussian distributions, then $e_{\hat{y}}$ calculated from them is the standard deviation of a Gaussian distribution—but this does not apply if the function is non-linear. Nevertheless, the computational error incurred when eqn. (3.59) is used in non-linear situations is generally very small. This is demonstrated by the following. Suppose 100 observations were made of the diameter of a spherical object and, because of random irregularities and imprecision of measurement, these are normally distributed as shown:

Diameter, x:	7	8	9	10	11	12	13	$\bar{x} = 10$
Frequency, f:	1	6	24	38	24	6	1	$s_{\bar{x}} = 0{\cdot}107$

The weight of the object is required and, the density of the material of which it is composed being 5·24 g/cm^3, the dependence of weight, y, on diameter, x, is given by $y = x^3/10$. If each of the observed diameters is transformed into weight we find that the standard deviation of the resulting distribution, $s_y = 32{\cdot}8$, so $s_{\bar{y}} = 3{\cdot}28$. If we transform $s_{\bar{x}} = 0{\cdot}107$ we find $s_{\hat{y}} = 3\bar{x}^2 s_{\bar{x}}/10 = 3{\cdot}21$. Thus, the standard error from eqn. (3.60d) is of quite acceptable accuracy, even in this rather extreme case.

(5) *Tolerance Limits*

Prediction of the outcomes of future sampling operations (say of the values of the next few experimental observations on a given system) is not the normal occupation of experimentalists. Nevertheless, predictions are sometimes made on the basis of available information. For example, eqns. (3.57) and (3.59) might be used explicitly to put limits on a future value of the dependent variable as in the planning of a precision experiment of the kind described by Worthing and Geffner (1943). The problem also arises in the use of calibration curves, determined from past data, discussed in Chapter IV.

Given that a distribution is of Gaussian (or other known) form and that its mean and standard deviation are also known, limits embracing stated proportions of individuals, and therefore probabilities of a particular member being drawn within those limits, can be laid down. Thus, in the Gaussian case, we can say

$$P(\mu - 1{\cdot}96\,\sigma < x_i < \mu + 1{\cdot}96\,\sigma) = 0{\cdot}95$$

and

$$P(\mu - 1{\cdot}64\,\sigma < x_i < \mu + 1{\cdot}64\,\sigma) = 0{\cdot}90$$

In other words, the limits $\mu \pm 1{\cdot}96\,\sigma$ and $\mu \pm 1{\cdot}64\,\sigma$ will embrace 95% and 90%, respectively, of individual items to be drawn from the universe. In the terminology of industrial statistics these are called *tolerance limits* and are applied to expectations with regard to future production on the assumption that the universe of products remains unchanged.

When $\bar{x}$ and s have to be used in place of μ and σ we have to make allowance for error and can no longer be certain that limits like $\bar{x} \pm 1{\cdot}96s$ embrace 95% of individuals. In repeated sampling this *population coverage*, *P*, will sometimes be greater than 95% and sometimes less. To ensure that on the average it is at least 95% with given confidence, 1 − P (the *confidence coefficient*), the factor 1·96 must be replaced by a new factor, *l*, which will be greater. Furthermore, the higher the degree of confidence required of the statement "that $\bar{x} \pm ls$ embraces 95% of individual items in the universe", the higher must the value of *l* be. However, for very large sample sizes (μ and σ being virtually known) *l* tends to 1·96 for $P = 95\%$ regardless of the requirements of 1 − P. Values of *l* for various values of *P* and 1 − P are given in Appendix Table VIII. Their use is discussed and examples (both one-and two-sided) given by Hald (1952) and Bowker and Lieberman (1959).

Notation of Chapter III

α	probability of wrongly rejecting H_0 (P in our notation)
A'	a quality control chart factor
β	probability of wrongly accepting H_0 ($1-\beta$ = the power of a test)
c	a function of degrees of freedom (eqn. 3.45)
C	a function of standard errors (eqn. 3.41 b)
χ^2	a test statistic
δ, d	true and estimated difference, generally between means
d_2	factor relating w to σ
D'	a quality control chart factor
e	an error in general, standard error, half confidence interval, specific or maximum error
e_i	absolute error in x_i
$e_{\hat{y}}$	error in a value, $\hat{y}$, computed from other values
e	an error estimated from differences between sample means

E	expected frequency
E	e.m.f.
f	degrees of freedom; frequency. f_i = frequency of items of kind i
F	a test statistic (variance ratio); cumulative frequency
γ	surface tension
γ_1, γ_2	measures of skewness and excess of kurtosis
h	value of a discrete variable, generally the number of "heads" in coin tossing
H_0	null hypothesis
H_a	alternative hypothesis or hypotheses
i	(generally subscript) individual item in a sample or kind of item
j	individual sample in a set
I, J	kinds of events
k	number of samples in a set; number of classes
l	factor used to find tolerance limits
m	sample median
M	test statistic used in comparing variances; universe median
MS	mean square = SS/f
μ	true universe mean
$\mu_{\bar{x}}$ etc.	true mean of sampling distribution of sample means etc.
μ_0	hypothetical value of μ
n	size of sample
N	size of universe; total number in a set of samples
O	observed frequency
p, sometimes p_i	an estimated relative frequency, proportion or probability
$\bar{p}$	an average proportion, the mean of sample values
P, sometimes P_i	true relative frequency, proportion or probability of items of kind i; probability; population coverage
P_0	a hypothetical true proportion
P	level of significance; the level of confidence or confidence coefficient, as opposed to level of significance, being indicated by $1-\text{P}$
pH	$-\log\,[\text{H}^+ \text{ activity}]$
q	studentized extreme deviate from the sample mean

Q	probability of items of kind other than i
r	Dixon's r, a statistic used in testing outliers
s	estimate of σ given by eqn. (2.17b)
$s_{\bar{x}}$ etc.	estimated standard error (deviation) of sampling distribution of sample means etc.
s_p	pooled estimate of σ
SS	sum of squares of deviations
σ	true universe standard deviation
σ_0	hypothetical value of σ
$\sigma_{\bar{x}}$ etc.	true standard error of sampling distribution of means etc.
t	deviate in estimated standard error units
t	time
φ	a function
u	the standardized deviate
v	coefficient of variation
w, $\bar{w}$	sample and mean sample range; weight of an observation
x, sometimes x_i	an individual observed value
x_1, x_n	lowest and highest values in a sample
$\bar{x}$, $\bar{\bar{x}}$	sample and grand means
X, sometimes X_i	the deviation $x_i - \bar{x}$
y	a dependent variable; $\hat{y}$ = best estimate of μ_y computed from x_1, x_2 etc.
1, 2	used generally (as subscripts) to differentiate samples
$\lvert \ldots \rvert$	without regard to sign—used in expressions for test statistics when the H_a are two-sided

CHAPTER IV

THE RELATIONSHIP BETWEEN TWO VARIABLES—SIMPLE LINEAR RELATIONSHIPS

MUCH of the data from experimental observation consist of pairs of associated observations: temperature and pressure, degrees of chemical advancement with time, voltage and current, etc. The purpose of investigations yielding such information is (a) to test the applicability of models (the "calculus of exploration" in the terminology of Hogben (1957)), (b) to test hypotheses concerning those models or to evaluate their parameters (the "calculus of judgments" or the "calculus of errors") or (c) to carry out operations (smoothing, differentiation, extrapolation etc.) in which (a) and (b) are implied even if the usual calculations are not performed. For example, smoothing of tabulated data and drawing a freehand curve through graphed points are simply methods of obtaining values freed as far as possible of random error without going to the trouble of selecting a suitable type of relationship and fitting an equation.

The primary information may be graphed for the sake of pictorial representation but the operations listed above are, in general, carried out on tabular forms. The basic treatment of linearly related data in so far as it involves equation fitting is the subject of the present chapter and modifications for the handling of non-linear relationships and linear relationships involving more than two variables are dealt with in the next. The methods of treatment presented in these two chapters are mainly statistical in the sense that they are largely based on the principle of least squares and on the supposition of normally distributed errors of observation. The treatment of tabulated data, depending in the main on successive differences between values, is the subject of Chapter VI.

4.1. Regression, Correlation and Confluence

The method of analysis appropriate to the investigation of relationships between two variables depends on their nature and on the nature of the relationship. Broadly, there are three possibilities corresponding to the three basic situations or models applying to the universe of which any set of observed pairs of values (x_1, y_1; $x_2, y_2 \dots x_n, y_n$) constitutes a sample.

(a) *Correlation Analysis*

For investigating the relationship between random variables, particularly when normally distributed and linearly connected. Consider the classic example in which x_i is the height of a father and y_i is the height of his son. Error in the measurement of the values is unimportant by comparison with their natural range of variation and may be ignored. The problem is to determine the degree of association and perhaps to estimate the frequency contours of the bivariate universe (analogous to the frequency curve in the univariate case) or to find equations for representing the relationship of x to y or y to x.

(b) *Regression Analysis*

For investigating the association of a random or "dependent" variable with one whose values are known exactly or can be selected ("independent"). For example, y_i may be the amount of substance found by an analytical method or an instrument reading and x_i an exactly known amount of substance taken. The problem is to find the equation relating mean values of y at particular values of x to the values of x in order to provide estimates of the constants of the true equation or for purposes of prediction.

(c) *Confluence Analysis*

For investigating the association between two variables, the observed values of each consisting of a random component (say error of measurement) superimposed on a not precisely known true vlaue or "structural" component. For example, y_i might be the compressive strength and x_i the bending stress strength of a cement test piece, both measurements being about equally subject to error. The problem here is to determine the relationship between the structural components.

All these models may be complicated by non-linearity or additional variables or by special constraints. The confluence model is perhaps the most generally applicable to physico-chemical data but in most cases the values of one variable are controllable or known with sufficient relative precision for it to be regarded as independent with respect to the other so that the simpler methods of regression analysis may be applied. It is to this kind of situation that we now turn.

4.2. The Simple Regression Line

Consider an independent variable x and a related, dependent variable y whose values are subject to a normally distributed error of measurement, unaffected by x, and whose true means at various points on the x scale fall on a straight line. Such a model may be assumed to apply to the data of Table 4.1, i.e. to represent the universe from which these data come as a random sample. Here,

Table 4.1. Results of Tests on an Analytical Method

Concentration taken, x (arbitrary units)	Concentration found, y (arbitrary units)
9	12·2
18	24·0
27	27·7
36	35·4
45	47·5
54	50·2
63	62·0

x is the known amount or concentration of a substance taken for testing purposes and y is the result found by an analytical method. The fact that the y values are given to the first decimal place implies that computational or instrumental reading error is of the order of 0·1 units but this takes no account of errors due to manipulation etc.

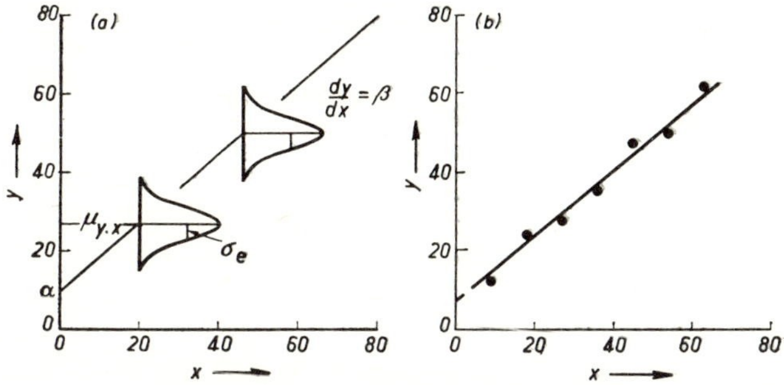

FIG. 4.1. Tests of an Analytical Method.

The kind of universe from which the sample of Table 4.1 may be presumed to have come is illustrated in (a). The data themselves are plotted in (b) with the least-squares line calculated in Table 4.2 drawn through them.

The situation may be represented graphically as in Fig. 4.1. The relevant equations are

$$\mu = \alpha + \beta x; \quad \hat{y} = a + bx \qquad (4.1\text{a, b})$$

$$y = \mu + e = \alpha + \beta x + e; \quad y = \hat{y} + \hat{Y} = a + bx + \hat{Y} \quad (4.2\text{a, b})$$

$$\sigma_e^2 = \Sigma(y - \mu)^2/N; \; s_e^2 = \Sigma(y - \hat{y})^2/(n-2) = \Sigma\hat{Y}^2/(n-2) \quad (4.3\text{a,b})$$

where n is the number of observed points, $\hat{y}$ estimates μ (more precisely $\mu_{y \cdot x}$, the true mean of all possible y values at a given point on the x scale) and a, b and s_e estimate the parameters of the model α, β and σ_e (the intercept, the slope or *regression coefficient*, and *standard error of estimate* respectively). To simplify the notation, subscripts indicating particular pairs of values or positions on the x scale are omitted unless strictly necessary to avoid ambiguity and the points are taken individually so the f of the equations in Tables 2.6 and 2.7 are also omitted. As before, the universe is considered finite (size N) in eqn. (4.3a), the equation taking the form of a ratio of integrals in the infinite case.

Now the required line, $\hat{y} = a + bx$, might be obtained in a number of ways:

(a) By drawing by eye through the graphed points, the positioning being carried out for preference with the aid of a device which enables the drawer to see all the points at the same time, string or a transparent plastic ruler having a line engraved down its centre being recommended.

(b) By the "method of selected points" using two points at or towards the ends of the range of x.

(c) By the "method of averages" whereby the points are divided in the x direction to give two equal terminal groups each of which are averaged for x and y.

(d) By minimizing algebraically the sum of squares of the deviations, $\Sigma \hat{Y}^2$, the "least squares" line so found being the two dimensional equivalent of the arithmetic mean. (When the x values are equally spaced this may be done by a summation method.)

(e) By a graphical method (for reasonably scattered points) which also gives the least squares line.

The algebraic methods must provide two simultaneous equations solvable for slope and intercept. In the least squares case these are called the *normal equations* and are readily derived as follows. It is required to find values of a and b for which $\Sigma \hat{Y}^2$ is a minimum and, therefore, for which $\mathrm{d}\Sigma \hat{Y}^2/\mathrm{d}a$ and $\mathrm{d}\Sigma \hat{Y}^2/\mathrm{d}b$ are zero. Let us for the moment distinguish the individual pairs of values by subscript, $i = 1, 2, \ldots, n$. Now for any individual value of y, y_i, we have

$$\hat{Y}_i = y_i - \hat{y}_i = y_i - a - bx_i$$

and

$$\hat{Y}_i^2 = y_i^2 - 2ay_i + 2abx_i - 2bx_iy_i + b^2x_i^2 + a^2$$

Adding the squares for the sample of n pairs of values gives

$$\Sigma \hat{Y}_i^2 = \Sigma y_i^2 - 2a\Sigma y_i + 2ab\Sigma x_iy_i - 2b\Sigma x_iy_i + b^2\Sigma x_i^2 + na^2 \quad (4.4)$$

Differentiating this with respect to a and equating to zero we obtain

$$\mathrm{d}\Sigma \hat{Y}_i^2/\mathrm{d}a = -2\Sigma y_i + 2b\Sigma x_i + 2na = 0$$

whence

$$a = \Sigma y_i/n - b\Sigma x_i/n = \bar{y} - b\bar{x} \quad (4.5)$$

Differentiating with respect to b and equating to zero gives

$$\mathrm{d}\Sigma \hat{Y}_i^2/\mathrm{d}b = 2a\Sigma x_i - 2\Sigma x_iy_i + 2b\Sigma x_i^2 = 0$$

from which, after substituting for a in terms of eqn. (4.5) and putting $n\bar{x}$ for Σx_i etc. where necessary, we find

$$b = (\Sigma x_iy_i - n\bar{x}\bar{y})/(\Sigma x_i^2 - n\bar{x}^2)$$

or, using the transformation provided by eqn. (2.31 a),

$$b = \Sigma X_i Y_i / \Sigma X_i^2. \tag{4.6}$$

By substituting these expressions for a and b back into eqn. (4.4) we find

$$\Sigma \hat{Y}_i^2 = \Sigma Y_i^2 - (\Sigma X_i Y_i)^2/\Sigma X_i^2 = \Sigma Y_i^2 - b\Sigma X_i Y_i = \Sigma Y_i^2 - b^2 \Sigma X_i^2 \tag{4.7}$$

In other words, the residual variation of the y values about the line is equal to their total variation (i.e., the sum of squares of their deviations about their mean, ΣY_i^2) less an amount, given by $b\Sigma X_i Y_i$ or an equivalent expression, which is accounted for by the regression itself.

The least squares line is also characterized by the fact that the algebraic sum of the deviations about it is zero, i.e.

$$\Sigma \hat{Y}_i = \Sigma(y_i - a - bx_i) = \Sigma y_i - na - b\Sigma x_i = 0 \tag{4.8}$$

The method of averages may be regarded as an attempt to apply this criterion within parts of the data. To obtain two simultaneous equations let us divide the values into equal groups, $x_1, y_1 \ldots x_j, y_j$ and $x_{j+1}, y_{j+1} \ldots x_n, y_n$ such that j is as close as possible to $n/2$ and let us call the sums of the values in the first group $\Sigma x'$ and $\Sigma y'$ and in the second $\Sigma x''$ and $\Sigma y''$. Adapting eqn. (4.8) we have

$$\Sigma y' - na/2 - b\Sigma x' = 0 \qquad \text{and} \qquad \Sigma y'' - na/2 - b\Sigma x'' = 0$$

By subtraction of the first from the second of these equations

$$\Sigma y'' - \Sigma y' - b\Sigma x'' + b\Sigma x' = 0$$

whence

$$b = (\Sigma y'' - \Sigma y')/(\Sigma x'' - \Sigma x') = (\bar{y}'' - \bar{y}')/(\bar{x}'' - \bar{x}') \tag{4.9}$$

and

$$a = \bar{y}' - b\bar{x}' = \bar{y}'' - b\bar{x}'' \tag{4.10}$$

where $\bar{x}'$, $\bar{y}'$ and $\bar{x}''$, $\bar{y}''$ are the within-group means. Other groupings can be adopted and the first and last thirds are held to give, on the average, a better estimate of slope (by neglecting the relatively non-contributing central values) than the first and second halves. Equations (4.9) and (4.10) still apply. An application is given in Section 4.12 below.

Of the various methods described, only (d) provides an inbuilt means of evaluating the sum of squares of the deviations for the estimation of the error variance. The number of independent differences, $\hat{Y}$, provided by the data after fitting the line is $n - 2$ and the division of $\Sigma\hat{Y}^2$ by this number of degrees of freedom gives an unbiased and most efficient estimator, s_e^2, of σ_e^2, from which confidence intervals for the parameters can be found. (The estimator, $s_e = \sqrt{s_e^2}$, of σ_e is of course slightly biased.) Equivalent estimates may be obtained from lines fitted in other ways but only by calculation or graphical measurement of the individual deviations or by indirect means. For example, an extension of the graphical method (e) gives the average deviation which is related approximately to the standard error by the factor 1·25.

When applied to the derivation of a linear relationship the method of least squares is not unduly tedious—hardly more so than the method of averages—and the labour of computation may be much reduced if the x values are equally spaced. In the actual fitting of the line no more assumptions are made in this than in any other method—all being based on the proposition that a linear relationship applies over the range of observation. The grounds for our preference for the method of least squares, however, are efficiency (in the sense that lines so determined are on the average closer to the true line than those obtained by other methods) and objectivity. Thus, if we are dealing with very scattered points the method is indicated because subjective sources of error are obviated while if the data are good they deserve adequate treatment, particularly if it yields useful confidence intervals for the parameters of the model.

4.3. Fitting the Least Squares Line

The three methods of fitting the least squares line described above will now be illustrated using the data of Table 4.1.

(1) *General Method*

We require the sums of squares of the deviations of the x values from their mean and of products of the x and y deviations. The sum of squares of the y deviations is not needed for fitting but is required for the evaluation of the standard error of estimate, s_e.

These quantities are best found from the sums of squares and products of the actual values. All necessary calculations are shown in Table 4.2.

TABLE 4.2. BASIC COMPUTATIONS FOR THE LEAST SQUARES LINE

x	y	x^2	y^2	xy
9	12·2	81	148·84	109·8
18	24·0	324	576·00	432·0
27	27·7	729	767·29	747·9
36	35·4	1296	1253·16	1274·4
45	47·5	2025	2256·25	2137·5
54	50·2	2916	2520·04	2710·8
63	62·0	3969	3844·00	3906·0

$\Sigma x = 252$; $\Sigma y = 259{\cdot}0$; $\Sigma x^2 = 11{,}340$; $\Sigma y^2 = 11{,}365{\cdot}58$; $\Sigma xy = 11{,}318{\cdot}4$
$\bar{x} = 36$; $\bar{y} = 37{\cdot}0$; $-n\bar{x}^2 = -9072$; $-n\bar{y}^2 = -9583{\cdot}00$; $-n\bar{x}\bar{y} = -9324{\cdot}0$
$n = 7$; $\Sigma X^2 = 2268$; $\Sigma Y^2 = 1782{\cdot}58$; $\Sigma XY = 1994{\cdot}4$

Slope (eqn. 4.6), $b = \Sigma XY/\Sigma X^2 = 1994{\cdot}4/2268 = 0{\cdot}879365$

Intercept (eqn. 4.5), $a = \bar{y} - b\bar{x} = 37{\cdot}0 - 31{\cdot}65714 = 5{\cdot}34286$

The fitted line is $\hat{y} = 5{\cdot}343 + 0{\cdot}879365\,x$

$\Sigma \hat{Y}^2 = \Sigma Y^2 - b\,\Sigma XY = 1782{\cdot}58 - 1753{\cdot}885528 = 28{\cdot}7744$

$s_e^2 = \Sigma \hat{Y}^2/(n-2) = 28{\cdot}774/5 = 5{\cdot}755$; $s_e = 2{\cdot}399$; $f = n - 2 = 5$

In machine calculation it is convenient to set x on the left-hand side of the setting register and y on the right and to square. This operation, which on some machines is automatic, gives $x^2 \ldots 2xy \ldots y^2$ and the accumulation in the products register is $\Sigma x^2 \ldots 2\Sigma xy \ldots \Sigma y^2$ with $\Sigma x \ldots \Sigma y$ in the revolutions register if equipped with tens carry-over. Some care is needed in establishing the capacity of the machine for the number of values of the size being processed. For example, a large machine with ten places in the setting register, eleven in the revolutions register and twenty-one in the product register can handle less than twenty pairs of three digit numbers averaging 500 although its capacity for two digit numbers is very large. When the capacity of the machine for accumulating sums of squares is exceeded because of the size of the numbers or of the sample or of the machine itself the operation can be carried out in two stages: Σx, Σx^2 and Σxy in one and Σy, Σy^2 and Σxy in the next. In the single operation the appearance of the correct Σx and Σy in the revolutions counter is a good check on the accuracy of the summations; in the split operation the agreement of the two Σxy is a certain test. If the columns must be summed separately the best check is provided by forming $q = x + y$ and finding Σq^2 equal to $\Sigma x^2 + \Sigma y^2 + 2\Sigma xy$. The later, and indeed all, stages of the calculation can be checked approximately by a graphical comparison of the fitted line with the points or absolutely by finding the individual deviations from the line and comparing the sum of their squares with the result of formula (4.7).

When calculating without the aid of a machine it is desirable to reduce the number of significant figures in the sample data by coding. For example, the x may be simplified by subtracting a constant close in value to $\bar{x}$ and the y may be treated likewise or may be simplified even further by subtracting a function of x, say $A + Bx$, where A and B are guessed approximations to a and b. Reconstruction of the equation in terms of y and x from that relating the coded values is a simple enough matter but a graphical check (at least) is always desirable.

The number of significant figures which should be retained in calculated constants depends to some extent on the uses to which they are to be put but the rule is to retain a superfluity of figures and to round off only at the end of a series of computations. In the example b correct to five significant figures still gives $\Sigma \hat{Y}^2$ correct to between three and four significant figures. The sums of squares and products of differences must themselves be accurate to justify the retention of many significant figures and, although equations of the form $\Sigma x^2 - n\bar{x}^2$ have been used for simplicity in setting out the example, the better computational forms given in Section 2.5 above should be used where necessary.

(2) *Summation Method*

This is applicable to the example because the x values are equally spaced. It is simply the appropriate modification of the method for fitting orthogonal polynomials described in Section 5.2 below and the notation of that section is used here. We require the sum and the first cumulative sum of the y values. The x must be coded so as to advance in steps of unity. The calculations are shown in Table 4.3.

A comparison with Table 4.2 shows that the line itself is fitted more easily by the summation method and there is a further gain if the cumulative sum is computed by the second orthogonal polynomial method of Section 5.2, rather than by the successive additions used in the table below. The gain is not so considerable when $\Sigma\hat{Y}^2$ and ΣX^2 must be obtained for later calculations although ΣX^2 is readily found from the corresponding value for the first n natural numbers. For $n = 7$ this is 28 and $\Sigma X^2 = 9^2 \times 28 = 2268$ as in Table 4.2.

TABLE 4.3. SUMMATION COMPUTATION FOR THE LEAST SQUARES LINE

x	$\varkappa$	y	$y(2)$	y^2
9	1	12·2	12·2	148·84
18	2	24·0	36·2	576·00
27	3	27.7	63.9	767.29
36	4	35·4	99·3	1253·16
45	5	47·5	146·8	2256·25
54	6	50·2	197·0	2520·04
63	7	62·0	259·0	3844·00

$$\Sigma y = S_1 = 259{\cdot}0;\ S_2 = 814{\cdot}4; \qquad 11{,}365{\cdot}58 = \Sigma y^2$$
$$\bar{y} = S_1/n = 37{\cdot}0; \quad \bar{\varkappa} = 4; \quad \varkappa = x/9$$
$$B = 6\bar{y}/(n-1) - 12S_2/(n^3 - n) = 37{\cdot}0 - 12(814{\cdot}4)/336 = 7{\cdot}914286$$
$$\hat{y} = \bar{y} + B(\varkappa - \bar{\varkappa}) = 37{\cdot}0 + 7{\cdot}914286(\varkappa - 4) = 5{\cdot}343 + 7{\cdot}914286\varkappa$$
$$\text{Substituting } x/9 \text{ for } \varkappa\text{:} \quad \hat{y} = 5{\cdot}343 + 0{\cdot}879365x$$
$$\Sigma\hat{Y}^2 = \Sigma y^2 - n\bar{y}^2 - (n^3 - n)\,B^2/12$$
$$= 11{,}365{\cdot}58 - 9583 - 336(7{\cdot}914286^2)/12 = 28{\cdot}7742$$
$$s_e^2 = \Sigma\hat{Y}^2/(n-2) = 28{\cdot}7742/5 = 5{\cdot}755; \quad s_e = 2{\cdot}399; \quad f = n - 2 = 5$$

(3) *Graphic Method*

Askovitz (1957, 1959) has described graphic methods for fitting the least squares line to equally and unequally spaced observations and for altering an already fitted line to take account of further observations. These techniques work best when the points are reasonably scattered (i.e. the y scale must be adequately opened up) and when the slope is low. Only one of Askovitz' methods will be illustrated here and for the example we take the data of Table 4.1 but use the analytical error, $z = y - x$, as the dependent variable. The least squares line, $\hat{z} = a + b'x$, has the same intercept and error variance as that fitted above but the slope $b' = b - 1$. The procedure is demonstrated in Fig. 4.2(a), the points being plotted so that thirds of the x spacings are clearly marked. No standard

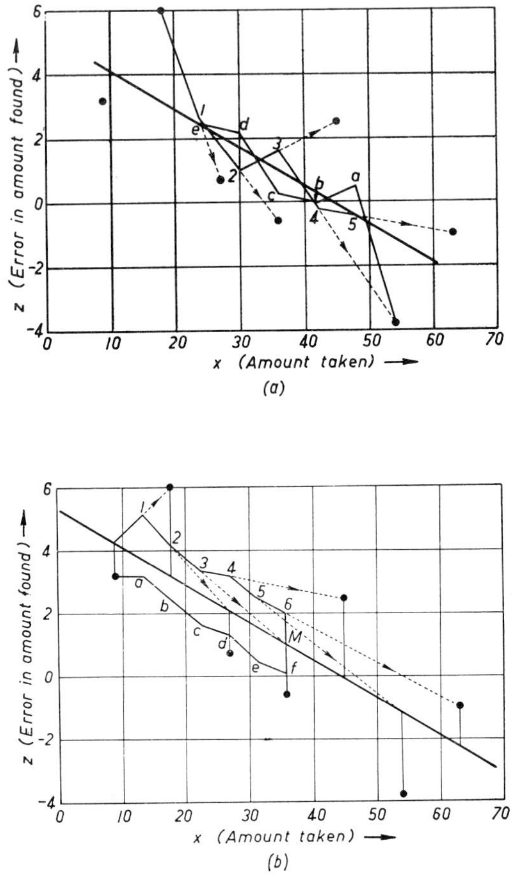

FIG. 4.2. Askovitz' Methods of Fitting and Finding the Average Deviation.

(a) Fitting the least-squares line. Starting from the second point, advance two-thirds of a plotting unit (6 scale units in this instance) towards the third point to 1, then two-thirds of a unit towards the fourth point to 2, and so on until 5 is reached. Starting from the second last point, advance two-thirds of a unit towards the third last to a, and so on until e is reached. The line through $e5$ is the least-squares line. (b) Finding the average deviation. Draw in lines representing the deviations of the points from the least-squares line. Starting from the top of the first line, advance half a plotting unit towards the top of the second to 1, then half a unit towards the top of the third to 2, and so on until 6 is reached. Then $6M$ (where M is a point on the line half-way between x_1 and x_n) is half the average deviation. As a check, starting from the bottom of the first line, advance in half units as before to f. Then $6f$ is the average deviation.

error of estimate is provided but a means of finding the fairly efficient average deviation is illustrated in Fig. 4.2(b) and this, multiplied by 1·25 (eqn. (2.24a)), gives the standard error. Here, the graphically determined average deviation is about 1·90 which leads to an estimated standard deviation of 2·38 compared with 2·40 obtained by direct calculation.

4.4. Confidence Intervals for the Parameters of the Simple Linear Model

A line may be fitted to a sample of associated data simply for the purpose of condensation of those data with the additional possibility that the equation so found may be used for occasional or repeated interpolation or even extrapolation without bothering about the scatter of the points. In general, however, we wish to make fuller application of the model and (making use of the postulate of normality) to exploit the estimate of error. We may wish to establish a confidence interval for one or other parameter—say for the true energy of activation as slope in a plot of logarithm of reaction rate against reciprocal temperature, or for a function of molecular weight of a gas as the intercept when applying the limiting density method. Alternatively a joint confidence region for slope and intercept, i.e. limits for the true line as a whole, may be required. Again, in problems of interpolation and extrapolation, we may wish to estimate the true value of y (the point μ on the true line), to put limits on an individual value of y at a particular value of x or, reversing the last procedure in calibration problems, to put limits on x given one or more additional observations of y. We will consider first the parameters.

(1) *Tests and Confidence Intervals for Slope and Intercept*

It can be shown that the sampling distributions of a and b obtained by repeated random sampling from a universe of the type being considered are normal (with standard errors related to the error variance) and unbiased (i.e. having α and β as true means). We may therefore apply tests of significance and calculate confidence intervals in the same way as is illustrated in Section 3.5 above. Thus, the slope,

$$b = \Sigma XY/\Sigma X^2 = X_1Y_1/\Sigma X^2 + X_2Y_2/\Sigma X^2 \ldots X_2Y_2/\Sigma X^2$$

Applying the result of eqn. (3.59) and remembering that s_Y^2 is s_e^2 in our present notation,

$$s_b^2 = (X_1/\Sigma X^2)^2\, s_e^2 + \ldots = (s_e/\Sigma X^2)^2\, (X_1^2 + X_2^2 + \ldots) = s_e^2/\Sigma X^2$$

so

$$s_b = s_e/\sqrt{\Sigma X^2} \qquad (4.11\,\text{a})$$

$$t = (b - \beta_o)/s_b \qquad (4.11\,\text{b})$$

$$100(1 - \text{P})\%\ \text{confidence limits of}\ \beta = b \pm t_{\text{P},f}\, s_b \qquad (4.11\,\text{c})$$

and for the intercept,

$$s_a = s_e\sqrt{(1/n + \bar{x}^2/\Sigma X^2)} = s_e\sqrt{(\Sigma x^2/n\Sigma X^2)} \qquad (4.12\text{a})$$

$$t = (a - \alpha_o)/s_a \qquad (4.12\text{b})$$

$$100\,(1 - \text{P})\,\%\ \text{confidence limits of}\ \alpha = a \pm t_{\text{P},f}\, s_a \qquad (4.12\text{c})$$

where s_e is found from $\Sigma \hat{Y}^2$ by eqn. (4.3 b) and is based on $f = n - 2$ degrees of freedom, α_o and β_o are hypothetical values under test and the other symbols have their usual significance.

Applying these equations to the example of Tables 4.1 and 4.2 we have, from

eqn. (4.11 a), $s_b = 2{\cdot}399/47{\cdot}624 = 0{\cdot}0537_4$

eqn. (4.12a), $s_a = 2{\cdot}399\sqrt{(0{\cdot}1429 + 36^2/2268)} = 2{\cdot}028$

eqn. (4.11c), 95% CL$\beta = b \pm t_{\cdot 05,5}\, s_b = 0{\cdot}879_4 \pm 2{\cdot}571(0{\cdot}0537)$

$= 0{\cdot}879 \pm 0{\cdot}138 = 0{\cdot}74$ to $1{\cdot}02$

eqn. (4.12c), 95% CL$\alpha = a \pm t_{\cdot 05,5}\, s_a = 5{\cdot}343 \pm 2{\cdot}571(2{\cdot}028)$

$= 5{\cdot}343 \pm 5{\cdot}214 = 0{\cdot}13$ to $10{\cdot}56$

The confidence intervals will serve to answer tests of significance or eqns. (4.11 b) and (4.12 b) may be used. Generally such tests are applied to see if there are reasonable grounds for supposing there to be any association at all (i.e. $\beta_o = 0$ on the null hypothesis) or to see if the true line might pass through the origin (i.e. $\alpha_o = 0$ on the null hypothesis).

An alternative to the t test for true zero slope is provided by an analysis of variance of the regression by which the total variability of the values about their mean is split into the component explained by the fitted line and a residual. The procedure, illustrated in Table 4.4, is not very instructive in the simple case but will be found useful in later sections. The probability of obtaining the observed ratio of mean squares (305) in random sampling is the same as that of getting $t_5 > |b|/s_b = |0{\cdot}8794|/0{\cdot}0537 = 16{\cdot}4$; it is extremely low.

TABLE 4.4. ANALYSIS OF VARIANCE OF REGRESSION

Source of variation	SS	f	MS	Ratio
Due to regression (line about mean)	$b\Sigma XY = 1753{\cdot}9$	1	1753·9	305
Residual (points about line)	$\Sigma Y^2 - b\Sigma XY = 28{\cdot}7$	$n-2=5$	5·74	$(F_{\cdot05,1/5} = 6{\cdot}61)$
Total (points about mean)	$\Sigma Y^2 = 1782{\cdot}6$	$n-1=6$		

(2) *A Joint Confidence Region for Slope and Intercept*

The results given above are valid independent interval estimates for α and β at the 95% level but we cannot say with this same degree of confidence that α and β might *at the same time* be at the lower or upper ends of the intervals calculated for them, i.e. 0·741 and 0·13 or 1·017 and 10·56 respectively. It may be inferred from an inspection of Fig. 4.1 that if (when the true line has positive slope) a sample turns out a relatively low estimate of slope it is likely to give a high value for the intercept and vice versa. Actually, if a is plotted against b for repeated random samples the points will be found to fall elliptically about the true centre (α, β) and, conversely, any confidence interval for the true combination of α *and* β will take the form of an elliptical region about the best estimates (a, b) as centre. While in general a sample giving a correct interval for α will also give a correct interval for β, there will be some cases in repeated sampling where α is covered by eqn. (4.12c) but not β by eqn. (4.11c) and vice versa. Consequently the extremes of the 100(1 − P)% joint interval are a little greater than the independent 100(1 − P)% limits for the parameters. The factor, $t_{P,f}$ (two-sided P) in eqns. (4.11c) and (4.12c) must be replaced by $\sqrt{(2F_{P,2/f})}$ (one-sided P) which, for 95% confidence, is 30% larger at $n = 10$ and 25% larger in the limit of very large sample size. For $n = 7$ and $f = 5$ as in the sample the values are $t = 2{\cdot}571$ and $\sqrt{(2F)} = 3{\cdot}403$, the latter being 32% greater in this case.

The equation for the 100(1 − P)% joint ellipse may be written

$$(\alpha - a)^2 + 2\bar{x}(\alpha - a)(\beta - b) + (\Sigma x^2/n)(\beta - b)^2 = 2F_{P,2/f}\, s_e^2/n \tag{4.13a}$$

This can be transposed in various ways but the most convenient form for plotting is

$$\alpha^2 + 2(\bar{x}\beta - \bar{y})\alpha + (\Sigma x^2/n)\beta^2 - 2(b\Sigma X^2/n + \bar{x}\bar{y})\beta + \bar{y}^2 + b^2\Sigma X^2/n - 2Fs_e^2/n = 0 \tag{4.13b}$$

which may be solved for α by the quadratic formula,

$$\alpha = (\bar{y} - \bar{x}\beta) \pm \sqrt{[-(\Sigma X^2/n)(\beta - b)^2 + (2Fs_e^2/n)]} \tag{4.13c}$$

The first term of eqn. (4.13c) is the line of slope $-\bar{x}$ passing through the point (a, b) whose projected length in the β direction is the overall interval for β, i.e.

$$b \pm \sqrt{(2F)}\, s_b \tag{4.13d}$$

The extreme limits in the α direction are those of the corresponding α interval,

$$a \pm \sqrt{(2F)}\, s_a \tag{4.13e}$$

and by applying the quadratic formula to eqn. (4.13b) and solving for β it may be found that the line passing through these extremes and the centre has a slope of

$$(\Sigma x^2/\Sigma x) \tag{4.13f}$$

A few values of the second term of eqn. (4.13c) added to the line given by the first term enable the ellipse to be sketched satisfactorily. Mandel and Linnig (1957) describe an approximate method of construction based on the extreme limit lines and two tangents passing through the points

$$b \pm e_b' \quad \text{and} \quad a \pm e_a' \tag{4.13g,h}$$

where

$$e_b' = \sqrt{(2F)}\, s_e \sqrt{[2n/(\Sigma x \sqrt{n} \sqrt{\Sigma x^2} + n\Sigma x^2)]} \qquad (4.13i)$$

and

$$e_a' = e_b'(s_a/s_b) \qquad (4.13j)$$

Substituting the appropriate value for F and the other relevant quantities given in Table 4.2 into eqn. (4.13c) we find for our example

$$\alpha = (37 - 36\beta) \pm \sqrt{[-324\,(\beta - 0{\cdot}8794)^2 + 9{\cdot}514]}$$

for the joint confidence ellipse. To aid in the construction of this we compute the other quantities from the expressions above.

Centre, $(a, b) = (5{\cdot}343, 0{\cdot}879)$, $\sqrt{(2F)} = 3{\cdot}403$, $\sqrt{(2F)}\, s_e = 8{\cdot}161$

Extreme limits in the β direction (eqn. 4.13d)

$$= 0{\cdot}879_4 \pm 3{\cdot}403(0{\cdot}0536) = 0{\cdot}708 \text{ to } 1{\cdot}051$$

on a line of slope (eqn. 4.13c) $= -36$

Extreme limits in the α direction (eqn. 4.13e)

$$= 5{\cdot}343 \pm 3{\cdot}403(2{\cdot}028) = -1{\cdot}56 \text{ to } 12{\cdot}23$$

on a line of slope (eqn. 4.13f) $= -11{,}340/252 = -45$

Tangent point coordinates from eqns. (4.13g) and (4.13i)

$$= 0{\cdot}879_4 \pm 8{\cdot}161 \sqrt{[14/(252 \sqrt{7} \sqrt{11{,}340} + 7(11{,}340))]}$$

$$= 0{\cdot}879_4 \pm 0{\cdot}078_7$$

$$= 0{\cdot}801 \text{ and } 0{\cdot}958 \text{ on the } \beta \text{ scale}$$

and from eqns. (4.13h), (4.13i) and (4.13j)

$$= 5{\cdot}343 \pm 0{\cdot}0788(2{\cdot}028/0{\cdot}05374)$$

$$= 2{\cdot}371 \text{ and } 8{\cdot}315 \text{ on the } \alpha \text{ scale}$$

The ellipse and its construction with these aids is illustrated in Fig. 4.3, but if it is desired to test only one or two pairs of possible values the F test of Section 4.6 is much less time consuming.

That individual and joint tests can differ in their results may be seen by comparing the figures here with the independent confidence intervals calculated above. It will be noted, for example, that it would not be unreasonable to suppose that the sample represents a true line of zero intercept and unit slope, i.e. $\mu = x$, since the point (0, 1) falls within the 95% confidence ellipse. However, had

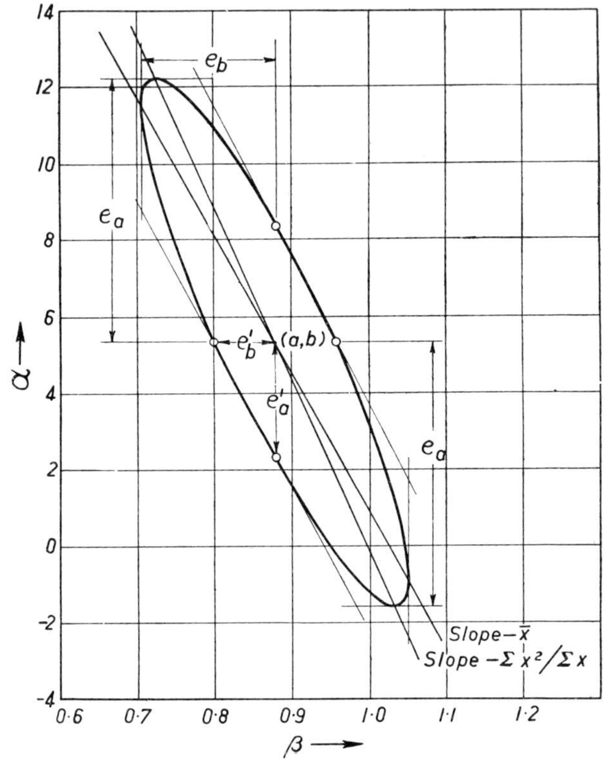

FIG. 4.3. The Joint Confidence Region for α and β.

The ellipse is the 95% confidence region for the true slope and intercept estimated from the data of Table 4.1. The 95% interval for α alone is $a \pm e_a$, and for β alone is $b \pm e_b$. The point representing a true intercept of zero and true slope of unity falls within the joint confidence region for these data.

the data been collected purely for testing the hypothesis that the true intercept were zero this would have been rejected by a test at the usual (5%) level of significance since the individual confidence interval for α does not embrace zero and, from eqn. (4.12b),

$$t = (a - 0)/s_a = 5{\cdot}343/2{\cdot}028 = 2{\cdot}635$$

which exceeds the critical value, 2·571, for this level and 5 degrees of freedom. The greater precision in the sense of a narrower interval estimate for the parameter by itself may be regarded as the reward for restricting the purposes of the investigation.

(3) *Interval Estimates for Points on the True Line*

The distinction just made between individual and joint estimates for α also applies in the more general case of estimation of points on the true line. The standard error of an estimated point may be obtained from the application of eqn. (3.59) to the variable quantities of the least squares eqn. (4.1 b) in the form

$$\hat{y} = \bar{y} + b(x - \bar{x}) = \bar{y} + bX,$$

i.e.

$$s_{\hat{y}}^2 = s_{\bar{y}}^2 + s_b^2 X^2 = s_e^2/n + s_e^2 X^2/\Sigma X^2$$

so

$$s_{\hat{y}} = s_e \sqrt{(1/n + X^2/\Sigma X^2)} \qquad (4.14\text{a})$$

Hence, $s_{\hat{y}} = 2{\cdot}399\sqrt{[0{\cdot}1429 + (x - 36)^2/2268]}$ for the example. For any given value of x, intervals for a point on the true line (i.e. the true mean of the y at that value) will be given by

$$100\,(1 - \text{P})\%\ \text{confidence limits of}\ \mu = \hat{y} \pm t_{\text{P},f}\, s_{\hat{y}} \qquad (4.14\text{b})$$

but, with the given degree of confidence, the true line may have any combination of values of α and β which lie within the joint confidence region given by the ellipse of eqn. (4.13 a), and the corresponding region for the line which allows for this joint unreliability of the estimates is a band bounded by

$$100\,(1 - \text{P})\%\ \text{confidence limits of}\ \mu = \hat{y} \pm \sqrt{(2F_{\text{P},\,2/f})}\, s_{\hat{y}} \qquad (4.14\text{c})$$

Thus, if we wish to find intervals for μ at several different values of x or, in other words, to use the fitted line for purposes of repeated interpolation with the same degree of confidence, then eqn. (4.14c) is the formula required.

As may be seen from Fig. 4.4(a), which shows eqns. (4.14b) and (4.14c) applied to the data of the example (the shaded and dotted bands respectively), they represent hyperbolae which are closest at the middle of the range of observation, $\bar{x}$, at which point X is zero and $s_{\hat{y}}$ reduces to $s_e/\sqrt{n}$. When we seek to make estimates of μ at greater and greater distances from $\bar{x}$ we have to contend with the increasing effect of the unreliability of the estimated slope and the intervals widen accordingly. At $x = 0$, $X = \bar{x}$ and eqn. (4.14a) reduces to eqn. (4.12a) for s_a.

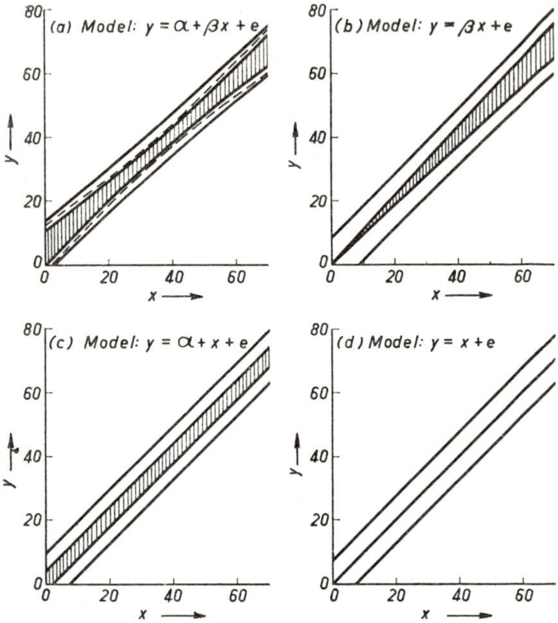

FIG. 4.4. Confidence Bands for Various Regression Models.

The inner shaded regions represent the 95% confidence bands for the true means of the y values at given x. The outer bands are for single values of y. In cases (b), (c) and (d), the bands are valid over the whole range of x. In the case of (a) the solid bands are valid only for particular values of x but the broken bands are valid over the whole range and correspond to the shaded bands in the other figures.

The process of interval estimation described here is simply refined interpolation or extrapolation—an additional risk naturally lying in extrapolation since the data themselves can give no clue as to the validity of the model outside the range of observation. This additional risk does not apply to the estimation of α as a parameter of the model but it could matter considerably if an attempt were made to actually use the regression equation at $x = 0$. For example, in fitting the Arrhenius Equation in the form

$$\log \mathrm{k} = \log \mathrm{A} + b(1/\mathrm{T})$$

we seek a parameter, the Arrhenius Constant, and have no illusion that this is in fact the experimental velocity constant, k, at infinite temperature ($1/T = 0$), but in fitting an equation to the data of Table 4.1 we must distinguish between α as a parameter of the model which we assume to apply over the observed range of x and α as $\mu_{y.0}$, the true mean y value at zero x or "analytical blank".

4.5. Predictions Based on a Fitted Line

The interval estimates described in the last section are subject to the general considerations of Section 3.5. The stated degree of confidence (say 95%) applies to the interval itself in the sense that 95% of intervals so calculated, from repeated sampling and fitting of lines, embrace the object of estimation. Such intervals may be the desired end of most physico-chemical investigations but sometimes the problem may be to find "prediction" intervals or tolerance limits by using the line as fitted or in reverse. In other words, instead of looking for limits of α and/or β based on measured values of the variables, we may wish to go a stage further and put limits on other (e.g. future) values of the variables based on the available estimates of α and β.

In the more concrete terms of our example, after fitting the line $\hat{y} = 5{\cdot}343 + 0{\cdot}8794x$, we may, for example, ask

(a) "Within what limits should we find the mean, $\dot{y}$, of m values of y, all determined on mixtures having a particular true concentration, say $x = 40$?"

(b) "What limits could we place on the unknown value of x which has given rise to $\dot{y}$, the mean of m determinations made at that x value?"

(c) "Within what limits should we find the means, $\dot{y}_1$, etc., each of m values of y determined on mixtures having various values of x?" or

"What limits could we place on the x values which have given rise to various observed $\dot{y}$ values?"

An extensive discussion of these and related problems may be found in Acton (1959) and other books. Only a brief outline is given here.

If such questions are asked of only one $\dot{y}$ or corresponding x value then confidence intervals can be found, i.e. intervals for which our "confidence" depends on the behaviour of all such which could be found by repeated sampling and fitting. The uncertainty in the mean, $\dot{y}$, of m observations at given x will be compounded of the variability of such means *per se* (measured by the variance, s_e^2/m) and of the uncertainty of the parameter they estimate (measured by the variance, $s_e^2(1/n + X^2/\Sigma X^2)$, from eqn. (4.14a)). These variances are independent and we may add them and take their square root (by virtue of eqn. (3.59)) to obtain a standard error

$$s_{\dot{y}} = s_e \sqrt{(1/m + 1/n + X^2/\Sigma X^2)} \qquad (4.15a)$$

Since the means, $\dot{y}$, would be distributed about the true mean of all values, μ, of which our best estimate is $\hat{y}$, we expect

$$100\ (1 - \mathrm{P})\%\ \text{confidence limits of}\ \dot{y} = \hat{y} \pm t_{\mathrm{P},f} s_{\dot{y}} \qquad (4.15b)$$

to represent an interval which embraces the mean value of y with the given degree of confidence. The way in which such intervals vary over the range of x is shown in Fig. 4.4a (outermost bands) for the case of single values ($m = 1$). The curvature and closeness of the bands increases with m until, in the limit of $m = \infty$, we reach those given by eqn. (4.14b) (innermost bands). If intervals valid for all values of x are required then we use

$$100\ (1 - \mathrm{P})\%\ \text{confidence limits of}\ \dot{y} = \hat{y} \pm \sqrt{(2F_{\mathrm{P},2/f})}\, s_{\dot{y}} \qquad (4.15c)$$

which in the limit of $m = \infty$ gives the same (dotted) bands as eqn. (4.14c).

When using the regression equation and its standard error in reverse, the precision of estimates of x depends on the reliability of the fitted line and also on the precision of the value of y involved. Approximately

$$s_{\hat{x}} = (s_e/b) \sqrt{[1/m + 1/n + (\dot{y} - \bar{y})^2/b^2\Sigma X^2]} \qquad (4.16a)$$

while

$$100\ (1 - \mathrm{P})\%\ \text{confidence limits of}\ x = \hat{x} \pm t_{\mathrm{P},f} s_{\hat{x}} \qquad (4.16b)$$

gives an interval valid for one point on the range, and

$$100\ (1 - \mathrm{P})\%\ \text{confidence limits of}\ x = \hat{x} \pm \sqrt{(2F_{\mathrm{P},2/f})}\, s_{\hat{x}} \qquad (4.16c)$$

gives intervals valid over the whole range of observation. In these equations $\hat{x}$ is the point estimate from the transformed equation

$$\hat{x} = (-a/b) + (1/b)\,\dot{y} = \bar{x} + (\dot{y} - \bar{y})/b \qquad (4.17)$$

As before, $\dot{y}$ is the available value of y (a single observation or a mean based on m observations) and the other quantities are the various properties of the sample by which the formula for the fitted line was found. $s_{\hat{x}}$ is not, of course, the standard deviation of a distribution of x values (since these are not random) but of values of x got from randomly varying y values.

The derivation of eqn. (4.16a) depends on the assumption that the variable quantities in eqn. (4.17) ($\dot{y}$, b, and a or $\bar{y}$) are normally distributed in repeated sampling (which is true with respect to the model) and independent (which is not). The resulting approximation, which by eqn. (4.16c) gives symmetrical intervals of x, is fair provided b is reliable. An exact relationship in terms of the model gives asymmetrical intervals (see, for example, Davies (1957)) and why this should be so can be seen from the shape of the curves in Fig. 4.4(a). In the limit of y being given or known exactly (in practice this would mean a vary large value of m) single or repeated values for x can be obtained graphically from the inner bands plotted in the figure. Limits of x from single values of y ($m = 1$) can be obtained from the curve derived from eqn. (4.15b).

The discrepancy between the exact or equivalent graphical solution and that obtained by eqn. (4.16b) will be the greater the further is x from $\bar{x}$ or $\dot{y}$ from $\bar{y}$ and the lower the value of b. Consider the use of the line fitted to the data of the example to find point and interval estimates of x from *given* values of y (i.e. $1/m = 0$). Equation (4.17) takes the form

$$\hat{x} = (-5{\cdot}343/0{\cdot}8794) + (1/0{\cdot}8794)\,\dot{y} = -6{\cdot}076 + 1{\cdot}1372\,\dot{y}$$

and $s_{\hat{x}} = (2{\cdot}399/0{\cdot}8794)\sqrt{(0{\cdot}1429 + (\dot{y} - 37)^3/1754)}$ from eqn. (4.16a). Using $t_{.05,5} = 2{\cdot}571$, we find the 95% interval for x at $y = 58$ to be $59{\cdot}9 \pm 4{\cdot}4$ or $55{\cdot}5$ to $64{\cdot}3$ whereas careful reading of the inner curves of Fig. 4.4(a) indicates an interval of $56{\cdot}0$ to $64{\cdot}9$ or $59{\cdot}9 - 3{\cdot}9$ to $59{\cdot}9 + 5{\cdot}0$. In the centre of the range of observation the difference between the two methods is minimal. Here, at $y = \bar{y} = 37$, the equations give $36{\cdot}00 \pm 2{\cdot}65$ or $33{\cdot}35$ to $38{\cdot}65$ and the graphical solution is indistinguishable from this. In the limit of the calibration line being known exactly, $s_{\hat{x}}$ reduces to $s_e/\beta\sqrt{m}$, which gives symmetrical intervals over the whole range.

Questions of type (c) are those most likely to arise in connection with prediction or calibration. If we wish to put many limits on values of $\dot{y}$ or x, i.e. to use a fitted line repeatedly in time (instead of or in addition to repeated use along the range of x) then we must use tolerance intervals analogous to those treated in Section 3.6(5) for the one-variable case. The ideal situation is one in which the estimates of the line and its standard error are so precise that any error can be ignored in relation to the scatter of the values themselves and limits can be placed on y values or, by inversion, on x values giving rise to particular y values, by using the appropriate value of the standard deviate, u. In practice this degree of precision could only be obtained by carrying out a very extensive calibration and in any case it must be remembered that even the most careful calibration confers no validity on the future use of an analytical method or instrument if the parameters of the model alter by reason of ageing reagents, instrumental wear or other causes.

In general, being content with less precisely fitted lines, we can use for any particular point on the scale

$$100\,P\%\text{ tolerance lts. of }\dot{y} = \hat{y} \pm l s_{\hat{y}}\text{ with }100\,(1 - \mathrm{P})\%\text{ confidence} \tag{4.18}$$

or

$$100\,P\%\text{ tolerance lts. of }\hat{x} \doteq \hat{x} \pm l s_{\hat{x}}\text{ with }100\,(1 - \mathrm{P})\%\text{ confidence} \tag{4.19}$$

where l is the appropriate factor from Appendix Table VIII. Better results than formula (4.19) can give might be obtained graphically from the tolerance bands of eqn. (4.18) or by using more accurate expressions for $s_{\hat{x}}$ referred to above.

The more realistic requirement is that of finding tolerance limits valid over the whole range of the fitted equation but, to quote Acton (1959), "although it is easy to formulate this problem, the necessary numerical integration seems laborious, and a useful approximation in terms of existing functions is as yet unknown". However, one might suppose that the tolerance interval bands valid over all values of x will be related to those valid for a single value in the way that the bands of eqns. (4.14b) and (4.15b) are related to those of eqns. (4.14c) and (4.15c). In other words, the tolerance limits required should be approximated by the expressions

$$100\,P\%\text{ tolerance lts. of }\dot{y} = \hat{y} \pm l' s_{\hat{y}}\text{ with }100\,(1 - \mathrm{P})\%\text{ confidence} \tag{4.20}$$

and

$$100\,P\%\text{ tolerance lts. of } x = \hat{x} \pm l's_{\hat{x}} \text{ with } 100\,(1-\text{P})\%\text{ confidence} \tag{4.21}$$

where the new factor is given by

$$l' = l[\sqrt{(2F_{\text{P},\,2/f})}/t_{\text{P},f}]. \tag{4.22}$$

4.6. Constrained Models and their Comparison with the Simple Regression Model

The characteristics of the model described above are (a) α and β are both unknown and must be estimated, and (b) $\Sigma\hat{Y}^2$ is reduced to an absolute minimum in the fitting process. If special knowledge is available, however, it should be taken into account by appropriate changes in the derivation of the least squares eqns. (4.4) to (4.7) above. For example, if we know α then the model presumed to fit the data still be represented by eqn. (4.1a) but the fitted equation takes the form

$$y_i = \alpha + bx + \hat{Y}_i \tag{4.23}$$

so

$$\hat{Y}_i^2 = (y_i - \alpha - bx_i)^2$$

and summing the n individual sets of values gives

$$\Sigma\hat{Y}_i^2 = \Sigma y_i^2 - 2\alpha\Sigma y_i + 2\alpha b\Sigma x_i - 2b\Sigma x_i y_i + b^2\Sigma x_i^2 + n\alpha^2 \tag{4.24a}$$

which, if $\alpha = 0$, reduces to

$$\Sigma\hat{Y}_i^2 = \Sigma y_i^2 - 2b\Sigma x_i y_i + b^2\Sigma x_i^2 \tag{4.24b}$$

By differentiating with respect to b and equating to zero:

$$\mathrm{d}\Sigma\hat{Y}_i^2/\mathrm{d}b = -2\Sigma x_i y_i + 2b\Sigma x_i^2 = 0$$

whence

$$b = \Sigma x_i y_i/\Sigma x_i^2 \tag{4.25}$$

Substituting this result back into eqn. (4.24b) gives

$$\Sigma\hat{Y}_i^2 = \Sigma y_i^2 - (\Sigma x_i y_i)^2/\Sigma x_i^2 = \Sigma y_i^2 - b\Sigma x_i y_i \tag{4.26}$$

The number of independent differences, $\hat{Y}_i$, in this $\Sigma\hat{Y}_i^2$ is equal to $n - 1$ so, since y is normally distributed with mean μ and standard deviation σ_e and $\hat{y}$ is an unbiased estimator of μ,

$$s_e^2 = \Sigma(y_i - \hat{y})^2/(n - 1) = \Sigma\hat{Y}_i^2/(n - 1) \tag{4.27}$$

is an unbiased estimator of σ_e^2. Rewriting eqn. (4.25) in full gives a linear relationship

$$b = x_1y_1/\Sigma x^2 + x_2y_2/\Sigma x^2 + \ldots x_ny_n/\Sigma x^2$$

to which eqn. (3.59) may be applied, so

$$s_b^2 = (x_1/\Sigma x_i^2)^2\, s_e^2 + (x_2/\Sigma x_i^2)^2\, s_e^2 + \ldots (x_n/\Sigma x_i^2)^2\, s_e^2 = s_e^2/\Sigma x^2$$

or

$$s_b = s_e/\sqrt{\Sigma x^2} \tag{4.28}$$

and, since $\hat{y} = bx$,

$$s_{\hat{y}}^2 = s_b^2x^2 = s_e^2x^2/\Sigma x^2$$

or

$$s_{\hat{y}} = s_ex/\sqrt{\Sigma x^2} = s_bx \tag{4.29}$$

The estimated variance of means of m values of y will be compounded of the estimated variance of such means *per se*, $s_e^2/\sqrt{m}$, and that of $\hat{y}$, so

$$s_{\dot{y}} = s_e\sqrt{(1/m + x^2/\Sigma x^2)} \tag{4.30}$$

and for inverse use of the line to predict x from $\dot{y}$

$$s_{\hat{x}} \doteq (s_e/b)\sqrt{(1/m + \dot{y}^2/b^2\Sigma x^2)} \tag{4.31}$$

Since only one parameter, β, is estimated here, limits for an unknown x based on $s_{\hat{x}}$ or $\dot{y}$ based on $s_{\dot{y}}$ will be valid for one and all values of x.

If α is known but not zero similar equations based on eqn. (4.24c) may be derived but those given above can be conveniently used with a transformed variable, $y - \alpha$.

Similarly, if β is known for the model, we can fit

$$y = a + \beta x + \hat{Y} \tag{4.32}$$

Fitting is simplest when $\beta = 1$ and the equations for the least squares estimates and their standard errors in this case are

$$a = \bar{y} - \bar{x} \tag{4.33}$$

$$\Sigma\hat{Y}^2 = \Sigma Y^2 - \Sigma X^2 \tag{4.34}$$

$$s_e = \sqrt{[\Sigma\hat{Y}^2/(n-1)]} \tag{4.35}$$

$$s_{\hat{y}} = s_a = s_e/\sqrt{n} \tag{4.36}$$

$$s_{\dot{y}} = s_{\hat{x}} = s_e\sqrt{(1/m + 1/n)} \tag{4.37}$$

When β is known but not unity (as, for example, in a Debye–Hückel extrapolation of mean activity coefficient or equivalent conductance to zero ionic strength), general equations can be derived or those above can be used with the dependent variable transformed to y/β.

It should be noted that special models should be adopted only for adequate prior reasons. For example, it would not be proper to fit the unconstrained model and then, simply because a test of significance fails to defeat the null hypothesis that $\alpha = 0$, switch to the constrained model, $\mu = \beta x$. On the other hand, if a particular model be deemed appropriate beforehand but some doubt is cast on its validity when the data are collected and plotted, common sense enjoins the application of a suitable test. Care is all the more necessary in these regression cases because the assumptions, in their applications, might have considerable influence on calculated standard errors and confidence intervals as can be seen from the computations (relevant to our example) below. On the other hand, one might take the view (see Section 3.5) that if, say, $\alpha = 0$ is an acceptable null hypothesis at the 5% level, the real value, even if not zero, is probably too small to have any great effect on the prediction intervals. That this can be so over part of the range is shown by the similarity of the outer bands of Fig. 4.4(a) and (b). In general, the greater the number of special constraints applied, the more precise will the remaining estimates appear to be and this is only slightly offset by the progressive increase in $\Sigma\hat{Y}^2$ which also results from the constraining process.

If both α and β are known there is of course no fitting to be done at all but for purposes of prediction an estimate of σ_e is required. For the case ($\alpha = 0$, $\beta = 1$) to which any other may be reduced, we have from eqn. (4.4)

$$\Sigma \hat{Y}^2 = \Sigma y^2 - 2\Sigma xy + \Sigma x^2 \tag{4.38}$$

and

$$s_e = \sqrt{(\Sigma \hat{Y}^2/n)} \tag{4.39}$$

Limits can be set for means of m values of y at given x based on

$$s_{\dot{y}} = s_e/\sqrt{m} \tag{4.40}$$

and, in reverse, these enable x to be estimated from $\dot{y}$ using

$$s_{\hat{x}} = s_e/\sqrt{m} \tag{4.41}$$

The application of these various relationships is illustrated with respect to the example already discussed above, in Table 4.5 and Fig. 4.4.

TABLE 4.5. COMPARISON OF UNCONSTRAINED AND CONSTRAINED MODELS

Model	Unconstrained	$\alpha = 0$	$\beta = 1$	$\alpha = 0, \beta = 1$
Equation	$\hat{y} = 5{\cdot}343 + 0{\cdot}8794x$	$\hat{y} = 0{\cdot}9981x$	$\hat{y} = 1{\cdot}000 + x$	$\mu = x$
$\Sigma \hat{Y}^2$	28·77	68·74	61·78	68·78
f	$n - 2 = 5$	$n - 1 = 6$	$n - 1 = 6$	$n = 7$
s_e	2·399	3·385	3·209	3·135
$s_{\hat{y}}$	$s_e\sqrt{(0{\cdot}143 + X^2/2268)}$	0·0138x	1·123	Not applicable
s_y	$s_e\sqrt{(1{\cdot}143 + X^2/2268)}$	$s_e\sqrt{(1 + x^2/11{,}340)}$	3·209	3·198
s_a	2·028	Not applicable	1·123	Not applicable
s_b	0·0537	0·0318	Not applicable	Not applicable
95% CLα	0·0129, 10·557	Not applicable	−1·751, 3·751	Not applicable
95% CLβ	0·7412, 1·1076	0·920, 1·076	Not applicable	Not applicable

The reality of the regression (i.e. $\beta \neq 0$) is generally assumed in the chosen model and is frequently so obvious as not to need demonstration by t test or the analogous analysis of variance of regression test. The interesting questions relate to the appropriateness of the model chosen or to comparisons of models and, as far as comparisons are concerned, the unconstrained model is generally the standard. Consider the following, applied to the data of the example using the results of calculations in Table 4.5.

(a) *Comparison of Singly Constrained with Unconstrained Model*

We can examine the reduction in residual variability:

	SS	f	MS	Ratio
Residual ($\alpha = 0$)	68·74	6		
Residual (unconst.)	28·77	5	5·755	
Difference	39·97	1	39·97	6·94 (cf. $F_{.05,1/5} = 6{\cdot}61$)

Or we can apply the t test

$$t = (a - 0)/s_a = 5{\cdot}343/2{\cdot}028 = 2{\cdot}63_4 \text{ (cf. } t_{.05,5} = 2{\cdot}57) \qquad (4.42)$$

Note that although the F test is one-sided in itself it is exactly equivalent to the two-sided t test; i.e., $2{\cdot}63_4 = \sqrt{6{\cdot}94}$. We conclude from either test that $\alpha = 0$ is inappropriate with respect to the unconstrained model and the latter fits the data not only better (as a qualitative comparison of the error variances or residuals shows) but *significantly* better in the usual sense that the improvement of fit can hardly be due to chance. These tests are merely alternatives to the inspection of the confidence interval for α which does not embrace zero (see in Table 4.5 above). Similar tests or inspection of the confidence interval for β reveal that $\beta = 1$ would be accepted for the unconstrained model at the 5% level of significance.

(b) *Comparison of Doubly Constrained with Unconstrained Model*

We can proceed as before

	SS	f	MS	Ratio
Residual ($\alpha = 0, \beta = 1$)	68·78	7		
Residual (unconst.)	28·77	5	5·755	
Difference	40·01	2	20·005	3·48 (cf. $F_{.05,2/5} = 5{\cdot}79$)

or construct the joint (α, β) ellipse and see where the point (0, 1) falls (see Fig. 4.3). Either way, we accept the hypothesis at the 5% level.

(c) *Comparison of Doubly with Singly Constrained Models*

The comparison can be made by testing the mean square difference in residual with the estimated error variance for the singly constrained model, or by the corresponding t test ($t = (b - 1)/s_b$ for $\alpha = 0$, or $t = (a - 0)/s_a$ for $\beta = 1$) or by inspection of the confidence intervals. The F test is facilitated by a variation devised by Youden (1962). If we designate the sums of squares of residuals as $\Sigma \hat{Y}^2_{01}$ for the doubly constrained model and $\Sigma \hat{Y}^2_{0b}$ and $\Sigma \hat{Y}^2_{a1}$ for those constrained as to intercept and slope respectively, then the latter sums of squares will be significantly lower than the former if their models fit the data better. Appendix Table IX provides the critical (P = 0·05) percentage reductions in the sums of squares of residuals. Here, $\Sigma \hat{Y}^2_{0b}$ (68·74) is only 0·06% lower than $\Sigma \hat{Y}^2_{01}$ (68·78) and $\Sigma \hat{Y}^2_{a1}$ (61·78) is 10·2% lower. For $n = 7$, these reductions are much less than the 50% called for by Table IX. All the tests give the same result—in the case of the example, relaxation of a constraint does not improve the goodness of fit significantly.

Table IX of the Appendix can also be used to test the unconstrained model ($\Sigma \hat{Y}^2_{ab} = 28·77$) against the constrained. $\Sigma \hat{Y}^2_{ab}$ is 53·4% lower than $\Sigma \hat{Y}^2_{a1}$ and 58·1% lower than $\Sigma \hat{Y}^2_{0b}$. In this case the table is entered at $n - 1 = 6$ and the critical reduction is 56·9%. Thus, the results (acceptance of the null hypothesis in the first test and rejection in the second) are the same as those indicated by the more formal procedures (a) and (b) above. Youden (1962) should be consulted for a further discussion from the point of view of analytical procedure.

Comparisons such as those described are based on the prior assumption of the unconstrained model (or of a singly constrained model in (c)) and the procedures simply examine possible parameter values. In so doing, hypotheses must be set up prior to testing and the apparent ambiguity (with respect to $\alpha = 0$) in the results of (a) and (b) arise out of the nature of these and of the data. (The ambiguity disappears when only one such test is made —either we are interested in α, or β, or α and β—and only one result obtained.) Apart from the valid object of examining possible theoretical parameter values, such comparisons might be made with the view that a constrained model is satisfactory only if it compares favourably with the unconstrained. However, when it is reasonable to select a particular model beforehand, any doubts which arise subsequently should be resolved without reference to other models and the test should depend on the nature of the residuals. If, for example, we adopt $y = \beta x + e$ with e distributed normally about zero with variance σ_e^2 and we fit $\hat{y} = bx$, then the $\hat{Y}$ should not be obviously inconsistent with the hypothetical e distribution. An analysis of the regression of $\hat{Y}$ on x would answer this question by revealing drift or non-normality.

One thing further merits consideration in a discussion of the comparison of these models and this relates to the optimum location of observations in controlled experimentation. In the case of the unconstrained model it is obvious that, since linearity is assumed, measurement at only two values of the independent variable need be made and eqns. (4.11) and (4.12) indicate that these should be located at the extreme ends of the range since this will increase the quantity ΣX^2 relative to that of any other arrangement and so reduce the standard errors of a and b. Consider for example four values of x equally spaced (1, 2, 3, 4) and then grouped in pairs (1, 1, 4, 4). In the first case we have $\Sigma x^2 = 30$ and $\Sigma X^2 = 5$ and in the second 34 and 9 respectively. Since s_e is the estimate of a constant which is unaffected by location, the second arrangement leads to a reduction of s_b by a factor of $\sqrt{5}/\sqrt{9}$ or 25%. s_a is reduced by a factor of $\sqrt{(30/20)}/\sqrt{(34/36)}$ or 16%, a result which improves the further the x values lie from the origin. In the case of the model constrained as to intercept it is obvious that all measurements should be located at the extreme end of the range of observation. Thus, taking four points as before and considering the application of eqn. (4.28) we have $s_b = s_e/\sqrt{30}$ from equal spacing but $s_e/\sqrt{64}$ from grouping at $x = 4$, a reduction of 32% in favour of the latter arrangement. This result, however, does not improve as the range of observation moves from the origin and for the arrangement (14, 14, 14, 14) by comparison with (11, 12, 13, 14) the reduction is only 10%. In the case of the line constrained as to slope it is immaterial with respect to s_a where the points are situated. In general we lack the courage of our convictions in this matter but, taking the unconstrained model as that most commonly adopted, three sets of observations (equal in number and equally spaced) will not only give more precise estimates of slope and intercept than equal or random spacing but will still serve to indicate and permit the evaluation of quadratic curvature should the linear model appear on examination to be inappropriate. Furthermore, such an arrangement is more economical from the experimental point of view. In most physico-chemical work (for example, regressions on 1/T °K to determine energy of activation or heat of reaction) there is a sufficient margin for safety in such an arrangement. Computations appropriate to values so grouped are discussed in the next section.

4.7. Computations in Simple Regression with Replication

If several observations are made at one or more of the values of x the form of computations of Table 4.2 can still be used but it is generally convenient to introduce some modification. For illustration we take seven points closely related to those of the previous example but grouped at only four values of x as shown in Table 4.6(a).

TABLE 4.6. COMPUTATION OF THE LEAST SQUARES LINE FROM REPLICATED DATA

(a) Computations based on totals

x	y	T	w	wx	wx^2	y^2	Σy^2	T^2/w	xT
10	12	12	1	10	100	144	144	144	120
	24					576			
25	28	81	3	75	1875	784	2201	2187	2025
	29					841			
45	48	48	1	45	2025	2304	2304	2304	2160
	58					3364			
61	60	118	2	122	7442	3600	6964	6962	7198
		259	7	252	11,442		11,613	11,597	11,503

$\bar{x} = \Sigma wx/n = 252/7 = 36;\quad \bar{y} = \Sigma T/n = 259/7 = 37$

$\Sigma wX^2 = \Sigma wx^2 - (\Sigma wx)^2/n = 11{,}442 - 9072 = 2370$

$\Sigma\Sigma Y^2 = \Sigma\Sigma y^2 - (\Sigma T)^2/n = 11{,}613 - 9583 = 2030$

$\Sigma w\dot{Y}^2 = \Sigma(T^2/w) - (\Sigma T)^2/n = 11{,}597 - 9583 = 2014$

$\Sigma wX\dot{Y} = \Sigma xT - \Sigma wx\Sigma T/n = 11{,}503 - 9324 = 2179$

Eqn. (4.45), $b = \Sigma wX\dot{Y}/\Sigma wX^2 = 2179/2370 = 0{\cdot}919409$

Eqn. (4.5), $a = \bar{y} - b\bar{x} = 37 - 33{\cdot}0987 = 3{\cdot}9013$

The fitted line is $\hat{y} = 3{\cdot}901 + 0{\cdot}91941x$

Eqn. (4.48a), $\Sigma\Sigma\hat{Y}^2 = \Sigma\Sigma Y^2 - (\Sigma wX\dot{Y})^2/\Sigma wX^2 = 2030 - 2179^2/2370 = 2030 - 2003{\cdot}4 = 26{\cdot}6$

Eqn. (4.48b), $s_e^2 = \Sigma\Sigma\hat{Y}^2/(n-2) = 26{\cdot}6/5 = 5{\cdot}32$

(The second last column is needed only to split the overall SS, $\Sigma\Sigma\hat{Y}^2$, into the within-sets component, 2030 − 2014 = 16, and the between-sets or means-about-line component, 2014 − 2003·4 = 10·6.)

TABLE 4.6 *(cont.)*

(b) Computations based on means

x	$\dot{y}$	w	wx	$w\dot{y}$	wx^2	Σy^2	$w\dot{y}^2$	$wx\dot{y}$
10	12	1	10	12	100	144	144	120
25	27	3	75	81	1875	2201	2187	2025
45	48	1	45	48	2025	2304	2304	2160
61	59	2	122	118	7442	6964	6962	7198
		7	252	259	11,442	11,613	11,597	11,503

$\bar{x} = \Sigma wx/n = 252/7 = 36; \quad \bar{y} = \Sigma w\dot{y}/n = 259/7 = 37$

$\Sigma wX^2 = \Sigma wx^2 - n\bar{x}^2 = 11{,}442 - 9072 = 2370$

$\Sigma\Sigma Y^2 = \Sigma\Sigma y^2 - n\bar{y}^2 = 11{,}613 - 9583 = 2030$

$\Sigma w\dot{Y}^2 = \Sigma w\dot{y}^2 - n\bar{y}^2 = 11{,}597 - 9583 = 2014$

$\Sigma wX\dot{Y} = \Sigma wx\dot{y} - n\bar{x}\bar{y} = 11{,}503 - 9324 = 2179$

(For subsequent computations see part (a) of the table. The individual y values whose squares form Σy^2 are given only in part (a).)

(c) Analysis of variance of regression

Source of Variation	Sum of squares, SS	f	MS	Ratio
Line about mean	$\Sigma w(\hat{y} - \bar{y})^2 = b\Sigma wX\dot{Y} = 2003{\cdot}4$	1	2003·4	
Means about line	$\Sigma w(\dot{y} - \hat{y})^2 = \Sigma w\dot{Y}^2 - b\Sigma wX\dot{Y} = 10{\cdot}6$	$k-2=2$	5·3	1·0
Within sets	$\Sigma\Sigma(y - \dot{y})^2 = \Sigma\Sigma Y^2 - \Sigma w\dot{Y}^2 = 16{\cdot}0$	$n-k=3$	5·3	
Total	$\Sigma\Sigma Y^2 = \Sigma\Sigma y^2 - n\bar{y}^2 = 2030{\cdot}0$	$n-1=6$		$(F_{.05,\,2/3} = 9{\cdot}55)$

Formally, we can designate the x values and sets by $i = 1, 2 \ldots k$ where $k = 4$ in the example and the individual y observations within each set by $j = 1, 2 \ldots w_i$. Thus, y_{ij} is the jth value of the ith set and the mean of the w_i such values is designated $\dot{y}_i$. Deviations of individual values and set means from corresponding points on the line, $\hat{y}_i$, and from the grand mean, $\bar{y}$, may be distinguished and we shall use the notation $Y_{ij} = y_{ij} - \bar{y}$, $\dot{Y}_i = \dot{y}_i - \bar{y}$, and $\hat{Y}_{ij} = y_{ij} - \hat{y}_i$, the latter being made up of the deviation of a value from its set mean, $y_{ij} - \dot{y}_i$, and of the mean from the line, $\dot{y}_i - \hat{y}_i$. For simplicity the subscripts are omitted from equations and column headings, Σ is used to designate summation over both the k sets

($\overset{k}{\Sigma}$) and the w values within a set ($\overset{w}{\Sigma}$) so $\Sigma\Sigma$ indicates summation within and over all sets and T denotes the sum of the y within a particular set.

Two methods of computation are illustrated—that in Table 4.6(a) makes use of the totals of the y values and is the better for avoiding rounding errors but the method of Table 4.6(b) reveals the true nature of the procedure, a calculation in which the values of the dependent variable are weighted according to the number of observations on which they are based. Broadly, the formulae for fitting are the same as those given above for non-replicated data except that the weighting factor, w, now appears in each, and most can be written in terms of sums or of means.

Thus, for the same model and form of fitted line, $\hat{y} = a + bx$, as for non-replicated data, we now have the formulae

Means,

$$\bar{x} = \Sigma wx/\Sigma w = \Sigma wx/n \tag{4.43}$$

$$\bar{y} = \Sigma w\dot{y}/\Sigma w = \Sigma w\dot{y}/n = \Sigma T/n \tag{4.44}$$

Slope,

$$b = \Sigma wX\dot{Y}/\Sigma wX^2 \tag{4.45}$$

Intercept (as before),

$$a = \bar{y} + b\bar{x} \tag{4.5}$$

"*Within sets*" *sum of squares* (squared deviations of values within sets from set means),

$$\begin{aligned}\Sigma\Sigma(y - \dot{y})^2 &= \Sigma\Sigma y^2 - \Sigma\Sigma w\dot{y}^2 = \Sigma\Sigma y^2 - \Sigma(T^2/w) \\ &= \Sigma\Sigma Y^2 - \Sigma w\dot{Y}^2\end{aligned} \tag{4.46a}$$

There are $w - 1$ independent differences of this kind in each set so the total or appropriate number of degrees of freedom is $\Sigma(w - 1) = n - k$, and σ_e^2 can be estimated by

$$s_e^2 = \Sigma\Sigma(y - \dot{y})^2/(n - k) \tag{4.46b}$$

"*Means about line*" *sum of squares* (squared deviations of set means from the line),

$$\Sigma w(\dot{y} - \hat{y})^2 = \Sigma w\dot{Y}^2 - b\,\Sigma wX\dot{Y} = \Sigma w\dot{Y}^2 - (\Sigma wX\dot{Y})^2/\Sigma wX^2 \tag{4.47a}$$

There are k such differences and $k - 2$ of them are independent so

$$s_e^2 = \Sigma w(\dot{y} - \hat{y})^2/(k - 2) \tag{4.47b}$$

Squared deviations of all values from the line can be obtained by adding the last two sums of squares and the corresponding "pooled" estimate of σ_e^2 by dividing this total sum of squares by the total degrees of freedom, $(n - k) + (k - 2) = n - 2$. Thus

Overall sum of squares of deviations

$$\Sigma\Sigma\hat{Y}^2 = \Sigma\Sigma(y - \dot{y})^2 + \Sigma w(\dot{y} - \hat{y})^2 = \Sigma\Sigma Y^2 - (\Sigma wX\dot{Y})^2/\Sigma wX^2 \tag{4.48a}$$

Error variance

$$s_e^2 = \Sigma\Sigma\hat{Y}^2/(n - 2) \tag{4.48b}$$

The total sum of squares of deviations of the n values from their grand mean is

$$\Sigma\Sigma(y - \bar{y})^2 = \Sigma\Sigma Y^2 = \Sigma\Sigma y^2 - n\bar{y}^2 = \Sigma\Sigma y^2 - (\Sigma T)^2/n \tag{4.49}$$

Since the model assumes a constant error variance, a comparison of the two independent estimates of this provides a test of its validity. If the implied null hypothesis were to fail (i.e. $\Sigma w(\dot{y} - \hat{y})^2/(k - 2)$ significantly greater than $\Sigma\Sigma(y - \dot{y})^2/(n - k)$) we would conclude that the variation of set means about the line could not be explained in terms of the "within-set" variability and the overall error variance and fitted line would both be meaningless. The most likely cause of such failure of the model is curvilinearity (an example of an analogous graphical test showing this is the quality control chart of Fig. 3.6) but a similar effect might be observed if the within-set variability were to vary significantly over the range (in which case both test and model would be invalid) or if the y values within one set were not fully equivalent (by reason of more constant conditions of measurement perhaps) to those in others. If the two independent estimates of σ_e^2 are comparable the model is justified and the overall or pooled estimate is used in all subsequent calculations. It is this estimate which would be obtained if the ordinary computational method illustrated in Table 4.2 were to be applied to the data. A formal test procedure analogous to that of Table 4.4 is given in Table 4.6(c). Since, in the case of the example, it shows pooling to be justified, it could be followed by a test for the reality of the regression involving comparison of the "line-about-mean" mean square with the pooled error variance. In the case of the example the reality of the regression is obvious.

4.8. The Comparison of Several Regression Lines

If several samples are available it may be a matter of interest to see if they have been drawn from the same universe or from universes having the same slope etc. and to what extent the data they provide can be combined to estimate the appropriate parameters. For this purpose the sums of squares and products of deviations of individual values from their means and the sums of squares of their deviations from the fitted lines are computed in the usual way. Formally, we designate the individual values within a sample or set by $i = 1, 2 \ldots n$ and the sets themselves by $j = 1, 2 \ldots k$. The totals of the k "within-sets" sums (designated Σ') taken away from the overall sums (calculated from the grand means and designated Σ^0, e.g. $\Sigma^0 X^2 = \sum^k \sum^n (x_{ij} - \bar{\bar{x}})^2$) leaves the "between-sets" quantities (designated Σ'') which reflect the difference from set to set. The overall slope (i.e. of all the values taken as a single sample) is given by

$$b^0 = \Sigma^0 XY / \Sigma^0 X^2 \tag{4.50}$$

the slope of the k mean points by

$$b'' = \Sigma'' XY / \Sigma'' X^2 \tag{4.51}$$

and the weighted average of the k sample slopes is given by

$$b' = \Sigma' XY / \Sigma' X^2 \tag{4.52}$$

The individual sample slopes, here, are as usual

$$b_j = \sum^n X_i Y_i / \sum^n X_i^2$$

The analysis of the total sum of squares of deviations of the individual values from the assumed overall regression line depends on the k error variances being comparable and these should be compared by Bartlett's test (Section 3.5) or some equivalent procedure if there is any doubt.

For the example we take the data from runs of a chemical reaction carried out at different temperatures and ionic strengths to see if the energy of activation is affected by the latter factor. It is

j	1		2		3		4		5	
	(I = 0·020)		(I = 0·018)		(I = 0·016)		(I = 0·014)		(I = 0·012)	
i	x	y	x	y	x	y	x	y	x	y
1	3956	924	4352	792	4590	157	4495	675	4602	237
2	3613	2304	3921	1741	3875	1869	3784	2248	4106	1165
3	2971	3617	3400	3092	3389	2909	3234	3560	3807	2028
4	2499	4686	2960	3988	2927	3923	2765	4417	3378	3045
5	1636	6595	2563	4654	2510	5011	2331	5480	2787	4406
6			2331	5353	1918	6180	1959	6503	2289	5539
7			2010	6222	1387	7513			1666	7084
Σ′	2935	3625	3077	3692	2942	3937	3095	3814	3237	3358

j = sample number; i = observation within sample; Σ′ = sample sum; I = ionic strength; T = temperature (°K); k = velocity constant; $x = (1/\text{T} - 0{\cdot}003)\,10^7$; $y = (\log \text{k} + 2)\,10^4$

(b) Sums of squares and products of differences, slopes and error variances

	$\Sigma X^2/10$	$\Sigma Y^2/10$	$\Sigma XY/10$	b_j	$\Sigma \hat{Y}^2/10$	f_j	$s_e^2/10$
Set 1	338,092	1,898,713	−797,427	−2·35861	17,892	3	5964
Set 2	441,518	2,274,928	−1,000,566	−2·26619	7449	5	1490
Set 3	743,965	3,859,430	−1,693,967	−2·27689	2361	5	472
Set 4	443,713	2,274,008	−1,003,540	−2·26170	4317	4	1079
Set 5	653,233	3,615,490	−1,536,086	−2·35153	3365	5	673
Within sets	2,620,521	13,922,569	−6,031,586	$-2{\cdot}30168 = b'$	35,384	22	(′)
Between sets	39,492	131,364	−54,038	$-1{\cdot}36833 = b''$		4	(″)
Total	2,660,013	14,053,933	−6,085,624	$-2{\cdot}28782 = b^0$		30	(⁰)

$N = \Sigma' n = 5 + 7 + 7 + 6 + 7 = 32;\quad k = 5$

TABLE 4.7 *(cont.)*

(c) The analysis of variance of regression

Source of variation	Sums of squares, SS/10	Degrees of freedom	Mean squares MS/10
Deviation within each set from the set line	$\Sigma'\hat{Y}^2 = S_1 = 35{,}384$	$N-2k = 22$	$1608 = s_1^2$
Differences among the regression coeffs., b_j	(by difference) $S_2 = 4422$	$k-1 = 4$	$1105 = s_2^2$
Deviation within sets from linear reg., b'	$\Sigma'Y^2 - b'\Sigma'XY = S_1 + S_2 = 39{,}806$	$N-k-1 = 26$	$1531 = s_3^2$
Deviation within sets from linear reg., b''	$\Sigma''Y^2 - b''\Sigma''XY = S_3 = 57{,}423$	$k-2 = 3$	$19{,}141 = s_4^2$
Difference between coefficients, b' and b''	(by difference)† $S_4 = 33{,}892$	1	$33{,}892 = s_5^2$
Total deviations from overall linear reg.	$\Sigma^0 Y^2 - b^0\Sigma^0 XY = 131{,}121$	$\hat{N}-2 = 30$	
† $S_4 = (b' - b'')^2 \Sigma' X^2 \Sigma'' X^2 / \Sigma^0 X^2$	$S_2 + S_3 + S_4 = 95{,}737$	$2k - 2 = 8$	$11{,}967 = s_6^2$

assumed that the temperatures (although not pre-set) are known accurately so that there is no error in the independent variable, 1/T °K, and that the dependent variable, the logarithm of the velocity constant, is distributed normally with constant variance. This assumption of normality may not be strictly correct but it is unlikely to be so far from the truth as to seriously invalidate the unconstrained regression model adopted in analysing the data. The data are set out in coded form in Table 4.7(a), the results of the initial computations in Table 4.7(b) and the analysis (following the form of Willsdon (1934)) in Table 4.7(c). For convenience the sums of squares etc. have been rounded to one less significant figure than provided by the data and the entries in Table 4.7(c) have been given special symbols. The actual basic computations made by the method of Table 4.2, which provide the link between Tables 4.7(a) and 4.7(b), are not shown. The differences between the three estimated regres-

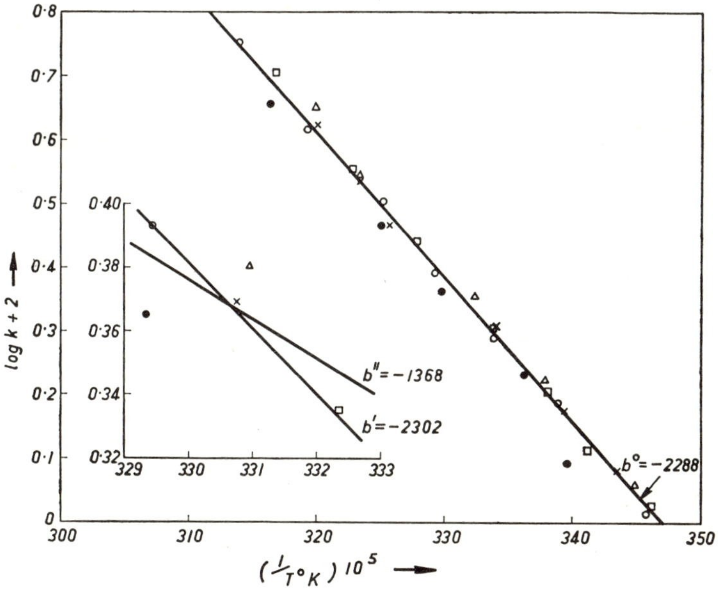

FIG. 4.5. The Comparison of Several Regression Lines.

The data plotted are those of Table 4.7(a). Set 1 is represented by ●, set 2 by ×, set 3 by ○, set 4 by △ and set 5 by □. b^0 is the overall regression coefficient and the line having this slope drawn through the individual points is the overall regression line. The inset diagram shows the set mean points and b'' is their regression coefficient. The within-sets regression coefficient is b'. The individual set regression coefficients are not shown.

sion coefficients and the relationships of these to the data are illustrated in Fig. 4.5.

The tests which may be made are

(a) *For the suitability of a single overall regression line,*

$$s_6^2/s_1^2 = 11{,}967/1608 = 7{\cdot}44 \quad \text{which is greater than} \quad F_{.05,\,8/22} = 2{\cdot}40$$

The departure of the individual sets from the overall regression line is significant. We may now test to find wherein lies the discrepancy.

(b) *For differences between the five regression coefficients,*

$$s_2^2/s_1^2 = 1105/1608 = 0{\cdot}69 \quad \text{which is less than} \quad F_{.05,4/22} = 2{\cdot}82$$

There are no significant differences between the regression coefficients and their mean, b', may be taken as the representative slope. The failure of the first test must be accounted for otherwise.

(c) *For linearity of the relationship between the five set means,*

$$s_4^2/s_3^2 = 19{,}141/1531 = 12{\cdot}5 \quad \text{which is greater than} \quad F_{.05,3/26} = 2{\cdot}98$$

This indicates a highly significant departure from linearity. The slope, b'', does not, therefore, satisfactorily represent the relationship between the mean points of the sets. However, it may be tested.

(d) *For the difference between the mean slope, b', and slope of the means, b'',*

$$s_5^2/s_3^2 = 33{,}892/1531 = 22{\cdot}1 \quad \text{which is greater than} \quad F_{.05,1/26} = 4{\cdot}23$$

The difference between $b' = -2{\cdot}302$ and $b'' = -1{\cdot}368$ is highly significant.

The failure of an overall regression line to fit the data appears to lie, therefore, in the differences between the mean points. These differences do not appear to be related to ionic strength but might be due to unsuspected errors in reagent concentrations. However, the energy of activation is unaffected and may be found from b' whose standard error is

$$s_{b'} = \sqrt{(\Sigma'\hat{Y}^2/\Sigma'f\Sigma'X^2)} = s_{ep}/\sqrt{\Sigma'X^2} \qquad (4.53)$$

where s_{ep} is a pooled standard error of estimate ($\sqrt{s_1^2}$ from Table 4.7c).

In the example this is 0·0248 and from it we obtain

$$95\% \text{ confidence limits of } \beta = b' \pm t_{.05,22}s_{b'} = 2{\cdot}302 \pm 0{\cdot}051.$$

In the original units of log k and 1/T these limits become −2302 ± 51 and the corresponding energy of activation lies within 10,532 ± 233 calories with 95% confidence.

When only two sample regressions have to be compared it is easier to proceed by t test. Suppose our information is confined to the data of sets 2 and 4 of Table 4.7(a). These we shall now designate 1 and 2 and record the results of the basic computations in Table 4.8, the sums of squares etc. being rounded as before.

TABLE 4.8. THE COMPARISON OF TWO REGRESSION LINES

Set	$\Sigma X^2/10$	$\Sigma Y^2/10$	$\Sigma XY/10$	b	$\Sigma \hat{Y}^2/10$	f	$s_e^2/10$	$\bar{x}$	$\bar{y}$
1	441,518	2,274,928	−1,000,566	−2·26619	7449	5	1490	3077	3692
2	443,713	2,274,008	−1,003,540	−2·26170	4317	4	1079	3095	3814
Σ′	885,231		−2,004,106		11,766	9			
Δ				+0·00449				−18	−122

(a) *Comparison of the error variances,*

$$s_{e_1}^2/s_{e_2}^2 = 1490/1079 = 1{\cdot}38 \text{ which is less than } F_{.05,\text{2-sided},5/4} = 9{\cdot}36.$$

Since the error variances are comparable they may be pooled:

$$s_{ep}^2 = \Sigma'\hat{Y}^2/\Sigma'f = 11{,}766/9 = 1307$$

and this value can be used in subsequent calculations.

(b) *Comparison of the regression coefficients,*

$$s_{\Delta b} = s_{ep}\sqrt{(1/\Sigma X_1^2 + 1/\Sigma X_2^2)} \tag{4.54}$$

so

$$s_{\Delta b} = \sqrt{[1307(1/441{,}518 + 1/443{,}713)]} = \sqrt{0{\cdot}00591} = 0{\cdot}0769$$

here and

$$t = \Delta b/s_{\Delta b} = 0{\cdot}00449/0{\cdot}0769 = 0{\cdot}06$$

which is less than $t_{.05,9} = 2{\cdot}26$.

This test is completely analogous to that for the comparison of two sample means, i.e. of their difference with zero, and the variance of the difference is simply the sum of the two regression coefficient variances based on the pooled error variance. Clearly, in this example the two regression coefficients may be considered to estimate the same true value and they can therefore be combined to get a better estimate,

$$b' = \Sigma' XY / \Sigma' X^2 = -2{,}004{,}106/885{,}231 = -2{\cdot}26394,$$

whose standard error is given by eqn. (4.53),

$$s_{b'} = s_{ep}/\sqrt{\Sigma' X^2} = \sqrt{(1307/885{,}231)} = 0{\cdot}0384.$$

With this combined estimate of slope we can replace the original fitted lines with two parallel lines represented by the equations

$$\hat{y}_1 = \bar{y}_1 + b'(x - \bar{x}_1) = 3692 - 2{\cdot}26394(x - 3077)$$
$$= 10{,}657 - 2{\cdot}26394x$$
$$\hat{y}_2 = \bar{y}_2 + b'(x - \bar{x}_2) = 3814 - 2{\cdot}26394(x - 3095)$$
$$= 10{,}820 - 2{\cdot}26394x$$

where 10,657 and 10,820 are new intercepts, a_1' and a_2'.

(c) *Comparison of the new equations* can be based on a consideration of slopes or of intercepts, the latter being the more informative. The displacement of the lines in the y direction and its standard error are

$$\Delta a' = a_1' - a_2' = \Delta\bar{y} - b'\Delta\bar{x} \quad (4.55)$$

and

$$s_{\Delta a'} = s_{ep}\sqrt{[1/n_1 + 1/n_2 + (\Delta\bar{x})^2/\Sigma' X^2]} \quad (4.56)$$

so, here, $t = |\Delta a'|/s_{\Delta a'} = -163/77 = 2{\cdot}12$ which is slightly less than the critical value for the 5% level of significance, $t_{.05,9} = 2{\cdot}26$. The same result is obtained for the ratio of the difference between b' and the slope of the means, $b'' = \Delta\bar{y}/\Delta\bar{x}$, divided by the standard error of this difference, $s_{(b'-b'')}$ or $s_{\Delta b'}$, which is given by

$$s_{(b'-b'')} = \sqrt{(s_{b'}^2 + s_{b''}^2)} = s_{ep}\sqrt{[1/\Sigma' X^2 + (1/n_1 + 1/n_2)/(\Delta\bar{x})^2]} \quad (4.57)$$

Here, $b'' = -122/-18 = 6{\cdot}78$ which differs from b' by 9·04, $s_{\Delta b'}$ is 4·26, and $t = 9{\cdot}04/4{\cdot}26 = 2{\cdot}12$, the same result as before. Since our adoption of the 5% level means that we choose to regard differences which have less chance of occurring than this as being accidental we could now proceed to combine the two sets of data and treat them as one.

The treatment above depends on step (a) showing that the error variances do not differ significantly. If this be not so the subsequent steps may still be carried out but

(i) if n_1 and n_2 are large,

$$s_{\Delta b} = (s_{e_1}^2/n_1 + s_{e_2}^2/n_2) \tag{4.58}$$

(ii) if n_1 and n_2 are small, the difference between b_1 and b_2 must be tested by an analogue of the Aspin–Welch test (Section 3.5) or its equivalent. If by the appropriate test the regression coefficients are found not to differ significantly a weighted mean can be derived

$$\bar{b} = (b_1/s_{b_1}^2 + b_2/s_{b_2}^2)/(1/s_{b_1}^2 + 1/s_{b_2}^2) \tag{4.59}$$

4.9. The Intersection of Two Regression Lines

The problem of locating the point of intersection of two lines is common enough in physico-chemical work (conductometric and analogous titration end-points come immediately to mind) but the establishment of confidence limits on the point of intersection seems generally to have been neglected. We consider only the case, dealt with by Fisher (1963) and Kastenbaum (1959), of two linear regression lines fitted to samples drawn from intersecting universes having a common and constant standard deviation. If the two fitted lines are

$$\hat{y}_1 = a_1 + b_1x_1 \quad \text{and} \quad \hat{y}_2 = a_2 + b_2x_2$$

based on n_1 and n_2 observations respectively, the point of intersection

$$x_I = (a_2 - a_1)/(b_1 - b_2) = \Delta a/\Delta b \tag{4.60}$$

estimates the true point

$$\chi = (\alpha_2 - \alpha_1)/(\beta_1 - \beta_2) \tag{4.61}$$

TABLE 4.9. THE SEMI-CONDUCTION OF PHTHALOCYANINE

Set 1				Set 2			
$10^3/T$	log R	x	y	$10^3/T$	log R	x	y
1·52	6·02	2	2	1·84	6·94	34	94
1·54	6·10	4	10	1·87	7·00	37	100
1·57	6·15	7	15	1·91	7·09	41	109
1·61	6·32	11	32	1·96	7·17	46	117
1·64	6·40	14	40	2·01	7·27	51	127
1·70	6·57	20	57	2·09	7·41	59	141
1·76	6·75	26	75	2·11	7·47	61	147
				2·17	7·62	67	162
				2·24	7·76	74	176
				2·30	7·87	80	187
$\Sigma x_1 = 84$; $\Sigma y_1 = 231$; $n_1 = 7$				$\Sigma x_2 = 550$; $\Sigma y_2 = 1360$; $n_2 = 10$			
$\Sigma X_1^2 = 454$; $\Sigma Y_1^2 = 4204$;				$\Sigma X_2^2 = 2240$; $\Sigma Y_2^2 = 9254$;			
$\Sigma X_1 Y_1 = 1379$				$\Sigma X_2 Y_2 = 4548$			
$\bar{x}_1 = 12$; $\bar{y}_1 = 33$				$\bar{x}_2 = 55$; $\bar{y}_2 = 136$			
$a_1 = -3{\cdot}449$; $b_1 = 3{\cdot}03744$				$a_2 = 24{\cdot}330$; $b_2 = 2{\cdot}03036$			
$\Sigma \hat{Y}_1^2 = 15{\cdot}3633$; $s_{e_1}^2 = 3{\cdot}073$				$\Sigma \hat{Y}_2^2 = 19{\cdot}9364$; $s_{e_2}^2 = 2{\cdot}492$			

Test for comparability of error variances:

$$F = s_{e_1}^2/s_{e_2}^2 = 3{\cdot}073/2{\cdot}492 = 1{\cdot}23$$

which is less than $F_{.05,\ \text{2-sided},\ 5/8} = F_{.025,\ \text{1-sided},\ 5/8} = 4{\cdot}82.$

Pooling of error variances:

$$s_{ep}^2 = (\Sigma \hat{Y}_1^2 + \Sigma \hat{Y}_2^2)/(n_1 + n_2 - 4) = 35{\cdot}2977/13 = 2{\cdot}7154$$

$$\Delta a = (a_2 - a_1) = 27{\cdot}779;\quad \Delta b = (b_1 - b_2) = 1{\cdot}0070;\quad \Delta a \Delta b = 27{\cdot}973$$

$$x_I = 27{\cdot}779/1{\cdot}0070 = 27{\cdot}586 \quad \text{on the coded scale.}$$

From eqn. (4.62b), $s_{\Delta a}^2 = 2{\cdot}7154(1/7 + 1/10 + 144/454 + 3025/2240) = 5{\cdot}1878$

From eqn. (4.62c), $s_{\Delta b}^2 = 2{\cdot}7154(1/454 + 1/2240) = 0{\cdot}007193$

From eqn. (4.62d), $s_{\Delta a \Delta b} = 2{\cdot}7154(12/454 + 55/2240) = 0{\cdot}13843$

Substituting in eqn. (4.62a) with $t_{.05,13} = 2{\cdot}1604$, $t^2 = 4{\cdot}6673$, for 95% confidence

$$(771{\cdot}673 - 24{\cdot}213) - 2\chi(27{\cdot}9757 - 0{\cdot}6461) + \chi^2(1{\cdot}014210 - 0{\cdot}33573) = 0.$$

Rearranging and solving by the quadratic formula:

$$0{\cdot}980637\chi^2 - 54{\cdot}6592\chi + 747{\cdot}460 = 0$$

95% *Confidence lts. of* $\chi = [54{\cdot}659 \pm \sqrt{(2987{\cdot}628 - 2931{\cdot}946)}]/1{\cdot}96127$
$= 24{\cdot}064$ to $31{\cdot}674$ on the coded scale.

and the confidence interval for the latter must satisfy

$$((\Delta a)^2 - t^2 s_{\Delta a}^2) - 2\chi(\Delta a \Delta b - t^2 s_{\Delta a \Delta b}) + \chi^2((\Delta b)^2 - t^2 s_{\Delta b}^2) = 0 \tag{4.62a}$$

Here,

$$s_{\Delta a}^2 = s_{ep}^2(1/n_1 + 1/n_2 + \bar{x}_1^2/\Sigma X_1^2 + \bar{x}_2^2/\Sigma X_2^2) \tag{4.62b}$$

$$s_{\Delta b}^2 = s_{ep}^2(1/\Sigma X_1^2 + 1/\Sigma X_2^2) \tag{4.62c}$$

and

$$s_{\Delta a \Delta b} = s_{ep}^2(\bar{x}_1/\Sigma X_1^2 + \bar{x}_2/\Sigma X_2^2) \tag{4.62d}$$

where s_{ep}^2 is the pooled estimate of error variance which is based on $n_1 + n_2 - 4$ degrees of freedom and t has the value appropriate to this number of degrees of freedom and the chosen level of confidence.

For illustration we take some data relating to semi-conduction in phthalocyanine. This changes from impurity semi-conduction at low temperatures (set 2) to intrinsic semi-conduction at higher temperatures (set 1). The independent variable is the reciprocal of absolute temperature, T °K, and the dependent variable is the logarithm of resistance, R. For convenience the data were coded as in Table 4.9 and all computed quantities are given in terms of the coded variables. They are illustrated in Fig. 4.6 whose axes are labelled in both coded and original units. The interval calculated for the true point of intersection given in the last line of the table is not symmetrical about x_I(27·584 − 3·520, 27·584 + 4·090) but when decoded (1·7406, 1·8167) and the reciprocals found the final interval is nearly symmetrical about $10^3/x_I$. Rounded to the nearest degree it is 563 ± 13°K. (It is assumed that the distribution of log R is sufficiently close to normal over the range of observation. That the error variances in the two samples estimate a common true variance is open to test.)

In many situations of the kind described, points in the vicinity of the intersection are difficult to assign or are suspect for theoretical reasons (hydrolysis, for example, causes discrepancies in conductimetric titration curves) but provided the samples are large enough the subjective rejection of such points is unlikely to affect greatly the calculated confidence intervals.

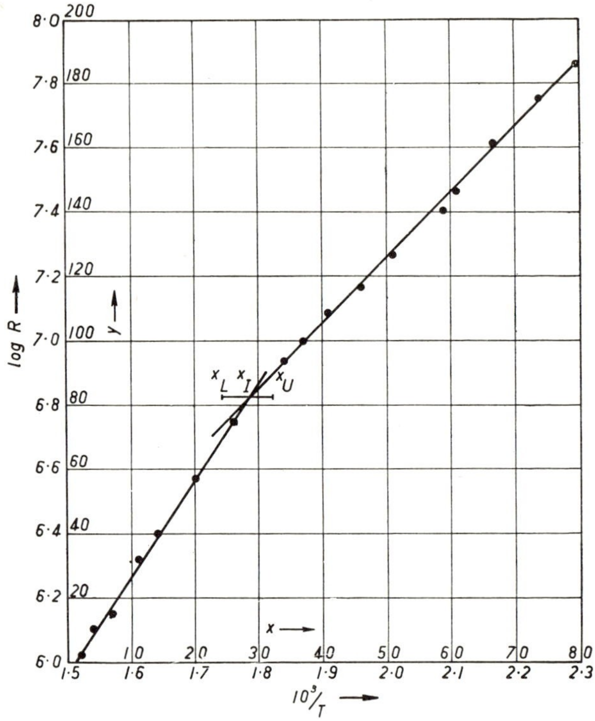

FIG. 4.6. Impurity and Intrinsic Semiconduction in Phthalocyanine.

The lower left-hand line represents intrinsic semiconduction at high temperatures and the upper right-hand impurity semiconduction at low temperatures. $x_L x_U$ is the 95% confidence interval for the true point of intersection of which x_I is the point estimate. The inner scales are marked in the coded units with which all the computations of Table 4.9 were made.

4.10. The Correlation Model

We turn now to the situation where we have two random variables whose values are in some way associated but which are on an equal footing in that neither can be regarded as independent like x in the simple regression model of Section 4.2 above. We assume that the variability of the values is "natural" in the sense that the values of y have some distribution for a particular value of x and vice versa but that errors of measurement are not a necessary part or cause of the spreads of these distributions. In general, of course, such distributions might be of any form and might alter in form or characteristic with changes in the value of the other variable and their means or other measures of central tendency might follow any path. Here, however, we shall consider only the normal bivariate universe model which is characterized by Gaussian distributions (of the y values for any x value and vice versa) having constant spread and whose means have linear loci or regression lines.

Such a universe is completely described by the overall means (μ_x, μ_y) and standard deviations (σ_x, σ_y) of the values of the two variables and by a dimensionless measure of association, the *correlation coefficient*, which is defined for the universe by the equation

$$\varrho = \sigma_{xy}/\sigma_x\sigma_y \tag{4.63}$$

where σ_{xy} is the covariance or mean product of all the pairs of values in the universe. The frequency density per unit area (out of a total number N) in the $x - y$ plane described by these five parameters is

$$f_{xy} = \left[\frac{N}{2\pi\sigma_x\sigma_y\sqrt{(1-\varrho^2)}}\right]\exp\left\{\left[-\frac{1}{2(1-\varrho^2)}\right]\left[\left(\frac{x-\mu_x}{\sigma_x}\right)^2 + \left(\frac{y-\mu_y}{\sigma_y}\right)^2 - \frac{2\varrho(x-\mu_x)}{\sigma_x}\frac{(y-\mu_y)}{\sigma_y}\right]\right\} \tag{4.64}$$

If frequency is represented at right angles to the $x - y$ plane, the surface given by this equation is a hump having a peak at the centre (μ_x, μ_y) and Gaussian cross-sections in any direction. The regression lines

$$\mu_{y \cdot x} = \alpha + \beta x; \quad \mu_{x \cdot y} = \alpha' + \beta' y \tag{4.65a, b}$$

intersect at the centre and their slopes and the standard deviations of individual values about them are related to the correlation coefficient and the overall standard deviations; thus

$$\varrho = \sqrt{(\beta\beta')}; \quad \beta = \varrho\sigma_y/\sigma_x; \quad \beta' = \varrho\sigma_x/\sigma_y \tag{4.66a, b, c}$$

and

$$\sigma_{y \cdot x} = \sigma_y\sqrt{(1-\varrho^2)}; \quad \sigma_{x \cdot y} = \sigma_x(1-\varrho^2) \tag{4.67a, b}$$

The two-dimensional representation of the normal bivariate universe is best achieved by a differential frequency or frequency density diagram which shows frequency contours (or their estimates) in the $x - y$ plane or by a cumulative frequency diagram which shows the corresponding boundaries of given proportions of the population taken from the centre (known or estimated) of the distribution. In either case the contours or boundaries form systems of ellipses. If there is no association between the variables, ϱ is zero and the axes of these ellipses are parallel to the co-ordinate axes and if, in addition, $\sigma_x = \sigma_y$ they degenerate into circles. Since lack of association implies that the means of the distributions of either variable are independent of changes in value of the other, in this situation the regression lines must be at right angles and coincide with the elliptic axes. If there is complete association between the variables the ellipses degenerate into a single line of non-zero slope which also represents both regression lines and the correlation coefficient takes one of its extreme values, $+1$ or -1, according to whether the association between x and y is direct or inverse. In intermediate cases the correlation coefficient is fractional and the ellipses are inclined to the reference axes; to the x axis, for example, by an angle θ given by

$$\tan 2\theta = 2\varrho\sigma_x\sigma_y/(\sigma_x^2 - \sigma_y^2) \tag{4.68}$$

and the regression lines (which pass through the extremities of the ellipses in the x or y directions) are inclined to one another by an angle φ given by

$$\tan \varphi = (1 - \varrho^2)\, \sigma_x\sigma_y/\varrho(\sigma_x^2 + \sigma_y^2) \tag{4.69}$$

These relationships can be applied, *mutatis mutandis*, to the appropriate estimations from sample data. The computations for the regression of y and x are identical with those applying in the simple regression situation and those for the regression of x on y are the same except that the roles of the two variables are reversed. To summarize:

$$\hat{y} = a + bx; \qquad \hat{x} = a' + b'y \tag{4.70a, b}$$

where

$$a = \bar{y} - b\bar{x}; \qquad a' = \bar{x} - b'\bar{y} \tag{4.71a, b}$$

and

$$b = \Sigma XY/\Sigma X^2; \quad b' = \Sigma XY/\Sigma Y^2 \tag{4.72a, b}$$

and

$$s_e = s_{y \cdot x} = \sqrt{[\Sigma \hat{Y}^2/(n-2)]}; \quad s_{e'} = s_{x \cdot y} = \sqrt{[\Sigma \hat{X}^2/(n-2)]} \qquad (4.73\text{a, b})$$

The correlation coefficient can be estimated in several ways but the most efficient estimator is the "product-moment" value:

$$r = s_{xy}/s_x s_y = \Sigma XY/\sqrt{(\Sigma X^2)}\sqrt{(\Sigma Y^2)} = \sqrt{bb'} \qquad (4.74\text{a, b, c})$$

where the estimated co-variance

$$s_{xy} = \Sigma XY/(n-1) \qquad (4.75)$$

If the standard errors of estimate are calculated from r by way of the appropriate modifications of eqns. (4.67a) and (4.67b), a correcting factor equal to $\sqrt{[(n-1)/(n-2)]}$ is involved due to the difference in degrees of freedom for ordinary and error variances.

The contour and boundary ellipses, and also confidence ellipses, are most easily fitted by transforming the variables to coordinates referred to the principal axes, so

$$\mathfrak{x} = x \cos\theta + y \sin\theta; \quad \mathfrak{y} = y \cos\theta - x \sin\theta \qquad (4.76\text{a, b})$$

θ is found by the sample form of eqn. (4.68) but the values of the transformed variables are not actually required. The equations of the ellipses thus referred to the principal axes are

$$\mathfrak{x}^2/s_{\mathfrak{x}}^2 + \mathfrak{y}^2/s_{\mathfrak{y}}^2 = c^2 \qquad (4.77)$$

where $s_{\mathfrak{x}}$ and $s_{\mathfrak{y}}$ are the (estimated) standard deviations of the transformed variables (i.e. of the normal distributions along the axial cross-sections) and $s_{\mathfrak{x}}c$ and $s_{\mathfrak{y}}c$ are the semi-axes. $s_{\mathfrak{x}}$ and $s_{\mathfrak{y}}$ are found by way of the relationships

$$s_{\mathfrak{x}}^2 + s_{\mathfrak{y}}^2 = s_x^2 + s_y^2; \quad s_{\mathfrak{x}} s_{\mathfrak{y}} = s_x s_y \sqrt{(1-r^2)} \qquad (4.78\text{a, b})$$

from which

$$s_{\mathfrak{x}} + s_{\mathfrak{y}} = \sqrt{(s_{\mathfrak{x}}^2 + s_{\mathfrak{y}}^2 + 2s_{\mathfrak{x}}s_{\mathfrak{y}})}; \quad s_{\mathfrak{x}} - s_{\mathfrak{y}} = \sqrt{(s_{\mathfrak{x}}^2 + s_{\mathfrak{y}}^2 - 2s_{\mathfrak{x}}s_{\mathfrak{y}})} \qquad (4.78\text{c, d})$$

The choice of c depends on whether contours, boundaries or confidence regions are required. If the first, then for the contour f_{xy},

$$f_{xy} = f_{xy\,(\max)} \exp\left(-\tfrac{1}{2}c^2\right) \qquad (4.79)$$

where, from the first term on the right-hand side of eqn. (4.64)

$$f_{xy\,(\max)} = n/2\pi s_x s_y \sqrt{(1 - r^2)} \tag{4.80}$$

so

$$c^2 = 4{\cdot}6052\,(\log f_{xy\,(\max)} - \log f_{xy}) \tag{4.81}$$

Obviously, contours may be found for f_{xy} relative to $f_{xy\,(\max)}$ by making the latter unity or relative to sample size n by substituting unity or 100(%) for that quantity. If cumulative frequency boundaries are required, c must be selected so that the required estimated proportion, P, of the universe values is embraced and the required value is

$$c^2 = \chi^2_{P,2} \tag{4.82}$$

If a confidence ellipse for the true mean point (μ_x, μ_y) is required the s_x^2 and s_y^2 of eqn. (4.77) must be replaced by the corresponding

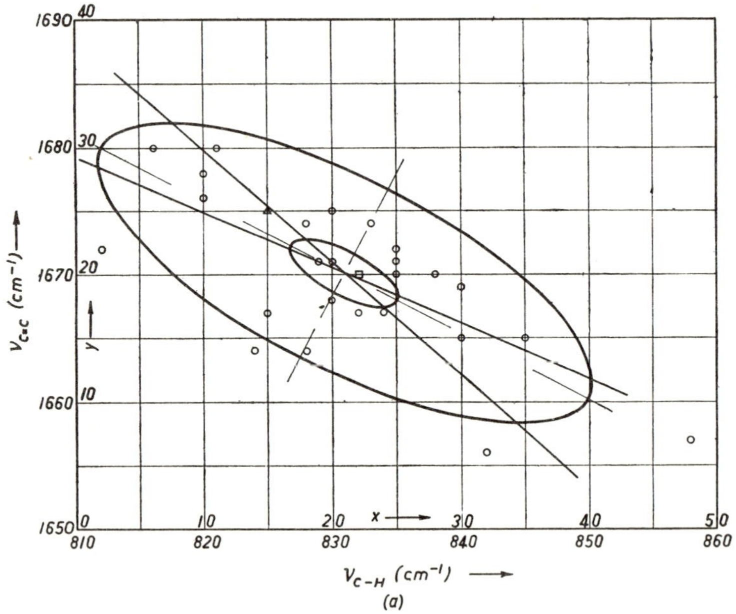

FIG. 4.7(a). Normal Bivariate Boundary and Confidence Regions.

The outer ellipse is the fitted 90% boundary for individual points. The inner ellipse is the 95% confidence region for the true mean point (μ_x, μ_y). The data plotted are those of Table 4.10, △ representing two points and □ four points. The regression lines of y on x and of x on y are shown full. The axes of the ellipses are shown broken. The inner scales are marked in the coded units used in the computations.

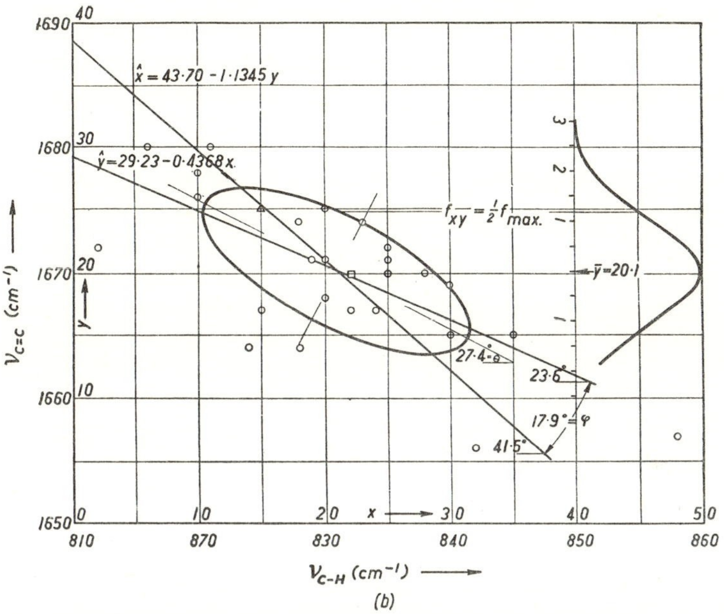

FIG. 4.7(b). A Normal Bivariate Frequency Contour.

The ellipse is the contour for half maximum frequency. The Gaussian curve on the right represents the shape of the normal bivariate surface along the $\bar{x}$ ordinate and is marked in standard error units. The angles made by the regression lines and the major axis to the x axis and the angle between the regression lines, all calculated in Table 4.10(b), are shown.

standard errors of the means, i.e. must be divided by n, and

$$c^2 = 2F_{\mathrm{P},2/f}(n-1)/(n-2) \tag{4.83}$$

(Alternatively, eqn. (4.77) may be used with c^2/n). This constant, sometimes known as Hotelling's T, is a generalized version of the one-dimensional statistic t, and has been used above in finding the joint confidence region for α and β in the simple regression model. If the true standard deviations and correlation coefficient were known it would be replaced by $2F_{\mathrm{P},2/\infty}$ which is $\chi^2_{\mathrm{P},2}$. In all cases, knowing the semi-axes, the ellipses are easily constructed by graphical methods. The geometrical relationships are illustrated, by application to the example below, in Figs. 4.7(a) and 4.7(b).

The data (in both original and coded form) in Table 4.10 (a) and the computations in Table 4.10(b) relate to frequency measurements on infrared absorption bands on some compounds of type $R_1R_2C{=}CR_3H$ in which isopropylidene-isopropenyl isomerism is possible. There appears to be an inverse relationship between the C=C stretching frequency of the isopropylidene groups (range,

TABLE 4.10. INFRARED ABSORPTION BANDS OF ISOPROPYLIDENE GROUPS

(a) Spectroscopic frequencies

No.	Compound	ν_{C-H}	x	$\nu_{C=C}$	y
1	Methyl heptanol	832	22	1670	20
2	β-Citronellol	830	20	1671	21
3	β-Citronellal	825	15	1675	25
4	2-Me-2-pentene	833	23	1674	24
5	3-Et-2-pentene	834	24	1667	17
6	3-Me-*cis*-2-pentene	825	15	1675	25
7	1-Me-1-cyclohexene	820	10	1676	26
8	Anymol	828	18	1674	24
9	Geraniol *(trans)*	838	28	1670	20
10	Nerol *(cis)*	832	22	1670	20
11	Dihydromyrcene	820	10	1678	28
12	Digeranyl	832	22	1670	20
13	Geraniolene	821	11	1680	30
14	Digeranyl ether	830	20	1668	18
15	Geranylamine HCl	835	25	1670	20
16	Geranyl chloride	845	35	1665	15
17	Linalool	835	25	1672	22
18	Isodigeranyl	832	22	1670	20
19	Geranyl acetate	830	20	1675	25
20	Polyisoprene	840	30	1665	15
21	2-Me-2-hexene	829	19	1671	21
22	2:6 Dime-2-octene	835	25	1671	21
23	3-Me-*trans*-2-pentene	812	2	1672	22
24	α-Pinene	842	32	1656	6
25	Terpinolene	816	6	1680	30
26	2:4:4 Trime-2-pentene	824	14	1664	14
27	Cholesterol	840	30	1669	19
28	Cholest-4-ene	858	48	1657	7
29	Cholest-5-ene	832	22	1667	17
30	Ergost-7-ene	825	15	1667	17
31	3β-hydroxyergost-7-ene	828	18	1664	14
32	24-Hydroxychol-9(11)-ene	827	17	1644	6

Note:

ν_{C-H} = out-of-plane hydrogen bending frequency (cm^{-1}), $x = \nu_{C-H} - 810$
$\nu_{C=C}$ = carbon–carbon stretching frequency (cm^{-1}), $y = \nu_{C=C} - 1650$

TABLE 4.10 *(cont.)*. (b) Calculation of the correlation coefficient, regression lines and contour, boundary and confidence ellipses

Basic quantities:

$n = 31$; $\Sigma x = 648$; $\Sigma y = 623$; $\bar{x} = 20{\cdot}903$; $\bar{y} = 20{\cdot}097$

$\Sigma x^2 = 15{,}978$; $\Sigma y^2 = 13{,}457$; $\Sigma xy = 11{,}960$

$(\Sigma x)^2/n = 13{,}545{\cdot}3$; $(\Sigma y)^2/n = 1250{\cdot}3$; $\Sigma x \Sigma y/n = 13{,}002{\cdot}7$

$\Sigma X^2 = 2432{\cdot}7$; $\Sigma Y^2 = 936{\cdot}7$; $\Sigma XY = -1062{\cdot}7$

$\div (n-1)$; $s_x^2 = 81{\cdot}090$; $s_y^2 = 31{\cdot}223$; $s_{xy} = -354{\cdot}233$

$s_x = 9{\cdot}005$; $s_y = 5{\cdot}589$; $s_x s_y = 50{\cdot}3289$

Calculation of the correlation coefficient and regression equations:

Eqn. (4.74), $r = -1062{\cdot}7/\sqrt{(2432{\cdot}7 \times 936{\cdot}7)} = -0{\cdot}7040$

$r^2 = 0{\cdot}4956$; $1 - r^2 = 0{\cdot}5044$; $\sqrt{(1 - r^2)} = 0{\cdot}7102$

Eqn. (4.72), $b = -1062{\cdot}7/2432{\cdot}7 = -0{\cdot}43684$

$b' = -1062{\cdot}7/936{\cdot}7 = -1{\cdot}13451$

Eqn. (4.71), $a = 20{\cdot}097 + 0{\cdot}43684 \times 20{\cdot}903 = 29{\cdot}228$

$a' = 20{\cdot}093 + 1{\cdot}13451 \times 20{\cdot}907 = 43{\cdot}703$

$\hat{y} = 29{\cdot}228 - 0{\cdot}43684x$; $\hat{x} = 43{\cdot}703 - 1{\cdot}13451y$

$\Sigma \hat{Y}^2 = 936{\cdot}7 - 0{\cdot}43684 \times 1062{\cdot}7 = 472{\cdot}5$

$\Sigma \hat{X}^2 = 2432{\cdot}7 - 1{\cdot}13451 \times 1062{\cdot}7 = 1227{\cdot}1$

Eqn. (4.73), $s_e^2 = 472{\cdot}5/29 = 16{\cdot}293$; $s_e = 4{\cdot}036$

$s_{e'}^2 = 1227{\cdot}1/29 = 42{\cdot}314$; $s_{e'} = 6{\cdot}505$

Eqn. (4.67), Corrected, Check $s_e = 5{\cdot}589 \times 0{\cdot}7102\sqrt{(30/29)} = 4{\cdot}036$

$s_{e'} = 9{\cdot}005 \times 0{\cdot}7102\sqrt{(30/29)} = 6{\cdot}505$

Calculation of angles (taken between 0° and 90° for convenience):

Eqn. (4.69), $\tan^{-1} |0{\cdot}5044 \times 50{\cdot}3289/-0{\cdot}7040(81{\cdot}090 + 31{\cdot}223)|$

$= \tan^{-1} 0{\cdot}3211 = 17{\cdot}9° = \varphi$

Check: $\tan^{-1} |0{\cdot}4368| = 23{\cdot}6°$; $\tan^{-1} |1/1{\cdot}13451| = 41{\cdot}5°$;

$41{\cdot}5° - 23{\cdot}6° = 17{\cdot}9°$

Eqn. (4.68), $\tan^{-1} |2 \times 0{\cdot}7040 \times 50{\cdot}3289/(81{\cdot}090 - 31{\cdot}223)| = 54{\cdot}8° = 2\theta$,

$\theta = 27{\cdot}4°$

Calculation of contour, boundary and confidence ellipses:

Eqns. (4.78a, b), $s_{x}^2 + s_{y}^2 = 81{\cdot}090 + 31{\cdot}223 = 112{\cdot}313$;

$s_x s_y = 50{\cdot}3289 \times 0{\cdot}7102 = 35{\cdot}922$

Eqns. (4.78c, d), $s_x + s_y = \sqrt{(112{\cdot}313 + 2 \times 35{\cdot}922)} = 13{\cdot}570$;

$s_x - s_y = \sqrt{(112{\cdot}313 - 2 \times 35{\cdot}922)} = 6{\cdot}361$

$s_x = (13{\cdot}570 + 6{\cdot}361)/2 = 9{\cdot}966$; $s_y = 13{\cdot}570 - 9{\cdot}966 = 3{\cdot}604$

Eqn. (4.80), $f_{xy\,(\max)} = 31/2\pi \times 50{\cdot}3289 \times 0{\cdot}7102 = 0{\cdot}1380$;

$\log f_{xy\,(\max)} = -0{\cdot}8601$

For half-maximum frequency:

$f_{xy} = 0{\cdot}0690$; $\log f_{xy} = -1{\cdot}1612$; $\log f_{xy\,(\max)} - \log f_{xy} = 0{\cdot}3011$

Eqn. (4.81), $c^2 = 4{\cdot}6052 \times 0{\cdot}3011 = 1{\cdot}3866$; $c = 1{\cdot}1775$

TABLE 4.10 *(cont.)*

Major semi-axis = $9{\cdot}966 \times 1{\cdot}1775 = 11{\cdot}735$;
Minor semi-axis = $3{\cdot}604 \times 1{\cdot}1775 = 4{\cdot}244$

For 90% boundary:

Eqn. (4.82), $c^2 = \chi^{2}_{.1,2} = 4{\cdot}6052$; $c = 2{\cdot}1460$

Major semi-axis = $9{\cdot}966 \times 2{\cdot}1460 = 21{\cdot}387$;
Minor semi-axis = $3{\cdot}640 \times 2{\cdot}1460 = 7{\cdot}735$

For 95% confidence region:

Eqn. (4.83), $c^2 = 2F_{.05,2/29} = 6{\cdot}8849$; $c = 2{\cdot}6239$; $c/\sqrt{(31)} = 0{\cdot}4713$

Major semi-axis = $9{\cdot}966 \times 0{\cdot}4713 = 4{\cdot}697$;
Minor semi-axis = $3{\cdot}604 \times 0{\cdot}4713 = 1{\cdot}1699$.

1650–1680 cm^{-1}) and the corresponding out-of-plane hydrogen bending frequency (range, 810–860 cm^{-1}).

The main intent of the calculations in this instance is to show the likely position of the distribution (given the Gaussian hypothesis) for diagnostic purposes or to check on suspected results. Confidence or tolerance intervals based on the regression equations and standard errors of estimate can be similarly used but the bands attached to one or other regression equation do not call attention to the likely bounds of the distribution which are a feature of the model. For example, the point for cholest-4-ene (no. 28, Table 4.10(a)) lies far from the estimated mean point and, in fact, is just outside the 99% boundary so that there might be grounds for suspicion that the two infrared frequencies are not ordinarily related in this case yet the point lies close to the y on x regression line and the orthogonal regression line (major axis) and would not be questioned on the basis of any test involving them. On the other hand the point for 24-hydroxychol-9(11)-ene (number 32) must be regarded as so highly exceptional as to make it reasonably certain that the values quoted in Table 4.10(a) are incorrect or that the compound has not the supposed structure. This conclusion is not affected when the point is included in the statistical analysis of the data as is commonly done when testing a suspected point on purely numerical grounds, but it is not, of course, desirable to include items in a sample when there is prior reason for believing that they do not belong with the remainder.

It must be remembered (especially when considering the rejection of single suspected points) that cumulative frequency boundaries like that in Fig. 4.7(b) are simply the result of fitting a normal bivariate distribution to the data and have the same status as the fitted curve on the right of Fig. 4.7(b) or a line drawn through points on probability paper. As such they take no account of sample fluctuations and do not represent tolerance regions like the intervals described in Section 3.6 above. Four, or rather more than 10%, of the thirty-one data points fall outside the fitted 90% boundaries of Fig. 4.7(a) but this must be regarded as fair agreement. Much more extreme sample fluctuations can be expected between observed and calculated differential frequencies and, in general, the latter are not of great interest, being worth computing only when the sample size is very great. The calculation in Table 4.10(b) is included only for purposes of demonstration. In this case the fitted maximum frequency is 0·14 points per square unit which means a density of 14 points per drawn square at the centre and 7 per square along the contour of Fig. 4.7(b). The sample points are more closely grouped about the centre than these values indicate. The effect of sampling error on the estimates of parameters and the application to them of tests of significance are described in the next section.

There is, however, one aspect of sampling error which may be considered here and this is the matter of "attenuation". The correlation coefficient calculated from sample data gives an overall measure of the relationship between the observed values and it estimates the true correlation in a universe of such values. If part of the variability of this universe is due to inherent errors of observation of the values, its correlation coefficient is lower than that of a universe consisting of similar items whose attached values are known without error. Specifically, if x' and y' are exactly known values (correlation coefficient, ϱ') and x and y are measured values (correlation coefficient, ϱ) subject to independent, random, normally distributed errors of measurement denoted by u and v respectively, then the relationship between the two coefficients is

$$\varrho = \varrho'/\sqrt{[(1 + \sigma_u^2/\sigma_x^2)(1 + \sigma_v^2/\sigma_y^2)]} \tag{4.84a}$$

Similar corrections apply to the regression coefficients; e.g.

$$\beta = \beta'/\sqrt{(1 + \sigma_u^2/\sigma_x^2)} \tag{4.84b}$$

where β is the true coefficient of the regression of y on x and β' in this instance is the "ultimate" value which would obtain if y and x were free of errors of measurement.

To return to the example, if ϱ, σ_x and σ_y were approximately as estimated (say $-0{\cdot}70$, $9{\cdot}0$ and $5{\cdot}6$ respectively) and σ_u and σ_v were $1{\cdot}0$ and $2{\cdot}0\ \text{cm}^{-1}$ (both plausible values) then the correlation between true error-free frequencies would be $-0{\cdot}75$. Conversely, if the true frequency values were completely correlated ($\varrho' = -1$) a coefficient of about $-0{\cdot}93$ would be expected from measured values. It follows from eqn. (4.84a) that if there are no errors of measurement the two coefficients are identical and the difference between them is negligible if errors of measurement are slight relative to the overall spread of the values.

4.11. Tests of Significance involving the Correlation Coefficient

One of the several uses of the sample correlation coefficient lies in the fact that it is a convenient single figure representing the apparent degree of association and as such can be used for testing hypothetical values of or finding confidence intervals for ϱ and for testing differences between samples. Unlike a and b, however, it is not normally distributed except in the important case where the universe value itself is zero. This exception simplifies the testing of a hypothesis that a sample yielding r has been drawn from a universe of zero correlation and we have here a simple test of association. For large samples from universes of low correlation the standard error of r,

$$\sigma_r = (1 - \varrho^2)/\sqrt{n} \tag{4.85}$$

may be used but for smaller samples convenient tables are available (see Appendix Table II). For testing r against other values of ϱ or for testing the difference between two or more values of r it is convenient to transform the correlation coefficient to another measure of association denoted by z for the sample (ζ for the universe) whose sampling distribution is close to normal regardless of sample size. The transformed statistic is

$$z = \ln\,[(1 + r)/(1 - r)] = \tanh^{-1} r \tag{4.86}$$

and its standard error is

$$\sigma_z = 1/\sqrt{(n - 3)} \tag{4.87}$$

The function is included in most collections of statistical tables and a convenient strip-chart is provided by Snedecor (1956). A brief table is included in the Appendix. Equation (4.87) is approximate. Kendall and Stuart (1961) give more elaborate equations appropriate to sample sizes of less than fifty.

Some examples of tests of significance related to the example of the last section and based on r and z follow.

(a) Is there reasonable evidence of association between the C=C and C—H frequencies of Table 4.10(a), r being $-0{\cdot}704$ from 31 observations?

For $n = 31$, whence $f = 29$, the most extreme values of r likely to arise (P $= 0{\cdot}05$) are $\pm 0{\cdot}306$ (Appendix Table II) so we reject the null hypothesis that $\varrho = 0$ in this case and prefer to believe that the data reveal some connection between the two frequencies.

(b) Could the true correlation coefficient be as high as $-0{\cdot}93$, a hypothetical value based, let us say, on the proposition that the frequencies are uniquely related but subject to certain errors of measurement?

To answer this question we must transform r to z and apply a one-sided u test. From Appendix Table X we have $z = -0{\cdot}877$ for the sample and a hypothetical ζ of $-1{\cdot}658$. Then

$$u = (z - \zeta)/\sigma_z = (-0{\cdot}877 + 1{\cdot}658)/0{\cdot}190 = +4{\cdot}11$$

This is much greater than the 5% one-sided critical value ($+1{\cdot}64$) so the proposition is clearly unlikely.

Alternatively, what is the 95% confidence interval for ϱ?

$$95\%\,\text{CL},\ \zeta = z \pm 1{\cdot}960\sigma_z = -0{\cdot}877 \pm 0{\cdot}372 = -0{\cdot}505 \text{ to } -1{\cdot}249.$$

Reconverting by Table X: 95% confidence limits for $\varrho = -0{\cdot}47$ to $-0{\cdot}85$.

This interval is not symmetrical about the best estimate, $-0{\cdot}704$.

Tests such as these assume that the sample is random and small relative to the universe. Now, we have no certain knowledge of randomness where the principle of collection is simply the availability of chemical compounds—but at least, in the case of the example, there was no conscious collection on the basis of the value of the properties under consideration (compare Hogben (1957)). Further, we can assume that, although the list of Table 4.10(a) may have exhausted almost all the available data, there exists or could exist a very large number of compounds of the same type.

TABLE 4.11. MANGANESE AND CHROMIUM CONTENT OF LOW CARBON IRON

No.	Mn contents, x	Cr contents, y	Product terms
$n_1 = 50$	$\bar{x}_1 = 148{\cdot}74$, $\Sigma X_1^2 = 107{,}834$	$\bar{y}_1 = 29{\cdot}96$, $\Sigma Y_1^2 = 4094$	$\Sigma X_1 Y_1 = 12{,}402$, $r_1 = 0{\cdot}590$
$n_2 = 49$	$\bar{x}_2 = 140{\cdot}12$, $\Sigma X_2^2 = 107{,}960$	$\bar{y}_2 = 32{\cdot}92$, $\Sigma Y_2^2 = 3507$	$\Sigma X_2 Y_2 = 12{,}830$, $r_2 = 0{\cdot}659$
$k = 2$	$t_x = +0{\cdot}92$, $\Sigma\Sigma X^2 = 215{,}794$	$t_y = -1{\cdot}66$, $\Sigma\Sigma Y^2 = 7601$	$\Sigma\Sigma XY = 25{,}232$, $r_p = 0{\cdot}623$

Tests for the comparison of two or more sample correlation coefficients are based on the z statistic and follow closely similar tests described in Chapter III. In their simple form, however, they can only indicate whether or not a number of samples have come from similarly *correlated* universes—a special test is required to see if they have come from the same or similarly *located* universes. Consider the following data which relate to random samples of pieces of iron analysed for manganese and chromium. The averages, sums of squares and products of deviations and correlation coefficients are given in Table 4.11.

(c) Could the two samples ($r_1 = 0{\cdot}590$, $r_2 = 0{\cdot}659$) have come from universes having the same degree of correlation?

Here we can apply a u test on the difference between two z values, the standard error of the difference (from the sum of the variances) being

$$\sigma_{\Delta z} = \left(\frac{1}{n_1 - 3} + \frac{1}{n_2 - 3}\right) \qquad (4.88)$$

Now, from Appendix Table X,

$$\Delta z = |z_1 - z_2| = |0{\cdot}678 - 0{\cdot}792| = 0{\cdot}114,$$

and

$$\sigma_{\Delta z} = \sqrt{(1/47 + 1/46)} = \sqrt{0{\cdot}04302} = 0{\cdot}2074,$$

so, testing the null hypothesis that the true $\Delta\zeta$ is zero against the two-sided alternative, we find

$$u = |\Delta z - 0|/\sigma_{\Delta z} = 0{\cdot}114/0{\cdot}207 = 0{\cdot}55 \quad (\text{cf. } u_{.05} = 1{\cdot}96)$$

This is clearly not significant and we may therefore say that the two samples have come from the same universe *as judged by the correlation coefficient*, i.e. that the sample values may be estimates of the same true ϱ. This being so a combined estimate of that ϱ can be obtained by weighting the z values according to sample size or inversely according to the variances (virtually the same thing),

$$\bar{z} = \Sigma nz/\Sigma n \quad \text{or} \quad \bar{z} = \Sigma[(n-3)\,z]/\Sigma(n-3) \qquad (4.89\text{a,b})$$

or by pooling the sums of squares and products of deviations. Either form of eqn. (4.89) gives an average z of 0·734 and the corresponding correlation coefficient is 0·626 which is very close to the "within-groups" value, 0·623, obtained directly from the data of Table 4.11. The standard error of the pooled z value is

$$\sigma_{\bar{z}} = \sqrt{[1/\Sigma(n-3)]} \tag{4.90}$$

and confidence intervals are found in the usual way:

$$95\%\ \text{CL},\ \zeta = \bar{z} \pm 1{\cdot}96\sigma_{\bar{z}} = 0{\cdot}734 \pm 1{\cdot}96(1/\sqrt{93}) = 0{\cdot}734 \pm 0{\cdot}203.$$

The corresponding interval for ϱ is 0·486 to 0·734.

The u test cannot of course be used to compare more than two randomly arrived-at correlation coefficients but a test similar to that of Bartlett for variances and based on χ^2 is available for larger numbers of estimates. Suppose a third sample of forty pieces of iron yielded a correlation coefficient of 0·760 and the question now becomes

(d) Could the three samples ($r_1 = 0{\cdot}590$, $r_2 = 0{\cdot}659$, $r_3 = 0{\cdot}760$) have come from universes having the same degree of correlation?

The null hypothesis is that all ϱ are the same and the alternative that some or all differ. We find χ^2 in the following way:

r	z	$n-3$	$(n-3)\,z$	$(n-3)\,z^2$
+0·590	+0·678	47	31·866	21·605
+0·659	+0·792	46	36·432	28·854
+0·760	+0·996	37	36·825	36·705
		130	105·150	87·164

$$\chi^2 = \Sigma(n-3)z^2 - (\Sigma(n-)\,z)^2/\Sigma(n-3) = 87{\cdot}164 - 85{\cdot}050 = 2{\cdot}114.$$

The critical value is that for the number of samples less one, i.e. $\chi_{\cdot 05,2} = 5{\cdot}99$, so we accept the null hypothesis. There being insufficient evidence to suggest that the sample coefficients differ significantly, we average them by eqn. (4.89b) which gives 105·15/130 = 0·809 for $\bar{z}$, whence $\bar{r} = 0{\cdot}668$. The standard error of $\bar{z}$ is given by eqn. (4.90), $1/\sqrt{130} = 0{\cdot}088$ in this instance.

Returning to the data of Table 4.11; we have seen that the two samples do not differ significantly in the degree of correlation between the Mn and Cr contents. Further, if the mean Mn contents and/or the mean Cr contents be compared by t test, these also do not differ significantly since the t values (0·92 and 1·66 given in the table) are well short of the critical value ($t_{\cdot 05,\Sigma n-2=97} = 1{\cdot}98$) for pairs of samples of these sizes. In applying ordinary t tests, however, no allowance is made for the correlation between x and y. The generalized form of t, Hotelling's T already referred to, takes this into account and the question of interest may be framed

(e) Could the two sample mean points ($\bar{x}_1, \bar{y}_1$ and $\bar{x}_2, \bar{y}_2$) be considered to estimate the same true mean point (μ_x, μ_y) of a bivariate universe?

To answer this question we use the pooled value of r given in the table,

$$r_p = \Sigma\Sigma XY/\sqrt{(\Sigma\Sigma X^2.\Sigma\Sigma Y^2)} = 0{\cdot}623,$$

and calculate first

$$T^2 = (t_x^2 - 2r_p t_x t_y + t_y^2)/(1 - r_p^2) \qquad (4.91)$$

$$= (0{\cdot}846 + 2 \times 0{\cdot}623 \times 0{\cdot}92 \times 1{\cdot}66 + 2{\cdot}7556)/0{\cdot}612 = 8{\cdot}89$$

and then

$$F = [(n_1 + n_2 - 2k + 1)/(n_1 + n_2 - k)\,k]\,T^2 \qquad (4.92)$$

$$= (96/97 \times 2)\,8{\cdot}99 = 4{\cdot}45$$

This value of F is to be compared with the critical value for $f_1 = k$ and $f_2 = n_1 + n_2 - 2k + 1$ degrees of freedom. Since this, $F_{\cdot 05,2/96}$, is only 3·09, we reject the null hypothesis and conclude that there is in fact a significant difference between the two samples with regard to overall quality of product. In view of the insignificance of the difference between the mean Mn and Cr contents taken separately this conclusion may appear paradoxical but the real situation is revealed by Fig. 4.8. In terms of t_x and t_y, the expression (4.91) is that of an ellipse having axes $2T\sqrt{(1 + r_p)}$ and $2T\sqrt{(1 - r_p)}$, at 45° to the coordinate axes, in a square of side $2T$. Substitution of the critical F in eqn. (4.92) gives 6·244 for T^2 and 2·50 for T. The ellipse in the Fig. 4.8 is constructed on the basis of this value. As can be seen, the sample point (t_x, t_y) is within the square of non-significant single values ($t_{\cdot 05,97} = 1{\cdot}98$) but is well beyond the critical region which takes account of the correlation.

The test procedure (and the example) is that of Hicks (1955) who also describes its extension to any number of variables but, as this number increases, the computation of T^2 increases in difficulty. On the other hand, for the comparison of a single sample ($k = 1$) with a hypothetical mean point (μ_x, μ_y), T^2 reduces to t^2 which is F for 1 and $n - 1$ degrees of freedom.

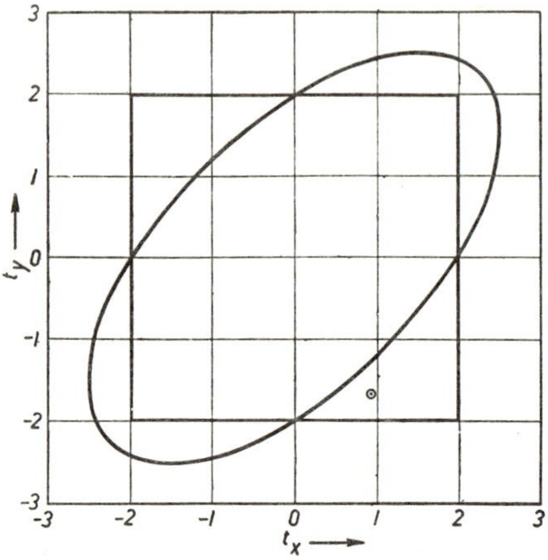

FIG. 4.8. A Test for the Comparability of Two Sample Points.

The square encloses values of t (for $f = 97$) which are not significant in single comparisons of sample means at the 5% level. The ellipse encloses values of T which are not significant in a joint comparison when the correlation coefficient is $+0{\cdot}623$. The point for the data of Table 4.11 falls outside the joint acceptance region.

4.12. THE APPROXIMATE TREATMENT OF ASSOCIATED DATA

The ordinary and sometimes rather tedious computations hitherto described can be avoided when the correlation coefficient or a regression equation are required with no great arithmetical precision. Grouping forms the basis of most approximate treatments and this may be done by way of an extension of the tabular methods illustrated in Chapter II or by semi-graphical procedure. Ranking methods can also be used.

(1) *The Grouping Method*

The grouping method is illustrated in Table 4.12 whose layout follows the form of the graphed data in Fig. 4.7. This form is convenient in that counts can be made on a scatter diagram and be directly transferred to the table and the usual visual checks for the sign of r or b are easy to make. More elaborate tables do however provide in-built checks on the arithmetic and a procedure for summation calculation is given by Elderton (1938). As far as

TABLE 4.12. INFRARED ABSORPTION BANDS OF ISOPROPYLIDENE GROUPS

ν_{C-H}		807·5–	817·5–	827·5–	837·5–	847·5–	857·5–				
x		−2	−1	0	1	2	3	f	fy	fy^2	fxy
$\nu_{C=C}$	y										
1682·5–	2	−4 1 −4	−2 2 −2					3	6	12	−8
1677·5–	1		−1 3 −1	0 3 0				6	6	6	−3
1672·5–	0	0 1 0		0 10 0	0 2 0			13	0	0	0
1667·5–	−1		+1 2 +2	0 3 0	−1 2 −2			7	−7	7	0
1662·5–	−2							0	0	0	0
1657·5–	−3				−3 1 −3		−9 1 −9	2	−6	18	−12
	f	2	7	16	5	0	1	31	−1	43	−23
	fx	−4	−7	0	5	0	3	−3			
	fx^2	8	7	0	5	0	9	29			

$$A_x = 832{\cdot}5; \quad h_x = 10; \quad A_y = 1675; \quad h_y = 5$$
$$\Sigma f X^2 = \Sigma f x^2 - (\Sigma f x)^2/n - n/12 = 29 - 0{\cdot}29 - 2{\cdot}58 = 26{\cdot}13$$
$$\Sigma f Y^2 = \Sigma f y^2 - (\Sigma f y)^2/n - n/12 = 43 - 0{\cdot}03 - 2{\cdot}58 = 40{\cdot}39$$
$$\Sigma f XY = \Sigma f xy - \Sigma f x \, \Sigma f y/n = -23 - 0{\cdot}10 = -23{\cdot}10$$
$$r = \Sigma f XY/\sqrt{\Sigma f X}\sqrt{\Sigma f Y} = -23{\cdot}10/\sqrt{26{\cdot}13}\sqrt{40{\cdot}39} = -0{\cdot}71$$

means and sums of squares of deviations go, the calculations in terms of the transformed variables follow the method laid down in Table 2.9 except that Sheppard's corrections have been made at an earlier stage than is usual. The only additional item is the calculation of the cross-product. This must be done cell by cell as shown (the figures are arranged [xy f fxy] in each cell) and the values summed by row or column or by both as a check on the addition. The calculations in the table are carried only as far as the correlation coefficient and since this is dimensionless the sums of squares and products need not be decoded. The correlation coefficient obtained by grouping is generally lower than that from exact calculation—an aspect of the attenuation effect described at the end of Section 4.10. It is therefore desirable that a reasonable number of class intervals be adopted in both scales but, when very large amounts of data are to be processed, extremely coarse grouping can be adopted.†

For regression calculations it is necessary to convert back to the original scale units by multiplying the squares and products of deviations by the appropriate power of the class interval width. Thus, in the example the sums of squares (SS) and products (SP) of deviations in the original units are

$$SS(\boldsymbol{\nu}_{C-H}) = h_x^2 \, \Sigma f X^2 = 10^2 \times 26{\cdot}13 = 2613$$

$$SS(\boldsymbol{\nu}_{C=C}) = h_y^2 \, \Sigma f Y^2 = 5^2 \times 40{\cdot}39 = 1010$$

$$SP(\boldsymbol{\nu}_{C-H} \, . \, \boldsymbol{\nu}_{C=C}) = h_x h_y \, \Sigma f XY = 10 \times 5 \times -23{\cdot}10 = -1155.$$

These values agree well with those found by exact calculation (from Table 4.10(b): 2432·7, 936·7 and −1062·7) and the results of the subsequent calculations are quite satisfactory; better, in fact, than might be expected in view of the rather coarse grouping adopted.

(2) *Three-group Analysis*

A useful semi-graphical method is described by Gibson and Jowett (1957) and called by them "three-group analysis" but it is in fact a form of the method of averages. The data are divided into three groups roughly in the proportion 1:2:1 (in the case of an approximately normal distribution), the central group is discarded, and the means found for the values in the first and last groups. If the division is made in the x direction the slope of the regression of y on x is

$$b = (\bar{y}'' - \bar{y}')/(\bar{x}'' - \bar{x}') \qquad (4.9)$$

If the division is made along the y axis the slope of the other regression is

$$b' = (\bar{x}'' - \bar{x}')/(\bar{y}'' - \bar{y}') \qquad (4.93)$$

† See, for example, some of the statistics described by Dixon and Massey (1957).

where (′) and (″) denote the first and last group respectively. The correlation coefficient can be calculated from the two regression coefficients by eqn. (4.74c). Gibson and Jowett recommend fitting the line through the grand mean point using eqn. (4.71a) or (4.71b). A fair value for the standard error of estimate can be found from the within-group ranges (taken parallel to the regression line) and the d_2 factors of Table V of the Appendix. It is convenient to split the central group into two for this purpose thus making four approximately equal groups. Since the use of the d_2 factors is not recommended for sample sizes in excess of twelve, this imposes a limit of less than fifty on the sample size treated unless one resorts to further subdivision. The standard error of the coefficient of the regression of y on x is given as

$$s_b = 0{\cdot}88 s_e \sqrt{[(1/n'' + 1/n')/(\bar{x}'' - \bar{x}')^2]} \tag{4.94}$$

in which the numerical factor takes into account a bias due to the 1:2:1 grouping of the normally distributed values. The way in which the data are grouped is not critical but the optimum proportions are (besides 1:2:1 for a normal) 1:1:1 for a rectangular, 2:1:2 for a U-shaped, 3:3:1 for a J-shaped and 4:5:2 for a fairly skewed distribution. In a simple regression situation the 1:1:1 split would be adopted and even when applied to a normal distribution this is almost as efficient as the recommended 1:2:1 split.

The application of the method to the infrared frequency example is illustrated in Table 4.13 and Fig. 4.9, the division being as close

TABLE 4.13. THREE-GROUP ANALYSIS APPLIED TO I.R. FREQUENCY DATA

Group	n	Σx	Σy	$\bar{x}$	$\bar{y}$	w	d_2	w/d_2	nw/d_2
Lower (′)	9	98	217	10·89	24·11	14·6	2·970	4·92	44·28
Middle	6 }	272	261			11·6	2·534	4·58	27·48
Middle	7 }					7·4	2·704	2·74	19·18
Upper (″)	9	278	145	30·89	16·11	13·1	2·970	4·41	39·69
	31	648	623						130·63

$\bar{x} = 648/31 = 20{\cdot}9$; $\bar{y} = 623/31 = 20{\cdot}1$; $\Delta\bar{x} = 20{\cdot}00$; $\Delta\bar{y} = -8{\cdot}00$

$s_e = 130{\cdot}63/31 = 4{\cdot}2$

Slope (eqn. 4.9), $b = (\bar{y}'' - \bar{y}')/(\bar{x}'' - \bar{x}') = -8/20 = -0{\cdot}400$

Intercept (eqn. 4.5), $a = \bar{y} - b\bar{x} = 20{\cdot}1 + 0{\cdot}400 \times 20{\cdot}9 = 28{\cdot}5$

Regression of y on x, $\hat{y} = 28{\cdot}5 - 0{\cdot}400x$

Std. error of b (eqn. 4.94), $s_b = 0{\cdot}88 \times 4{\cdot}22 \sqrt{[(1/9 + 1/9)/20^2]} = 0{\cdot}083$

to the optimum as possible. Only the regression of y on x is dealt with, the values being those of Table 4.10(a). The value of the regression coefficient is a little low by comparison with the exact calculated value. This is also true of the coefficient of the regression of x on y and the correlation coefficient. The latter (the calculations are not shown) comes out to be $-0{\cdot}62$ compared with the exactly computed $-0{\cdot}704$. On the other hand, the value for the standard error of estimate is better than might have been expected.

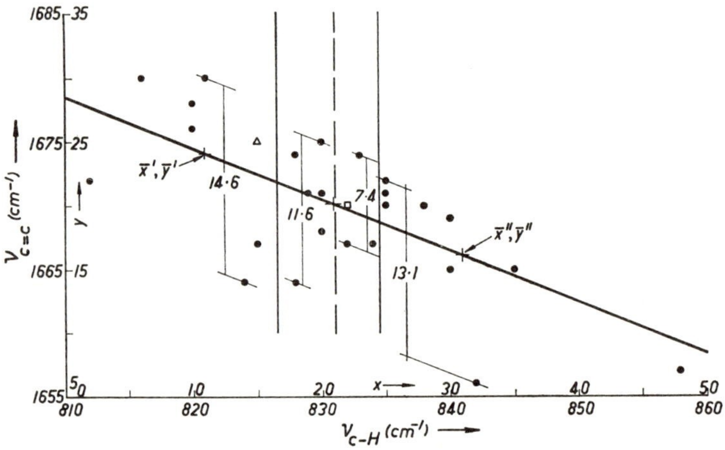

FIG. 4.9. "Three-group" Analysis of Associated Data.

The data plotted are those of Table 4.10, grouped according to Table 4.13. The full line is the approximate regression of y on x.

(3) *The Ranking Method*

Of the several statistics based on ranking of data, Spearman's correlation coefficient is most used. If the ranked values come from a normal bivariate universe this is an estimate of ϱ and therefore an approximation to the ordinary or "product-moment" coefficient, r. The usual computing equation is

$$r_S = 1 - 6\,\Sigma d^2/(n^3 - n) \tag{4.95a}$$

where d is a difference between ranks of x and y taken from 1 to n (low to high) in each case. When values tie in ranking, this should be taken into account unless the correction is very small. Kelly (1947) gives an equation which allows for such ties,

$$r_S = 1 - 6\,\Sigma d^2/\sqrt{[(n^3 - n - \Sigma(k_x^3 - k_x))\,(n^3 - n - \Sigma(k_y^3 - k_y))]} \tag{4.95b}$$

where k is the number of items in each tie. For a double tie $k = 2$, $k^3 - k = 6$; for a triple, $k = 3$, $k^3 - k = 24$ etc.

TABLE 4.14. RANK CORRELATION

Batch	Zn, % x	Ash, % y	Rank x	Rank y	Difference in rank, d	d^2	Corrections
1	5·33	17·97	2½	5	−2½	6·25	x:1 double tie
2	5·63	18·13	11	9	+2	4·00	$\Sigma(k^3 - k) = 6$
3	5·42	18·06	9	8	+1	1·00	$n^3 - n - 6 = 1314$
4	5·39	18·05	7	7	0	0·00	
5	5·45	18·16	10	10	0	0·00	y:2 double ties
6	5·33	17·75	2½	1½	+1	1·00	$\Sigma(k^3 - k) = 12$
7	5·36	17·75	4	1½	+2½	6·25	$n^3 - n - 12 = 1308$
8	5·37	17·92	5	3½	+1½	2·25	
9	5·39	18·01	7	6	+1	1·00	
10	5·39	19·02	7	11	−4	16·00	
11	5·30	17·92	1	3½	−2½	6·25	

$\Sigma d^2 = 44{\cdot}00$

Eqn. (4.95a), $r_s = 1 - 6\,\Sigma d^2/(n^3 - n) = 1 - 6 \times 44/1320 = 0{\cdot}800$

Eqn. (4.95b), corrected $r_s = 1 - 6 \times 44/\sqrt{(1314 \times 1308)} = 0{\cdot}799$

The example here relates to the percentages of zinc and ash in batches of a corrosion inhibitor. The data and the computations are given in Table 4.14. Litchfield and Wilcoxon (1955) give a convenient nomograph which permits r_s to be read off for n from 6 to 40 and Σd^2 from 2 to 20,000 (Fig. 4.10). They also give a table of approximate critical values of the latter which is reproduced in Table XI of the Appendix. In the case of the example Σd^2 is 44 and this is well below the lower critical values in the table, 88 for the 5% level and 58 for the 1% level. The rank correlation is therefore significantly high (positive) and it is most unlikely that the sample comes from an uncorrelated universe. The difference between the uncorrected value and that corrected for ties is negligible.

Rank correlation coefficients are calculated for significance testing rather than for descriptive purposes although it is sometimes convenient to rank a series of items in order of magnitude of properties

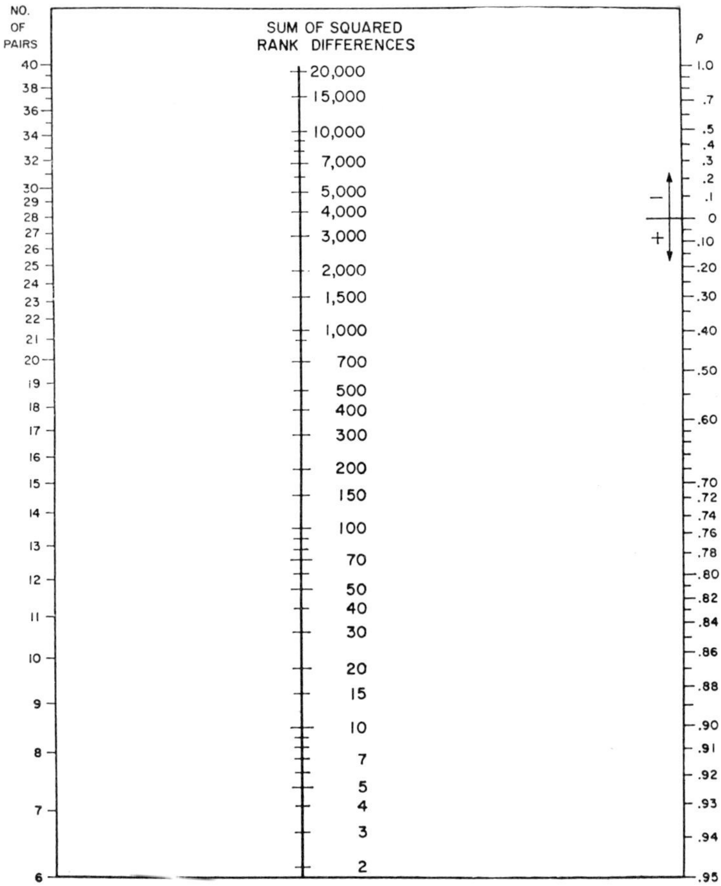

FIG. 4.10. A Nomograph for Spearman's Rank Correlation Coefficient.

A line connecting n and Σd^2 gives r_s on the right-hand scale.

Reprinted from *Analytical Chemistry*, 27, 299, 1955. Copright 1955 by the American Chemical Society and reprinted by permission of the copyright owner.

for which variable scales do not exist or would be difficult to devise and thus to characterize the degree of relationship between the properties.

(4) *A Control Chart Method*

Other measures can be used to detect significant association, particularly in the form of a trend of values in time or some other order of sampling. Quality control charts for single values are useful for such tests and also the number of runs above and below the median or runs up and down. Hald (1952) gives useful information about distributions of runs—the quality control chart procedure is illustrated here.

Consider the following data. An observer, measuring the surface tension of a solution under somewhat poorly controlled conditions noticed that his first four readings decreased in magnitude so he continued to take readings until he had twenty in all. His measurements, in the order in which they were made, were

63·8 (0·5) 63·3 (0·4) 62·9 (1·3) 61·6 (2·4) 64·0 (1·9) 62·1 (0·5) 61·6 (0·1)

61·7 (0·1) 61·8 (0·8) 61·0 (1·3) 62·3 (0·5) 62·8 (0·6) 62·2 (0·8) 61·4 (0·7)

60·7 (0·6) 61·3 (0·8) 62·1 (0·6) 61·5 (0·5) 62·0 (1·1) 60·9 dyn/cm,

the figures in parentheses being the differences between successive readings or *moving ranges*. Regression analysis is pointless in this instance, especially as the independent variable (time) is not properly characterized, but simple tests for trend can adequately answer the question of interest which is whether or not the true mean of the dependent variable is changing with time. The data of the example have a mean of 62·05 ($\bar{x}$) and a mean moving range of 0·816 ($\bar{w}'$).

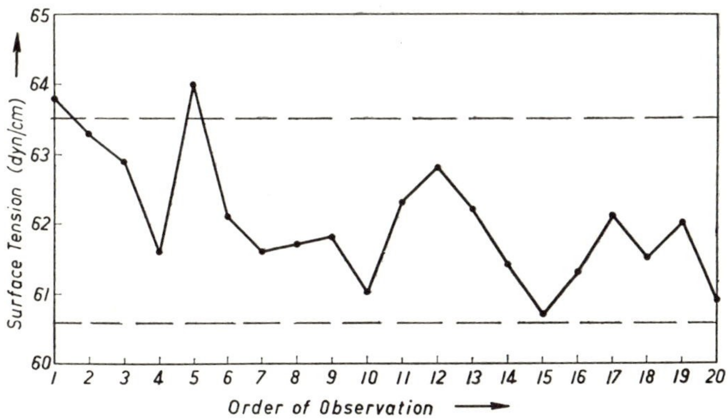

FIG. 4.11. Apparent Trend in Successive Observations.

The broken lines (based on the mean moving range) represent the limits within which 95% of individual points should fall in the absence of a significant trend.

The "control chart" limits are $62{\cdot}05 \pm 1{\cdot}74 \times 0{\cdot}816 = 60{\cdot}59$ to $63{\cdot}51$, the factor $1{\cdot}74$ being $1{\cdot}96/d_2$ (for samples of size two from Table V of the Appendix). Two early points fall outside these limits so it could be claimed that there is some evidence of drift (or, at least, of lack of control) on this basis. The data are plotted in Fig. 4.11. Tests like this are very insensitive but their usefulness when a quick, rough answer is all that is required is obvious enough.

Notation of Chapter IV

a, b	estimated intercept and regression coefficient (slope) of the line of regression of y on x
a', b'	estimated intercept and regression coefficient of the line of regression of x on y, intercepts and mean regression coefficient of averaged regression lines
$\alpha, \beta, \alpha', \beta'$	true intercepts and regression coefficients estimated by the above
α_0, β_0	hypothetical values of intercept and regression coefficient
A, B	guessed approximations to a and b
A_x, A_y	arbitrary means on the scales of x and y
$\boldsymbol{\beta}'$	coefficient of the regression of $\boldsymbol{y}'$ on $\boldsymbol{x}'$
CL	confidence limits
d	difference between ranks
d_2	factor relating range, w, to standard deviation, σ. (Table V of Appendix)
e	the actual error of an observed value
f	degrees of freedom; frequency
h_x, h_y	class interval widths on the x and y scales
i, j (generally subscripts)	individual values in a sample or set
I	ionic strength
k	number of samples or sets; number of items in a tie
k	velocity constant
l, l'	tolerance limit factors
m	number of observations at a particular value of x

μ or $\mu_{y \cdot x}$	true mean of y values at given x value
MS	mean square = SS/f
n	size of sample
N	size of universe
P	population coverage (in tolerance limits)
P	level of significance; 1 − P = level of confidence, confidence coefficient
ϱ, r	universe and sample correlation coefficient
r_p	pooled estimate of ϱ
ϱ'	coefficient of correlation between $\boldsymbol{x}'$ and $\boldsymbol{y}'$
r_S	Spearman's rank correlation coefficient
R	electrical resistance
σ_e, s_e	universe and sample standard error of estimate
σ_e^2, s_e^2	universe and sample error variance
s_{ep}	pooled estimate of σ_e
s_a, s_b, $s_{\hat{y}}$ etc.	standard errors of sampling distributions of a, b, $\hat{y}$ etc.
$s_{\Delta a'}$, $s_{\Delta b}$, $s_{\Delta b'}$	standard error of the difference between intercepts of regression lines averaged with respect to slope, regression coefficients, and the averaged regression coefficient and that of the sample means
SP	sum of products of deviations
SS, S	sum of squares of deviations
T	a total; Hotelling's statistic
T	temperature
θ	inclination of the major axis of a correlation ellipse to the x-axis
φ	the angle between regression lines
u, v	errors of measurement of x and y respectively in correlation
w	weight of an observation or number of observations in a set; range
w'	moving range (i.e. difference between two successive values)
x	a variable, generally independent
$\hat{x}$	a value of x estimated from an observed y value

χ, x_I	true and estimated points of intersection of regression lines
$\not{x}$, $\not{y}$	transformed or coded values of x and y
X, Y	deviations of x and y values from their means, $x - \bar{x}$ and $y - \bar{y}$
$\mathbf{x}'$, $\mathbf{y}'$	values of x and y free of experimental error
y	a variable, generally dependent
$\hat{y}$	estimate of $\mu_{y \cdot x}$
$\hat{Y}$	the deviation, $y - \hat{y}$, an estimate of e
$\dot{y}$	mean of a sub-set of m observations
$\dot{Y}$	the deviation, $\dot{y} - \bar{y}$
ζ, z	universe and sample transformed correlation coefficients
$'$, $''$	superscripts indicating lower and upper halves or thirds of a sample
$^\circ$, $''$, $'$	superscripts indicating overall, between and within samples sums of squares etc. or estimates

CHAPTER V

THE RELATIONSHIP BETWEEN TWO OR MORE VARIABLES—MULTIPLE, NON-LINEAR AND OTHER RELATIONSHIPS

THE principal models of Chapter IV (the simple regression and the correlation models) can be modified in three main ways to meet more complicated situations encountered in the treatment of experimental data. First, the number of variables can be increased beyond two without other alterations to the specification of the models. Secondly, the simple linear dependence of, say, y on x can be changed to accommodate curvilinear relationships and these, in turn, can be incorporated into equations involving a multiplicity of variables. Thirdly, the requirement of constancy of error in the dependent variable can be relaxed to allow for its variation over the range of measurement of the independent variable. In this chapter these and related matters will be treated in turn.

5.1. MULTIPLE LINEAR CORRELATION AND REGRESSION

It is convenient to consider first the extension of the linear correlation model of Section 4.10 to three variables. With the guidance of the more complicated model we can investigate the interrelationships between all three variables; in particular, we can estimate (a) the relationship or degree of association between the values of any two with the effect of the third removed, i.e. the *partial* regression equations and correlation coefficients, and (b) the relationship or degree of association between one variable and the combination of the other two, i.e. *multiple* regression equations and correlation coefficients. In the correlation case, where all the variables have a natural dispersion, the number of multiple regression equations is equal to the number of variables, each equation representing a least squares plane which best fits the points so as to minimize the deviations in the direction of one of the variable scales. In the regression case there is only one equation of interest, that which minimizes the deviations of the dependent variable.

In the correlation case, we can represent the situation by the multiple regression equations

$$\mu_{y \cdot xz} = \alpha + \beta x + \gamma z; \qquad \hat{y} = a + bx + cz \qquad (5.1\,\text{a,b})$$

$$\mu_{x \cdot yz} = \alpha' + \beta' y + \gamma' z; \qquad \hat{x} = a' + b'y + c'z \qquad (5.1\,\text{c,d})$$

$$\mu_{z \cdot xy} = \alpha'' + \beta'' x + \gamma'' y; \quad \hat{z} = a'' + b''x + c''y \qquad (5.1\,\text{e,f})$$

where, in eqns. (5.1 a) and (5.1 b) for example, $\mu_{y \cdot xz}$ is the true mean of the y values at particular values of x and z, α is an intercept on the y axis and β and γ are *partial regression coefficients* or slopes estimated by a, b and c. The latter are the constants of the equation of a plane, eqn. (5.1 b), fitted so that $\hat{y}$ is the best estimate of $\mu_{y \cdot xz}$ and the sum of the squares of the deviations of the actual y values from it is a minimum. If eqns. (5.1 a) and (5.1 b) are adapted to a particular value of z then they take the form

$$\mu_{y \cdot x(z)} = (\alpha + \gamma z) + \beta x; \quad \hat{y}_{(z)} = (a + cz) + bx \qquad (5.2\,\text{a,b})$$

which are the true and estimated partial regression equations having $(\alpha + \gamma z)$ and $(a + cz)$ as intercepts.

Similar meanings are attached to the terms of the other pairs of equations. The three regression planes, true or estimated, intersect at the mean points, (μ_x, μ_y, μ_z) and $(\bar{x}, \bar{y}, \bar{z})$ respectively. How closely the planes lie with respect to each other depends on the degree of association between the variables. If there is no association then the three true planes are at right angles to each other. If there is complete association they coincide. It should be noted that nothing is implied in the model, or in the analysis of data based on it, about causation. We cannot say in general that the values of one variable are caused by the other two although this might be a natural implication in some experimental situations.

Returning again to the first pair of equations; we can say in connection with their errors or residuals

$$y = \mu_{y\cdot xz} + e_y = \alpha + \beta x + \gamma z + e; \tag{5.3a}$$

$$y = \hat{y} + \hat{Y} = a + bx + cz + \hat{Y} \tag{5.3b}$$

and

$$\sigma^2_{y\cdot xz} = \Sigma(y - \mu_{y\cdot xz})^2/N = \Sigma e_y^2/N \tag{5.4a}$$

$$s^2_{y\cdot xz} = \Sigma(y - \hat{y})^2/(n-3) = \Sigma \hat{Y}^2/(n-3) \tag{5.4b}$$

where e and $\hat{Y}$ are the deviations of points from the true and estimated regression planes, $\sigma^2_{y\cdot xz}$ is the true error variance of the y values over the range of x and z values and $s^2_{y\cdot xz}$ is an unbiased estimator of this property. It is assumed in the model under discussion that the error variance is constant. The number of degrees of freedom, $n - 3$, appropriate to its estimate is the number of independent differences of the type, $y - \hat{y}$, having regard to the fact that the data must provide three constants (a, b and c) in order that $\hat{y}$ can be found. In eqn. (5.4a) we assume for convenience that the universe is of size N—strictly, a function of integrals replaces this in the case of an infinite universe.

Following the procedure of Section 4.2 (eqns. (4.4) to (4.6)), normal equations can be found whose solution gives the constants of the least squares equations. For eqn. (5.1b) these are

$$a = \bar{y} - b\bar{x} - c\bar{z} \tag{5.5a}$$

$$b\Sigma X^2 + c\Sigma XZ = \Sigma XY \tag{5.5b}$$

$$b\Sigma XZ + c\Sigma Z^2 = \Sigma YZ \tag{5.5c}$$

or, alternatively, in terms of the sums of squares and products of the original values,

$$\left.\begin{aligned} an + b\Sigma x + c\Sigma z &= \Sigma y \\ a\Sigma x + b\Sigma x^2 + c\Sigma xz &= \Sigma xy \\ a\Sigma z + b\Sigma xz + c\Sigma z^2 &= \Sigma yz \end{aligned}\right\} \tag{5.6}$$

It is convenient to solve the simultaneous eqns. (5.5b) and (5.5c) by finding

$$A = \Sigma XY/\Sigma XZ; \quad B = \Sigma X^2/\Sigma XZ \qquad (5.7\text{a, b})$$

$$C = \Sigma YZ/\Sigma Z^2; \quad D = \Sigma XZ/\Sigma Z^2 \qquad (5.7\text{c, d})$$

whence

$$b = (A - C)/(B - D) \quad \text{and} \quad c = A - Bb \qquad (5.7\text{e, f})$$

Product–moment correlation coefficients, as described in Section 4.10, can be calculated from the values of any pair of variables but the real measure of association between them is found by correcting for the influence of the third variable. These corrected or partial correlation coefficients can be found from the crude or overall values by equations of the form

$$r_{xy\cdot z} = (r_{xy} - r_{xz}r_{yz})/\sqrt{[(1 - r_{xz}^2)(1 - r_{yz}^2)]} \qquad (5.8)$$

The partial correlation coefficient, $r_{xy\cdot z}$, is that which would be obtained if a sample array of paired x and y values were available for which the z value was constant. For computational purposes, a term of the type $\sqrt{(1 - r^2)}$ can be found as the cosine of the angle of which r is the sine. A nomograph designed by Lees and Lord (1962) is of help in computing partial correlation coefficients.

A measure of association between the values of a particular variable on the one hand and those of the remaining variables on the other is provided by the multiple correlation coefficient. For the correlation of y with x and z we have

$$r_{y\cdot xz} = \sqrt{[(r_{xy}^2 + r_{yz}^2 - 2r_{xy}r_{yz}r_{xz})/(1 - r_{xz}^2)]}$$
$$= \sqrt{[1 - (1 - r_{xy}^2)(1 - r_{yz\cdot x}^2)]} \qquad (5.9\text{a, b})$$

Lord's (1955) nomograph enables $r_{y\cdot xz}$ to be found from the three zero-order coefficients.

By appropriate interchange of symbols, eqns. (5.8) and (5.9) can be adapted to the other partial and multiple coefficients, $r_{yz\cdot x}$, $r_{xz\cdot y}$, $r_{x\cdot yz}$, $r_{z\cdot xy}$, and to the true or parameter values, $\varrho_{xy\cdot z}$ etc. Partial correlation coefficients, like the overall ϱ_{xy} and r_{xy} measures, fall within the limits of ± 1 but the multiple coefficients only take positive values to $+1$.

Just as r_{xy} is not only a measure of association between paired x and y values but is also numerically equal to that between the actual and calculated values (i.e. $|r_{xy}| = r_{x\hat{x}} = r_{y\hat{y}}$) and, further, is related to the standard errors of estimate by the sample versions of eqns. (4.67a) and (4.67b), so is the multiple correlation coefficient connected with the corresponding partial and multiple regression equations. Thus, $r_{y \cdot xz}$ is the same as $r_{y\hat{y}}$, $\hat{y}$ being the calculated value from the multiple regression eqn. (5.1b). The standard error of eqn. (5.1b) is given in a general way by eqn. (5.4b) but, by a derivation similar to that of eqn. (4.7) we find

$$s_{y \cdot xz} = \sqrt{[(\Sigma Y^2 - b\Sigma XY - c\Sigma YZ)/(n-3)]} \qquad (5.10\text{a})$$

$$= s_y\sqrt{[(1 - r^2_{y \cdot xz})(n-1)/(n-3)]} \qquad (5.10\text{b})$$

The first of these equations shows how the total variability of the y values, ΣY^2, is reduced by contributions due to x and z to yield the variability of the y values from the regression plane, $\Sigma \hat{Y}^2$.

Following the forms of eqns. (4.11a) and (4.14a) we have for the standard errors of the partial regression coefficients and of $\hat{y}$

$$s_b = s_{y \cdot xz}/\sqrt{[\Sigma X^2(1 - r^2_{xz})]}; \quad s_c = s_{y \cdot xz}/\sqrt{[\Sigma Z^2(1 - r^2_{xz})]} \qquad (5.11\text{a,b})$$

$$s_{\hat{y}} = s_{y \cdot xz}\sqrt{(1/n + K)} \qquad (5.12\text{a})$$

where

$$K = [X^2/\Sigma X^2 + Z^2/\Sigma Z^2 - 2XZr^2_{xz}/\Sigma XZ)/(1 - r^2_{xz})] \qquad (5.12\text{b})$$

Since the intercept, a, is the value of $\hat{y}$ for $x = z = 0$, its standard error is given by eqns. (5.12a) and (5.12b) when the deviations from the means in the latter, X and Z, are replaced by $\bar{x}$ and $\bar{z}$. Confidence intervals for α, β, γ and $\mu_{y \cdot xz}$ can be found in the usual way (i.e. by eqn. (3.36b)), the value of t being that for $f = n - 3$ degrees of freedom. The 100 (1 − P)% limits for a single forecast by eqn. (5.1b) are given by $\hat{y} \pm t_{P,f}s_{y \cdot xz}\sqrt{(1 + 1/n + K)}$. A convenient procedure for finding the standard errors, when one is interested in only one of the regression equations, is described in connection with the pure regression model below.

A simple example of the application of the correlation model (subject to the reservations as to randomness etc. discussed in Section 4.10) is provided by certain of the compounds listed in Table 4.10(a). These are characterized by three spectroscopic frequencies, the two given in that table and a "skeletal" frequency (in the range 790–810 cm^{-1}) which appears when the isopropylidene group in the molecule forms part of a cyclohexenyl ring. Only the

TABLE 5.1. INFRARED ABSORPTION BANDS OF ISOPROPYLIDENE GROUPS

(a) Coded data and results of basic calculations

Compound	$\nu_{C-H} - 810$ x	$\nu_{C=C} - 1650$ y	$\nu_{sk} - 780$ z	Calculated quantities
Terpinoline	6	30	12	$\Sigma X^2 = 1320{\cdot}89$
1-Me-1-cyclohexene	10	26	17	$\Sigma Y^2 = 527{\cdot}56$
Ergost-7-ene	15	17	15	$\Sigma Z^2 = 296{\cdot}00$
3β-hydroxyergost-7-ene	18	14	12	$\Sigma XY = -686{\cdot}78$
Anymol	18	24	18	$\Sigma XZ = 392{\cdot}33$
Cholest-5-ene	22	17	17	$\Sigma YZ = -76{\cdot}67$
Cholesterol	30	19	20	$r_{xy} = -0{\cdot}823$
α-Pinene	32	6	9	$r_{xz} = +0{\cdot}627$
Cholest-4-ene	48	7	30	$r_{yz} = -0{\cdot}194$
	$\Sigma x = 199$ $\bar{x} = 22{\cdot}11$	$\Sigma y = 160$ $\bar{y} = 17{\cdot}78$	$\Sigma z = 150$ $\bar{z} = 16{\cdot}67$	

(b) Correlation coefficients and multiple regression equation

Partial correlation coefficient (eqn. 5.8):

$r_{xy \cdot z} = (-0{\cdot}823 + 0{\cdot}627 \times 0{\cdot}194)/\sqrt{[(1 - 0{\cdot}627^2)(1 - 0{\cdot}194^2)]} = -0{\cdot}918$

Similarly: $r_{xz \cdot y} = +0{\cdot}846$; $r_{yz \cdot x} = +0{\cdot}741$

Multiple correlation coefficient (eqn. 5.9a):

$r_{y \cdot xz} = \sqrt{(0{\cdot}823^2 + 0{\cdot}194^2 - 2 \times 0{\cdot}823 \times 0{\cdot}627 \times 0{\cdot}194)}/\sqrt{(1 - 0{\cdot}627^2)} = 0{\cdot}921$

Similarly: $r_{x \cdot yz} = 0{\cdot}954$; $r_{z \cdot xy} = 0{\cdot}852$

Regression of y on x and z (eqns. 5.7a, b, c, d):

$A = -686{\cdot}78/392{\cdot}33 = -1{\cdot}75052$; $C = -76{\cdot}67/296{\cdot}00 = -0{\cdot}25902$

$B = 1320{\cdot}89/392{\cdot}33 = 3{\cdot}36678$; $D = 392{\cdot}33/296{\cdot}00 = 1{\cdot}32544$

Eqn. (5.7e), $b = (-1{\cdot}75052 + 0{\cdot}25902)/(3{\cdot}36678 - 1{\cdot}32544) = -0{\cdot}7306$

Eqn. (5.7f), $c = -1{\cdot}75052 + 0{\cdot}7306 \times 3{\cdot}36678 = +0{\cdot}7094$

Eqn. (5.5a), $a = 17{\cdot}78 - 0{\cdot}7306 \times 22{\cdot}11 + 0{\cdot}7094 \times 16{\cdot}67 = 22{\cdot}11$

Whence $\hat{y} = 22{\cdot}11 - 0{\cdot}7306x + 0{\cdot}7094z$

Variation about the regression of y on x and z (eqn. 5.10a):

$\Sigma \hat{Y}^2 = 527{\cdot}56 - 0{\cdot}7306 \times 686{\cdot}78 + 0{\cdot}7094 \times 76{\cdot}67 = 527{\cdot}56 - 447{\cdot}39 = 80{\cdot}17$

$s_{y \cdot xz} = \sqrt{[80{\cdot}17/(9 - 3)]} = \sqrt{13{\cdot}36} = 3{\cdot}66$

Standard errors and 95% confidence limits of regression coefficients:

Eqn. (5.11a), $s_b = 3{\cdot}66/\sqrt{[1320{\cdot}89(1 - 0{\cdot}627^2)]} = 0{\cdot}129$ $(t_{.05,6} = 2{\cdot}447)$

95% confidence limits of $\beta = -0{\cdot}7306 \pm 2{\cdot}447 \times 0{\cdot}129 = -0{\cdot}73 \pm 0{\cdot}32$

Eqn. (5.11b), $s_c = 3{\cdot}66/\sqrt{[296{\cdot}00(1 - 0{\cdot}627^2)]} = 0{\cdot}273$

95% confidence limits of $\gamma = +0{\cdot}7094 \pm 2{\cdot}447 \times 0{\cdot}273 = +0{\cdot}71 \pm 0{\cdot}67$

Eqn. (5.12a), $s_{\hat{y}}$ (at $x = \bar{x}$, $z = \bar{z}$) $= 3{\cdot}66/\sqrt{9} = 1{\cdot}22$

TABLE 5.1 *(cont.)*

(c) The analysis of variance of regression

Source of variation	SS	f	MS	Ratio
Due to x, $(\Sigma XY)^2/\Sigma X^2$	357·08	1	357·08	26·72
Additional due to z, by difference	90·31	1	90·31	6·76
Due to regression on x and z, $b\Sigma XY + c\Sigma XZ$	447·39	2	223·68	16·74
Residual	80·17	6	13·36	($F_{\cdot05,1/6}$ = 5·99)
Total	527·56	8		($F_{\cdot05,2/6}$ = 5·14)

coded data and the results of the preliminary calculations are given in Table 5.1. The subsequent, illustrative calculations are confined to the regression of y on x and z. The basic calculations are carried out as shown in Table 4.2 and Table 4.10 (i.e. using relationships of the form of eqn. 2.31)). (Table 5.3 provides another example.) To check the necessary sums of squares and products of the actual values, it is convenient to find $q = x + y + z$, then $\Sigma q^2 = \Sigma x^2 + \Sigma y^2 + \Sigma z^2 + 2\Sigma xy + 2\Sigma xz + 2\Sigma yz$ contains all the quantities.

One expression of the relationship between x and y for the nine compounds considered here is the equation for the regression of y on x,

$$\hat{y} = 29{\cdot}27 - 0{\cdot}5199x \quad \text{(see Fig. 5.1)}$$

and the part of the total variability of the y values ($\Sigma Y^2 = 527{\cdot}56$) which is accounted for by this regression is given by $b\Sigma XY$ or $(\Sigma XY)^2/\Sigma X^2$ (b here being the overall regression coefficient) and is 357·08. The standard deviation of the y values calculated about their mean is 8·12 and this is reduced to 4·93 about the regression line. That these reductions are significant can be verified by an analysis of variance of the regression or by a test on the correlation coefficient, −0·827, which clearly is one not likely to have been observed had the variables been unrelated. Considered in the same way, the values of z do not, at first sight, appear to contribute to the relationship. The crude correlation coefficient between x and z, +0·627, is of doubtful significance and that between y and z, −0·194, could easily have arisen by chance since $r_{\cdot05,7} = 0{\cdot}666$. (Appendix Table II.)

The picture is changed, however, when we consider the three sets of values together. We find that z is, in fact, related to the other two variables—a relationship concealed in this instance when the effect or interdependence of either x or y is ignored—and from the point of view of calculation or prediction of y values it is, therefore, worthwhile including a z term in the equation. This conclusion can be expressed in various ways and verified by several equivalent tests of significance. (a) The partial correlation coefficients given at the top of Table 5.1 (b) are all numerically greater than the corresponding crude coefficients and all are significantly greater than zero.

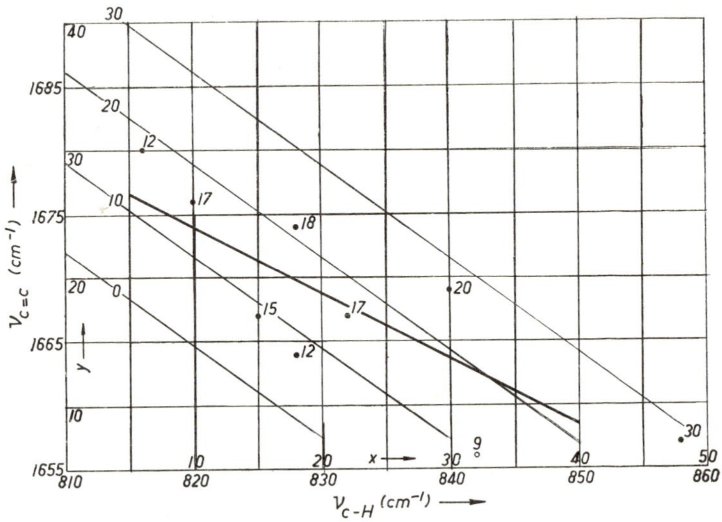

FIG. 5.1. The Data of Table 5.1.

The thin lines are the partial regression lines for four fixed values of z, 0, 10, 20 and 30. The heavy line represents the regression of y on x not taking the z values into account.

(They are tested for significance against a null hypothesis that the true value is zero by comparison with the critical values of Appendix Table II for $n - k$ degrees of freedom where k is the number of variables included in the analysis. Here k is 3 and n is 9 so the 5% critical value is 0·707.) (b) The multiple regression equation accounts for 447·39 of the total variability of the y values, leaving a residual of 80·17 which we attribute to error and unexplained sources of variation. If we deduct from 447·39 the 357·08 accounted for by the simple regression of y on x and test the difference, attributed to z, against the mean square residual or error variance, we find the z contribution to be significant at between the 5% and 2·5% levels. This is illustrated in Table 5.1 (c). To put the point in a different way; we can say that the multiple regression standard error of estimate, 3·66, is significantly lower than the simple regression standard error of estimate, 4·93. (c) As shown at the bottom of Table 5.1 (b), the confidence interval for γ is $+0{\cdot}04$ to $+1{\cdot}38$ and, although the limits are somewhat wide, they do not embrace zero and it is therefore unlikely that the true partial regression coefficient takes this value.

The third of the tests contained in Table 5.1 (c) is that of the multiple regression mean square against the residual and is, in fact, a test of the multiple correlation coefficient, 0·921, against a hypothetical true value of zero. The ratio in the table, 16·74, is also given by the expression $r^2(n-k)/(1-r^2)\,k$ (in which r is the multiple coefficient) and this can be calculated and compared with $F_{\mathrm{P},k/(n-k)}$ without extracting the sums of squares of deviations. The magnitude of the multiple correlation coefficient, of the multiple regression to residual mean square ratio, and the smallness of $s_{y \cdot xz}$

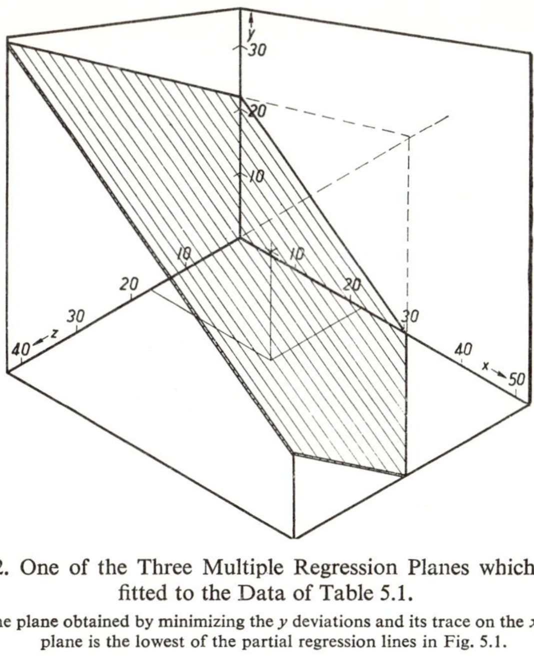

FIG. 5.2. One of the Three Multiple Regression Planes which can be fitted to the Data of Table 5.1.

This is the plane obtained by minimizing the y deviations and its trace on the xy ($z = 0$) plane is the lowest of the partial regression lines in Fig. 5.1.

in comparison with s_y all indicate the closeness of the experimental or observed points to the multiple regression plane. Similar results are obtained if the same tests are applied when $\Sigma\hat{X}^2$ and $\Sigma\hat{Z}^2$ are respectively minimized. The two regression planes so obtained are close to that of y on x and z whose position in space is illustrated in Fig. 5.2. Its trace on the xy ($z = 0$) plane is the lowest of the partial regression lines shown in Fig. 5.1.

The final (decoded) equations for all three regression planes are

$$\boldsymbol{\nu}_{\mathrm{C-H}} = 1874{\cdot}6 - 1{\cdot}1252\boldsymbol{\nu}_{\mathrm{C=C}} + 1{\cdot}0340\boldsymbol{\nu}_{\mathrm{sk.}}; \quad s_e = 4{\cdot}85\ \mathrm{cm}^{-1}$$

$$\boldsymbol{\nu}_{\mathrm{C=C}} = 1711{\cdot}2 - 0{\cdot}7306\boldsymbol{\nu}_{\mathrm{C-H}} + 0{\cdot}7094\boldsymbol{\nu}_{\mathrm{sk.}}; \quad s_e = 3{\cdot}66\ \mathrm{cm}^{-1}$$

$$\boldsymbol{\nu}_{\mathrm{sk.}} = -1019{\cdot}2 + 0{\cdot}6853\boldsymbol{\nu}_{\mathrm{C-H}} + 0{\cdot}7468\boldsymbol{\nu}_{\mathrm{C=C}}; \quad s_e = 3{\cdot}74\ \mathrm{cm}^{-1}$$

For purposes of diagnosis or prediction all three are equivalent and the one chosen for calculating a value for a particular variable is that for which the deviations of that variable have been minimized. In this particular example the intercepts have no physical significance and the true means, estimated by $\boldsymbol{\nu}_{\mathrm{C-H}}$ etc., are of little interest, this being confined to individual values. Approximate limits for an expected frequency of a single compound may be computed from the appropriate standard error of estimate, provided its other frequencies are close to the sample mean values but, when attempting to estimate far from these, the limits for individual or true mean values become very wide due to the influence of the K term in eqn. (5.12a).

In the pure regression case, only the relationship between the dependent variable on the one hand and the independent variables (whose values are known or preset) on the other is of interest and, although the other regression planes and partial correlation coefficients can be found from the associated data, they have no meaning in terms of the pure regression model. Apart from this, the regression analysis does not differ from the correlation analysis although it is convenient to modify the notation. If we denote the dependent variable by y, the independent variables by $x_1, x_2 \ldots x_j$, and the standard error of estimate by s_e, we can express the regression equation by

$$\hat{y} = a + b_1x_1 + b_2x_2 + \ldots b_jx_j \tag{5.13}$$

while

$$s_e = \sqrt{[\Sigma \hat{Y}^2/(n - k)]}$$
$$= \sqrt{[(\Sigma Y^2 - b_1 \Sigma YX_1 - b_2 \Sigma YX_2 - \ldots b_j \Sigma YX_j)/(n - k)]} \tag{5.14}$$

(Here, j is the number of independent variables and k is the total number.)

As before, the intercept is a linear function of the means:

$$a = \bar{y} - b_1\bar{x}_1 - b_2\bar{x}_2 - \ldots b_j\bar{x}_j \tag{5.15}$$

and the partial regression coefficients are found by solving a set of j simultaneous equations of the form of eqns. (5.5b) and (5.5c). These are

$$\left.\begin{array}{l} b_1 \Sigma X_1^2 + b_2 \Sigma X_1 X_2 + \ldots b_j \Sigma X_1 X_j = \Sigma YX_1 \\ b_1 \Sigma X_1 X_2 + b_2 \Sigma X_2^2 + \ldots b_j \Sigma X_2 X_j = \Sigma YX_2 \\ \cdots\cdots\cdots\cdots\cdots\cdots\cdots\cdots \\ b_1 \Sigma X_1 X_j + b_2 \Sigma X_2 X_j + \ldots b_j \Sigma X_j^2 = \Sigma YX_j \end{array}\right\} \tag{5.16}$$

(Again, eqns. (5.15) and (5.16) can be combined in a single set of equations in terms of the sums of squares and products of the original values like eqn. (5.6).)

The coefficients of the b in these equations form a square symmetric matrix and the coefficients of the inverse matrix, known as the *Gauss multipliers*, can be found by replacing the b in their columns and the column of products of deviations on the right-hand side of eqn. (5.16) by

$c_{11}, c_{12}, \ldots, c_{1j}$ and $1, 0, 0, \ldots$ in the first set,

$c_{21}, c_{22}, \ldots, c_{2j}$ and $0, 1, 0, \ldots$ in the second,

and so on to

$c_{j1}, c_{j2}, \ldots, c_{jj}$ and $0, 0, \ldots, 1$ in the last set.

For an account of the origin of these equations and of methods for their solution see, for example, Anderson and Bancroft (1952) and Dwyer (1951). Such methods can also be applied in multiple correlation analysis and they are necessary when the number of variables is large. Furthermore, they enable variables, found to be unimportant in the course of the analysis, to be fairly readily eliminated. When a large number of values is to be handled, an extension of the approximate "three-group" method of Section 4.12 (see Gibson and Jowett (1957)) effects a considerable reduction in calculation. The ordinary grouping method illustrated in Table 4.12 can also be applied, in which case the basic equations will contain frequency or weighting factors as shown in that table.

In the three variable regression case, we have

$$\left.\begin{aligned} b_1\Sigma X_1^2 + b_2\Sigma X_1X_2 &= \Sigma YX_1 \\ b_1\Sigma X_1X_2 + b_2\Sigma X_2^2 &= \Sigma YX_2 \end{aligned}\right\} \qquad (5.17\text{a})$$

$$\left.\begin{aligned} c_{11}\Sigma X_1^2 + c_{12}\Sigma X_1X_2 &= 1 \\ c_{11}\Sigma X_1X_2 + c_{12}\Sigma X_2^2 &= 0 \end{aligned}\right\} \qquad (5.17\text{b})$$

$$\left.\begin{aligned} c_{21}\Sigma X_1^2 + c_{22}\Sigma X_1X_2 &= 0 \\ c_{21}\Sigma X_1X_2 + c_{22}\Sigma X_2^2 &= 1 \end{aligned}\right\} \qquad (5.17\text{c})$$

A convenient way of solving for the constants in the three sets is to find

$$D = \Sigma X_1^2\Sigma X_2^2 - (\Sigma X_1X_2)^2 \qquad (5.18)$$

Then

$$b_1 = (\Sigma X_2^2\Sigma YX_1 - \Sigma X_1X_2\Sigma YX_2)/D \qquad (5.19\text{a})$$

$$b_2 = (\Sigma X_1^2\Sigma YX_2 - \Sigma X_1X_2\Sigma YX_1)/D \qquad (5.19\text{b})$$

and

$$c_{11} = \Sigma X_2^2/D; \quad c_{22} = \Sigma X_1^2/D; \quad c_{12} = c_{21} = -\Sigma X_1X_2/D \qquad (5.20\text{a, b, c})$$

The standard errors are

$$s_{b_1} = s_e\sqrt{c_{11}}; \quad s_{b_2} = s_e\sqrt{c_{22}} \qquad (5.21\text{a, b})$$

and

$$s_{\hat{y}} = s_e\sqrt{(1/n + c_{11}X_1^2 + c_{22}X_2^2 + c_{12}X_1X_2)} \qquad (5.22)$$

In applying this notation and these equations to the previous example, b becomes b_1, c becomes b_2, x becomes x_1, and z becomes x_2 and so on. From the calculated quantities in the right-hand column of Table 5.1 (a) we find:

from eqn. (5.18),

$$D = 1320{\cdot}89 \times 296{\cdot}00 - 392{\cdot}33^2 = 237{,}061$$

from eqn. (5.19a),

$$b_1 = (296{\cdot}00 \times -686{\cdot}78 - 392{\cdot}33 \times -76{\cdot}67)/237{,}061 = -0{\cdot}7036$$

from eqn. (5.19b),

$$b_2 = (1320{\cdot}89 \times -76{\cdot}67 - 392{\cdot}33 \times -686{\cdot}78)/237{,}061 = +0{\cdot}7094$$

from eqn. (5.20a),

$$c_{11} = 296{\cdot}00/237{,}061 = 0{\cdot}00124863; \quad \sqrt{c_{11}} = 0{\cdot}035336$$

from eqn. (5.20b),

$$c_{22} - 1320{\cdot}89/237{,}061 = 0{\cdot}00557194; \quad \sqrt{c_{22}} = 0{\cdot}074646$$

from eqn. (5.21a, b),

$$s_{b_1} = 3{\cdot}66 \times 0{\cdot}035336 = 0{\cdot}129; \quad s_{b_2} = 3{\cdot}66 \times 0{\cdot}074646 = 0{\cdot}273$$

The calculated values do not differ from those of Table 5.1 (b) by more than small rounding errors.

There are many other problems relating to multiple correlation and regression not dealt with here. Two may be mentioned for their possible relevance in some situations. The first is the problem of taking into account additional information about one or more of the coefficients of the regression equation which might be available from previous experiments of a similar nature or from a different source. Durbin (1953) discusses this matter. The other is the problem of finding joint confidence intervals for the true values of the coefficients. This has been discussed in Section 4.4 in relation to the intercept and slope of a line. The extension to the multiple regression case is treated by Durand (1954) and by Davies (1957).

5.2. The Fitting of Polynomials. Statistical Methods

The simplest and most flexible kind of curvilinear relationship between two variables is the polynomial for which the supposed "true" and actual fitted equation may be written

$$\mu = \alpha + \beta_1 x + \beta_2 x^2 + \beta_3 x^3 + \dots + \beta_j x^j \qquad (5.23\text{a})$$

$$\hat{y} = a + b_1 x + b_2 x^2 + b_3 x^3 + \dots + b_j x^j \qquad (5.23\text{b})$$

Equations such as (5.23 b) can be used to describe or *adjust* a wide variety of physico-chemical data, the molar heat capacity versus temperature relationships being perhaps the best known example. The fitting can be carried out by such approximate methods as those based on selected points or averages if the experimenter wants only a formula, or according to the principle of least squares if tests of significance are to be applied or confidence intervals for the coefficients of eqn. (5.23 a) required. In the latter situation it may be assumed that the variable y is dependent and subject to normally distributed random errors (having a constant standard deviation, σ_e) while the values of x are exactly known or preset. As before, the supposedly unvarying σ_e can be estimated from the deviations of the experimental points from the fitted line, $y - \hat{y} = \hat{Y}$, by

$$s_e = \sqrt{[\Sigma \hat{Y}^2/(n - k)]} \tag{5.24}$$

where n is the number of points and k the number of x terms in the equation (i.e. the power to which x is taken) plus one for the dependent variable itself.

The straight line, represented in its true and fitted forms by eqns. (4.1 a) and (4.1 b), is the first degree of this general power series type of relationship (the zeroth degree is represented by the horizontal lines, $\mu_{y \cdot x} = \alpha$, a constant, and $\hat{y} = a = \bar{y}$) and the methods of least squares fitting of equations of higher degree are merely the extension of those already described in Chapter IV and in Section 5.1 above. In effect, these methods amount to *rectification* (see Section 5.5) in which, for example, the equation for a parabola is reduced to that of a plane like eqn. (5.1 b) and so on. As a consequence, a quadratic can be fitted by the application of eqns. (5.5) to (5.7) or (5.17) to (5.19) in which x^2 is substituted for z or x_2 and s_e can be obtained by the appropriate modification of eqn. (5.10). This multiple regression method is of quite general application and can be varied to suit whatever degree of polynomial is required or to suit particular forms like the "square-root·parabola",

$$\hat{y} = a + b_1 x + b_2 x^{\frac{1}{2}}$$

which has more moderate curvature at high values of x. This equation is also used for adjusting molar heat data.

Equation (5.23 a) represents a whole series of models and it is, of course, possible to fit eqn. (5.23 b) to a set of points in sufficient degree so as to leave no residual. Thus, a straight line completely fits two points, a quadratic three, a cubic four and so on. Since y is subject to random error, however, there must be some residual variation and, in the absence of theoretical guidance, the process of fitting is only carried as far as a term in x whose contribution to the total variability of the y values is significant with respect to the error in y or, in other words, until an equation is obtained whose residuals show no significant trend with respect to x. In a sense, then, the data determine the model assumed to apply to them and greater than usual precautions are needed in interpreting results, particularly when extrapolating. In point of fact, the theoretical curvilinear relationships of physics and chemistry do not usually take a simple power series form and the polynomial model is only adopted as a local approximation when theoretical guidance is uncertain, unduly complicated or lacking.

When using the multiple regression method of fitting, eqns. (5.15) and (5.16) can be applied directly, only substituting x for x_1, x^2 for x_2 etc. Accordingly, for the evaluation of the intercept by eqn. (5.15) we have

$$a = \bar{y} - b_1\bar{x} - b_2\overline{x^2} - b_3\overline{x^3} - \ldots, \tag{5.25}$$

where $\overline{x^2}$ and $\overline{x^3}$ are the means of the squares and of the cubes of the x values. Similarly, in setting up the equations for the regression coefficients (eqns. (5.16)) and for the Gauss multipliers, ΣX_1^2 is ΣX^2, ΣX_2^2 is the sum of the squares of the x^2 values from their mean, $\Sigma X_1 X_2$ is the sum of products of the kind $(x_i - \bar{x})(x_i^2 - \overline{x^2})$ and so on. The equations can be solved conveniently by the method of elimination or by determinants if the required polynomial is of low degree (for example, eqns. (5.19a) and (5.19b) can be adapted to give the coefficients of a quadratic equation) or by the inverse matrix method if the number of terms required is large.

In their alternative form, the normal equations can be more simply expressed:

$$\left.\begin{array}{l} an + b_1\Sigma x \quad + b_2\Sigma x^2 + \ldots b_j\Sigma x^j \quad = \Sigma y \\ a\Sigma x + b_1\Sigma x^2 \quad + b_2\Sigma x^3 + \ldots b_j\Sigma x^{j+1} = \Sigma yx \\ \ldots\ldots\ldots\ldots\ldots\ldots\ldots\ldots\ldots\ldots \\ a\Sigma x^j + b_1\Sigma x^{j+1} + \ldots \qquad b_j\Sigma x^{2j} = \Sigma yx^j \end{array}\right\} \tag{5.26}$$

but their solution is, of course, one degree more difficult than in the case of the ordinary form.

The method is illustrated by application to the data of Table 5.2 (a repetition of Table 3.7) which shows the variation of some determinations of E^0 for the antimony electrode with pH (the latter

TABLE 5.2. E^0 VALUES FOR THE ANTIMONY ELECTRODE

pH $= x$	Mean $E^0 - 255$ (mV) $= y$	Duplicates; y_1, y_2
2·00	0·36	0·38, 0·34
3·00	0·12	0·12, 0·12
4·00	0·09	0·06, 0·12
5·00	0·01	0·02, 0·00
6·00	0·51	0·59, 0·43
7·00	0·83	0·93, 0·73
8·00	1·71	1·70, 1·72

assumed to be accurately known). In the first instance we need consider only the mean E0values (these being coded to simplify the arithmetic) and the calculations are set out in Table 5.3. Because the fitting of all the first three degrees of polynomial (indicated in the table by subscripts, l for the line, q for the quadratic and c for the cubic curve respectively) is shown, some computations appear which are superfluous to the fitting of the polynomial actually appropriate to the data. These, however, show the amount of extra work required to find successive terms and permit comparison with the methods next described. Likewise, we are not concerned for the time being with Σy^2 or measures of variability derived from it which are also included in the table. It should be noted that the normal equations in the quadratic and the cubic set are very similar in their coefficients because of the high correlation between the values of x and x^2, x^2 and x^3 etc., and it is necessary to carry a considerable number of significant figures, at least in the early stages of the computations.

Although the multiple regression method can always be used in treating curvilinear data some simplification is possible when the values of the independent variable are equally spaced. Several methods which take advantage of this situation have been put forward, some based simply on the sums of powers of numbers in equ ence (to which any series of equally spaced x values can be

TABLE 5.3. POLYNOMIAL FITTING BY THE MULTIPLE REGRESSION METHOD

y	x	y^2	x^2	x^3	x^4	x^5	x^6	yx	yx^2	yx^3
0·36	2	0·1296	4	8	16	32	64	0·72	1·44	2·88
0·12	3	0·0144	9	27	81	243	729	0·36	1·08	3·24
0·09	4	0·0081	16	64	256	1024	4096	0·36	1·44	5·76
0·01	5	0·0001	25	125	625	3125	15625	0·05	0·25	1·25
0·51	6	0·2601	36	216	1296	7776	46656	3·06	18·36	110·16
0·83	7	0·6889	49	343	2401	16807	117649	5·81	40·67	284·69
1·71	8	2·9241	64	512	4096	32768	262144	13·68	110·72	875·52
3·63	35	4·0253	203	1295	8771	61775	446963	24·04	172·68	1283·50
Σy	Σx	Σy^2	Σx^2	Σx^3	Σx^4	Σx^5	Σx^6	Σyx	Σyx^2	Σyx^3

$\bar{y} = 3{\cdot}63/7 = 0{\cdot}518571$; $\bar{x}_1 = \bar{x} = 35/7 = 5$; $\bar{x}_2 = \overline{x^2} = 203/7 = 29$;
$\bar{x}_3 = \overline{x^3} = 1295/7 = 185$

$$\Sigma Y^2 = \Sigma y^2 - (\Sigma y)^2/n = 4{\cdot}0253 - 3{\cdot}63^2/7 = 2{\cdot}1429$$

$$\left.\begin{aligned} \Sigma YX_1 &= \Sigma yx - \Sigma y \Sigma x/n \\ &= 24{\cdot}04 - 3{\cdot}63 \times 35/7 = 5{\cdot}89 \\ \Sigma X_1^2 &= \Sigma x^2 - (\Sigma x)^2/n \\ &= 203 - 35^2/7 = 28 \end{aligned}\right\} \quad \begin{aligned} b_{1l} &= 5{\cdot}89/28 = +0{\cdot}210357 \\ a_l &= 0{\cdot}519 - 0{\cdot}21036 \times 5 = -0{\cdot}533 \end{aligned}$$

The straight line of best fit: $\hat{y}_l = -0{\cdot}533 + 0{\cdot}21036x$

$$\left.\begin{aligned} \Sigma YX_2 &= \Sigma yx^2 - \Sigma y\Sigma x^2/n = 172{\cdot}68 - 3{\cdot}63 \times 203/7 = 67{\cdot}41 \\ \Sigma X_2^2 &= \Sigma x^4 - (\Sigma x^2)^2/n = 8771 - 203^2/7 = 2884 \\ \Sigma X_1X_2 &= \Sigma x^3 - \Sigma x\Sigma x^2/n = 1295 - 35 \times 203/7 = 280 \end{aligned}\right\}$$

$$\text{Eqn. (5.17a)}, \left.\begin{aligned} 28b_{1q} + 280b_{2q} &= 5{\cdot}89 \\ 280b_{1q} + 2884b_{2q} &= 67{\cdot}41 \end{aligned}\right\} \begin{aligned} b_{1q} &= -0{\cdot}802738 \qquad b_{2q} = +0{\cdot}101310 \\ a_q &= 0{\cdot}5186 + 0{\cdot}80274 \times 5 - 0{\cdot}10131 \\ &\quad \times 29 = 1{\cdot}594 \end{aligned}$$

The quadratic of best fit: $\hat{y}_q = 1{\cdot}594 - 0{\cdot}80274x + 0{\cdot}10131x^2$

$$\left.\begin{aligned} \Sigma YX_3 &= \Sigma yx^3 - \Sigma y\Sigma x^3/n = 1283{\cdot}50 - 3{\cdot}63 \times 1295/7 = 611{\cdot}95 \\ \Sigma X_3^2 &= \Sigma x^6 - (\Sigma x^3)^2/n = 446{,}963 - 1295^2/7 = 207{,}388 \\ \Sigma X_1X_3 &= \Sigma x^4 - \Sigma x\Sigma x^3/n = 8771 - 35 \times 1295/7 = 2296 \\ \Sigma X_2X_3 &= \Sigma x^5 - \Sigma x^2\Sigma x^3/n = 61{,}775 - 203 \times 1295/7 = 24{,}220 \end{aligned}\right\}$$

$$\text{Eqn. (5.16)}, \left.\begin{aligned} 28b_{1c} + 280b_{2c} + 2296b_{3c} &= 5{\cdot}89 \\ 280b_{1c} + 2884b_{2c} + 24{,}220b_{3c} &= 67{\cdot}41 \\ 2296b_{1c} + 24{,}220b_{2c} + 207{,}388b_{3c} &= 611{\cdot}95 \end{aligned}\right\} \begin{aligned} b_{1c} &= -0{\cdot}387183 \\ b_{2c} &= +0{\cdot}009643 \\ b_{3c} &= +0{\cdot}006111 \end{aligned}$$

$$\text{Eqn. (5.25)}, \quad a_c = 0{\cdot}5186 + 0{\cdot}3872 \times 5 - 0{\cdot}009643 \times 29 - 0{\cdot}006111 \times 185 = 1{\cdot}044$$

The cubic of best fit: $\hat{y}_c = 1{\cdot}044 - 0{\cdot}38718x + 0{\cdot}009643x^2 + 0{\cdot}006111x^3$

$$\text{Eqn. (5.14)}, \quad \Sigma\hat{Y}_l^2 = \Sigma Y^2 - b_{1l}\Sigma YX_1 = 2{\cdot}1429 - 0{\cdot}21036 \times 5{\cdot}89 = 0{\cdot}9039$$

$$\begin{aligned} \Sigma\hat{Y}_p^2 &= \Sigma Y^2 - b_{1p}\Sigma YX_1 - b_{2p}\Sigma YX_2 \\ &= 2{\cdot}1429 + 0{\cdot}80274 \times 5{\cdot}89 - 0{\cdot}10131 \times 67{\cdot}41 = 0{\cdot}0417 \end{aligned}$$

$$\begin{aligned} \Sigma\hat{Y}_c^2 &= \Sigma Y^2 - b_{1c}\Sigma YX_1 - b_{2c}\Sigma YX_2 - b_{3c}\Sigma YX_3 \\ &= 2{\cdot}1429 + 0{\cdot}38718 \times 5{\cdot}89 - 0{\cdot}00943 \times 67{\cdot}41 \\ &\quad - 0{\cdot}00611 \times 611{\cdot}95 = 0{\cdot}0337 \end{aligned}$$

reduced) and others on the properties of orthogonal polynomials (the term implying zero products of corresponding coefficients). Thus, Svensson's (1959) method is based on a coding of x to the arithmetic series, 0, ± 1, ± 2, $\pm 3, \ldots$, and he provides tables of sums of powers suitable for fitting up to fifty observations. Similar methods are described or references to them given by Worthing and Geffner (1943) and by Deming (1943). The orthogonal polynomial methods (or, rather, two main versions of the same method) were devised by Fisher and accounts of them can be found in the general references already cited. (See, for example, Fisher (1963), Anderson and Bancroft (1952) and Snedecor (1956).) Their great advantage lies in the fact that not only do they simplify the computations required to reach a polynomial of given degree but they also simplify the testing of successive terms against the residual variability and the adjustment of the polynomial accordingly.

The first orthogonal polynomial method depends on successive accumulation of the values of the dependent variable as shown in Table 5.4(a) and the number of such accumulations depends on the

TABLE 5.4. POLYNOMIAL FITTING BY THE ORTHOGONAL POLYNOMIAL METHODS

(a) Using accumulations of the y values

x	$\mathscr{x}$	y	$y(2)$	$y(3)$	$y(4)$
2	−3	0·36	0·36	0·36	0·36
3	−2	0·12	0·48	0·84	1·20
4	−1	0·09	0·57	1·41	2·61
5	0	0·01	0·58	1·99	4·60
6	+1	0·51	1·09	3·08	7·68
7	+2	0·83	1·92	5·00	12·68
8	+3	1·71	3·63	8·63	21·31
$\bar{x} = 5$	$h = 1$	$S_1 = 3·63$	$S_2 = 8·63$	$S_3 = 21·31$	$S_4 = 50·44$

Eqn. (5.27a), $A = 3·63/7 = 0·518571$

Eqn. (5.27b), $B = (6/6)(0·518571 - 2 \times 8·63/7 \times 8)$
$= 1(0·518571 - 0·308214) = 0·210357$

Eqn. (5.27c), $C = (30/6 \times 5)(0·518571 - 3 \times 0·308214 + 12 \times 21·31/7 \times 8 \times 9)$
$= 1(0·518571 - 0·924642 + 0·507381) = 0·101310$

Eqn. (5.27d), $D = (140/6 \times 5 \times 4)(0·518571 - 2 \times 0·924642 + 5 \times 0·507381$
$- 120 \times 50·44/7 \times 8 \times 9 \times 10)$
$= 1·166667(0·518571 - 1·849284 + 2·536905 - 1·200952)$
$= 0·006111$

TABLE 5.4 *(cont.)*

Eqn. (5.28), $\hat{y} = 0{\cdot}518571 + 0{\cdot}210357\varkappa_1 + 0{\cdot}101310\varkappa_2 + 0{\cdot}006111\varkappa_3$

Eqn. (5.31a), $\hat{y}_l = (0{\cdot}518571 - 5 \times 0{\cdot}210357) + 0{\cdot}210357x$
$= -0{\cdot}533214 + 0{\cdot}210357x$

Eqn. (5.31b), $\hat{y}_q = (-0{\cdot}533214 - (25 - 4)\,0{\cdot}101310) + (0{\cdot}210357 - 10 \times 0{\cdot}101310)x + 0{\cdot}101310x^2$
$= 1{\cdot}594296 - 0{\cdot}802743x + 0{\cdot}101310x^2$

Eqn. (5.31c), $\hat{y}_c = (1{\cdot}594296 - (125 - 35)\,0{\cdot}006111) + (-0{\cdot}802743 + (75-7)\,0{\cdot}006111)x + (0{\cdot}101310 - 15 \times 0{\cdot}006111)x^2 + 0{\cdot}006111x^3$
$= 1{\cdot}044296 - 0{\cdot}387187x + 0{\cdot}009643x^2 + 0{\cdot}006111x^3$

(b) Using the factors of Appendix Table XII

x	$\varkappa$	y	ξ_1	$y\xi_1$	ξ_2	$y\xi_2$	ξ_3	$y\xi_3$
2	1	0·36	−3	−1·08	+5	1·80	−1	−0·36
3	2	0·12	−2	−0·24	0	0·00	+1	0·12
4	3	0·09	−1	−0·09	−3	−0·27	+1	0·09
5	4	0·01	0	0·00	−4	−0·04	0	0·00
6	5	0·51	+1	0·51	−3	−1·53	−1	−0·51
7	6	0·83	+2	1·66	0	0·00	−1	−0·83
8	7	1·71	+3	5·13	+5	8·55	+1	1·71
Eqn. (5.32)		7) 3·63		28) 5·89		84) 8·21		6) 0·22

$\varkappa = x - 1$; $A = 0{\cdot}518571$; $B = 0{\cdot}210357$; $C = 0{\cdot}101310$; $D' = 0{\cdot}036667$

Eqn. (5.33) (Table XII),

$$\hat{y} = A + B(\varkappa - 4) + C(\varkappa^2 - 8\varkappa + 12) + (D'/6)(\varkappa^3 - 12\varkappa^2 + 41\varkappa - 36)$$

Whence: $b_{1l} = 0{\cdot}210357$; $b_{2q} = 0{\cdot}101310$; $b_{3c} = 0{\cdot}006111$

$a_l = 0{\cdot}518571 - 4 \times 0{\cdot}210357 = -0{\cdot}322857$; $\quad b_{1q} = 0{\cdot}210357 - 8 \times 0{\cdot}101310 = -0{\cdot}600123$

$a_q = -0{\cdot}322857 + 12 \times 0{\cdot}101310 = 0{\cdot}892863$; $\quad b_{1c} = -0{\cdot}600123 + 41 \times 0{\cdot}006111 = -0{\cdot}349567$

$a_c = 0{\cdot}892863 - 36 \times 0{\cdot}006111 = 0{\cdot}672861$; $\quad b_{2c} = 0{\cdot}101310 - 12 \times 0{\cdot}006111 = 0{\cdot}027977$

$\hat{y}_l = -0{\cdot}322857 + 0{\cdot}210357\varkappa = -0{\cdot}533214 + 0{\cdot}210357x$

$\hat{y}_q = 0{\cdot}892863 - 0{\cdot}600123\varkappa + 0{\cdot}101310\varkappa^2$
$= 1{\cdot}594296 - 0{\cdot}802743x + 0{\cdot}101310x^2$

$\hat{y}_c = 0{\cdot}672861 - 0{\cdot}349567\varkappa + 0{\cdot}027977\varkappa^2 + 0{\cdot}006111\varkappa^3$
$= 1{\cdot}044294 - 0{\cdot}387198x + 0{\cdot}009644x^2 + 0{\cdot}006111x^3$

degree of polynomial required. Thus, for a straight line we require the sum of the y values, S_1, and the first accumulation of these in the sense of increasing x values, S_2; to fit a quadratic we need S_3 (the accumulation of the values which go to form S_2) as well; to fit a cubic we need S_4 and so on. From these sums we can find the quantities

$$A = S_1/n = \bar{y} \tag{5.27a}$$

$$B = [6/(n-1)]\,[S_1/n - 2S_2/n(n+1)] \tag{5.27b}$$

$$C = [30/(n-1)(n-2)]\,[S_1/n - 6S_2/n(n+1) + 12S_3/n(n+1)(n+2)] \tag{5.27c}$$

$$D = [140/(n-1)(n-2)(n-3)]\,[S_1/n - 12S_2/n(n+1) + 60S_3/n(n+1)(n+2) - 120S_4/n(n+1)(n+2)(n+3)] \tag{5.27d}$$

which are the coefficients of an equation

$$\hat{y} = A + B\mathfrak{x}_1 + C\mathfrak{x}_2 + D\mathfrak{x}_3 \tag{5.28}$$

where

$$\mathfrak{x}_1 = \mathfrak{x}\bar{\mathfrak{x}} \tag{5.29a}$$

$$\mathfrak{x}_2 = (\mathfrak{x}\bar{\mathfrak{x}})^2 - (n^2 - 1)/12 \tag{5.29b}$$

$$\mathfrak{x}_3 = (\mathfrak{x}\bar{\mathfrak{x}})^3 - (3n^2 - 7)(\mathfrak{x}\bar{\mathfrak{x}})/20 \tag{5.29c}$$

where, in turn, $\mathfrak{x}$ is a coded form of the independent variable which goes in steps of unity. If the origin of the coded scale is chosen to be the centre of the x scale, then $\bar{\mathfrak{x}}$ is zero and

$$\mathfrak{x} = (x - \bar{x})/h \tag{5.30}$$

h being the interval between the different values of x. In this case we have for the successive stages of fitting (straight line, quadratic and cubic):

$$\hat{y}_l = [A - \bar{x}B/h] + [B/h]\,x = a_1 + b_{1l}x \tag{5.31a}$$

$$\hat{y}_q = [a_1 + C(\bar{x}^2/h^2 - (n^2-1)/12] + [b_{1l} + 2\bar{x}C/h^2]\,x + [C/h^2]\,x^2 = a_q + b_{1q}x + b_{2q}x^2 \tag{5.31b}$$

$$\hat{y}_c = [a_q - D(\bar{x}^3/h^3 - \bar{x}(3n^2-7)/20h)] + [b_{1q} + D(3\bar{x}^2/h^3 - (3n^2-7)/20h)]\,x + [b_{2q} - 3\bar{x}D/h^3]\,x^2 + [D/h^3]\,x^3 = a_c + b_{1c}x + b_{2c}x^2 + b_{3c}x^3 \tag{5.31c}$$

The equations for further terms can be devised from formulae given in the references cited.

Simple expressions can be derived for maxima and minima in terms of orthogonal coefficients. For example, the single turning point of a quadratic occurs at $\bar{\varkappa} - B/2C$. On this basis, Horton (1955) describes a computational procedure for locating symmetric peaks in spectrometric work.

The second orthogonal polynomial method does not require the successive accumulation of the y values but, instead, their multiplication by simple tabulated factors which are given in Table XII of the Appendix. Here, we have

$$A = \bar{y}, \qquad B = \Sigma y\xi_1/\Sigma\xi_1^2 \qquad (5.32\text{a,b})$$

$$C = \Sigma y\xi_2/\Sigma\xi_2^2, \quad D' = \Sigma y\xi_3/\Sigma\xi_3^2 \qquad (5.32\text{c,d})$$

as the coefficients of an equation

$$\hat{y} = A + B\xi_1 + C\xi_2 + D'\xi_3 \qquad (5.33)$$

which is identical with eqn. (5.28) except for the last term and in which ξ_1, ξ_2 and ξ_3 are simple functions of a coded variable, $\varkappa$, going in steps of unity from 1 to n. Consequently, for purposes of decoding, we have

$$(\varkappa - \bar{\varkappa}) = (x - \bar{x})/h \quad \text{or} \quad \varkappa = (x - \bar{x})/h + (n + 1)/2 \quad (5.34\text{a,b})$$

The values of $\Sigma\xi_i^2$ and the functions ξ_i for i from 1 to 3 (i.e. to the cubic stage) and for n from 3 to 12 are given in Table XII. More extensive tables, with a further description and modification of the method for larger numbers of observations, can be found in Fisher and Yates (1948) and Pearson and Hartley (1958). It should be noted that, as with the previous method, the coefficients of a given polynomial can be built up out of quantities already found in computing those of lower degree. Use has been made of this in setting out the computations for the example in Table 5.4.

The three methods just described lead to the same equations (apart from small rounding errors) for the variation of E^0 with pH (viz., those in the tables with the addition of 255 mV to the constant term) and the question which now arises is which is appropriate to the data. Still considering only the mean E^0 or y values and *regarding these as single determinations*, the procedure is to find the amount of the total variability of the y values accounted for by a given equation and the residual sum of squares of deviations left over by it and to compare the two by way of the appropriate F test. In the case of the multiple regression method the residuals can be found from the bracketed expressions in eqns. (5.10) or (5.14) and the results are given in the last lines of Table 5.3. When using the orthogonal polynomial method, the share of the total variability of the y values accounted for by a particular term can be found from the orthogonal coefficients. The expressions are:

due to the linear term,

$$n(n^2 - 1)\, B^2/12 = B^2 \Sigma \xi_1^2 \qquad (5.35\text{a})$$

due to the quadratic term,

$$n(n^2 - 1)\,(n^2 - 4)\, C^2/180 = C^2 \Sigma \xi_2^2 \qquad (5.35\text{b})$$

due to the cubic term,

$$n(n^2 - 1)\,(n^2 - 4)\,(n^2 - 9)\, D^2/2800 = D'^2 \Sigma \xi_3^2 \qquad (5.35\text{c})$$

From the residual left over from each of these contributions an error variance can be calculated based on the number of independent differences of the form $(y - \hat{y})$ entering into them, this being the number of residual degrees of freedom, $n - k$. Thus, each successive term accounts for one degree of freedom as appears in Table 5.5.

TABLE 5.5. TESTING FOR THE SIGNIFICANCE OF POLYNOMIAL TERMS

Component of variation	SS	f	MS	Ratio	$F_{\cdot 05}$
Total, $\Sigma Y^2 = \Sigma y^2 - (\Sigma y)^2/n$	2·1429	6			
Reduction due to linear term:					
eqn. (5.35a), $28 \times 0{\cdot}0442500$	1·2390	1	1·2390	6·85	6·61
Residual from the line, $\Sigma \hat{Y}_l^2$	0·9039	5	0·1808		
Reduction due to quadratic term:					
eqn. (5.35b), $84 \times 0{\cdot}0102637$	0·8622	1	0·8622	72·9	7·71
Residual from the quadratic, $\Sigma \hat{Y}_q^2$	0·0417	4	0·0104		
Reduction due to cubic term:					
eqn. (5.35c), $216 \times 0{\cdot}000037345$ $= 6 \times 0{\cdot}00134447$	0·0081	1	0·0081	0·72	10·1
Residual from the cubic curve, $\Sigma \hat{Y}_c^2$	0·0336	3	0·0112		

The results of the various tests in Table 5.5 show that the linear and the quadratic terms are significant, the latter highly so. The cubic term, on the other hand, accounts for a very small proportion of the total variability, less in fact than the residual remaining after fitting it, and it follows that the quadratic is the appropriate curve to fit the points. The corresponding standard error of estimate, from eqn. (5.24) and the table, is $\sqrt{0{\cdot}0104} = 0{\cdot}102$ mV and this is a measure of the experimental reliability of the E^0 values from which the equation has been calculated. Confidence intervals for the true E^0 at any particular pH within the range of observation or for the parameters of the model can be found by the methods of the preceeding section.

It should be noted that the linear term does not appear to have a marked effect in this instance and, had the points followed a strict U-shaped pattern, would have sunk into insignificance (by comparison with the residual left after fitting it) because of the symmetry of the E^0 values across the pH scale. It is conceivable that, in a situation like this, the practice of successively fitting terms and concurrently testing for their significance against the apparent residual might lead to the computations being stopped at too early a stage.†

When observations of the dependent variable have been repeated, the residual variability or estimate of measuring error can be found from the replicate values. Turning to the last column of Table 5.2, we have duplicates (i.e. observations which are presumed to be quite independent but have in common the fact that they were made at the same pH) from which a "within samples" sum of squares or average range can be obtained by the methods of Section 3.5(3). The equation for the total variation is eqn. (3.46) and the residual variation can be found as the difference between eqns. (3.46) and (3.47) which, in a notation more convenient to the present chapter, may be written

$$\Sigma\Sigma(y - \bar{y})^2 = \Sigma\Sigma y^2 - \Sigma T^2/r \qquad (5.36)$$

where $\bar{y}$ is the mean, T the total and r the number of individual values at any value of x and $\Sigma\Sigma$ indicates summation within and

† A visual appreciation along these lines appears to have misled the original observers, whose data have been slightly modified for the purposes of the example, into striking a mean E^0 for the whole of the pH range, 2 to 7. Youden (1947) and Mandel (1947) discuss the point.

over all the values of x. The degree of replication, r, is taken as constant in eqn. (5.36). When it is 2, as in the example, the residual is also half the sum of the squares of the ranges.

The computation of the total and residual variability is carried out in Table 5.6(a) and the analysis of variance of the regression in Table 5.6(b). The contributions of the various terms of the curvilinear regression are twice the quantities in Table 5.5 because the y values to which the curve has been fitted are now taken as the means of two observations. Table 5.6(b) shows that, judged by the experimental error, a quadratic expression is appropriate to

TABLE 5.6. THE SIGNIFICANCE OF POLYNOMIAL TERMS, DUPLICATE OBSERVATIONS

(a) Calculation of the total and residual variation

pH	$E^0 - 255$ (mV); y_1, y_2	y_1^2, y_2^2	T	T^2	w	w^2
2	0·38, 0·34	0·1444, 0·1156	0·72	0·5184	0·04	0·0016
3	0·12, 0·12	0·0144, 0·0144	0·24	0·0576	0·00	0·0000
4	0·06, 0·12	0·0036, 0·0144	0·18	0·0324	0·06	0·0036
5	0·02, 0·00	0·0004, 0·0000	0·02	0·0004	0·02	0·0004
6	0·59, 0·43	0·3481, 0·1849	1·02	1·0404	0·16	0·0256
7	0·93, 0·73	0·8649, 0·5329	1·66	2·7556	0·20	0·0400
8	1·70, 1·72	2·8900, 2·9584	3·42	11·6964	0·02	0·0004
	$\Sigma\Sigma y = 7·26$	$\Sigma\Sigma y^2 =$ 8·0864		2) 16·1012	7) 0·50	0·0716
$n = 14;$		$-(\Sigma\Sigma y)^2/14 = -3·7648$		8·0506	0·071 $= \bar{w}$	
$r = 2;$		$\Sigma\Sigma Y^2 =$ 4·3216				

$$\Sigma\Sigma(y - \bar{y})^2 = 8·0864 - 8·0506 = 0·0358 = \tfrac{1}{2}\Sigma w^2$$

(b) The analysis of variance of regression

Component of variation	SS	f	MS	Ratio	$F_{0·5}$
Line about mean, $\Sigma(\hat{y}_l - \bar{\bar{y}})^2$; $2 \times 1·2390$	2·4780	1	2·4780	485	5·59
Parabola about line, $\Sigma(\hat{y}_q - \hat{y}_l)^2$; $2 \times 0·8622$	1·7244	1	1·7244	337	5·59
Cubic about parabola, $\Sigma(\hat{y}_c - \hat{y}_q)^2$; $2 \times 0·0081$	0·0162	1	0·0162	3·17	5·59
Means about cubic, $\Sigma(\bar{y} - \hat{y}_c)^2$ (by difference)	0·0672	3	0·0224	4·38	4·35
Residual ("within pHs")	0·0358	7	0·005114		
Total	4·3216	13			

the data and a cubic not. There is a questionable contribution to the total variation which could be accounted for by higher terms but, if this is pooled with the non-significant cubic contribution, the mean square so obtained, (0·0162 + 0·0672)/4 = 0·0208, is no longer significant with respect to the residual. In any case, a quartic would not be fitted in practice unless there were a graver warning of its necessity than is provided by these experimental values.

As a practical measure when treating duplicate or more highly replicated observations, a quality control type chart based on the average range has much to commend it. Such charts have already been presented for the data of the example in Fig. 3.6 which shows the comparability of the ranges or differences between duplicates and the significant differences between the means of the duplicates. Given the average range, 0·071, the allowed 95% limits of variation of the means of pairs of values, $\pm 1{\cdot}229 \times 0{\cdot}071 = \pm 0{\cdot}09$, can be laid off not only from $\bar{\bar{y}}$, as was done in Fig. 3.6, but also from the fitted curves. The effect is demonstrated in Fig. 5.3 where the differences between the observed and calculated values, $\hat{Y}_l = y - \hat{y}_l$, $\hat{Y}_q = y - \hat{y}_q$, $\hat{Y}_c = y - \hat{y}_c$, are compared with the $\pm 0{\cdot}09$ limits. The systematic departure of the points from the straight line can be clearly seen to indicate some kind of curvilinear relationship and the just slightly significant discrepancies of the deviations from the calculated quadratic and cubic values are in accord with the results of the tests in Table 5.6(b) of which Fig. 5.3 provides, therefore, a good illustration.

Although our attention has been confined to single or equally weighted observations of the dependent variable, problems associated with larger numbers of values can be handled by the methods already given. When differing numbers of observations have been made at various values of x, the analysis of variance of regression resembles the procedure just described but the fitting must allow for the differing weights of the $\bar{y}$ values and is an extension of that described for the straight line in Section 4.7. When a large number of values is available and the values of x are scattered over their range (or when both x and y are natural variables so that the problem is one of curvilinear correlation) the points can be grouped to simplify the computations which again follow the method of Section 4.7.

When the intercept of a polynomial curve is known exactly (or at least very much more accurately than the other true y values) it is

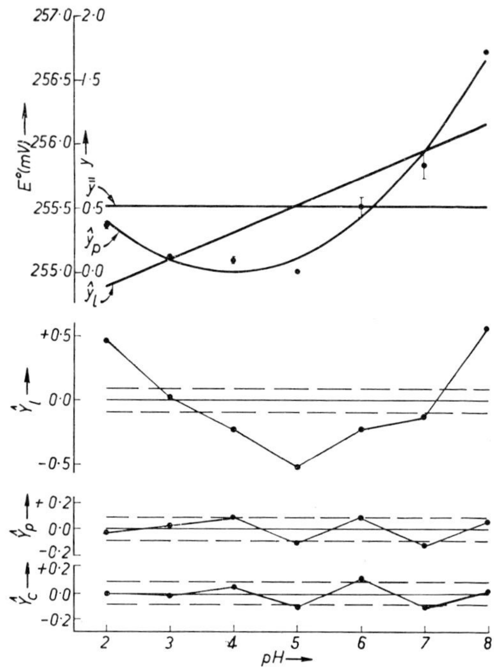

FIG. 5.3. Graphical Tests on Curvilinear Data.

The uppermost graph shows the mean data points of Table 5.2 and the best fitting horizontal line (mean), straight line and parabola through them. The lower graphs show the deviations of the mean data points from values given by the best fitting linear, quadratic and cubic equations compared with the amount of deviation attributable to random error as judged by the differences between duplicates.

tempting to rearrange eqn. (5.23b) so as to reduce it in degree. Thus the least squares quadratic

$$\hat{y} = \alpha + b_1x + b_2x^2 \tag{5.37}$$

can be transformed to

$$\hat{y}' = b_1 + b_2x \tag{5.38}$$

and fitted by regressing $y' = (y - \alpha)/x$ on x. This would, however, minimize the sum of the squares of the $(y' - \hat{y}')$ deviations and have the effect of weighting favourably the observations at the lower end of the x scale (see Section 5.5). On the other hand, to use the ordinary methods already described would be to discard the additional information altogether. Returning to the procedure described in Section 4.2 and deriving the normal equations, we find that for $z = y - \alpha$,

$$\left.\begin{aligned} b_1\Sigma x^2 + b_2\Sigma x^3 &= \Sigma zx \\ b_1\Sigma x^3 + b_2\Sigma x^4 &= \Sigma zx^2 \end{aligned}\right\} \tag{5.39}$$

in terms of the sums of squares etc. and products of the actual values. If the known intercept is zero, i.e. the quadratic is to be fitted through the origin, y and z are indentical.

5.3. The Fitting of Polynomials. Non-statistical Methods

When an equation only is desired, e.g. for reporting curvilinear data in a condensed form or for simple manipulations, the method of selected points or of averages is available. The latter is best in the case of more numerous and scattered data and simply involves grouping the values more or less equally along the scale of the independent variable and using the group means in the simultaneous equations required for finding the coefficients. Thus, three groups yielding three sets of means are needed to fit a quadratic, four for a cubic and so on. The method of selected points should only be applied when the values are relatively free of experimental error or after its influence has been reduced by smoothing as described in the next chapter.

When fitting equations by approximate methods such as these, the problem is how to decide, preferably beforehand, the degree of polynomial required. In the case of smoothed or otherwise good data, to which it is intended to apply the method of selected points, a test can be based on item differences. This, the *tabular suitability test*, depends on the fact that for an equation of the form

$$y = a + b_1x + b_2x^2 + \ldots b_jx^j$$

the jth derivative, d^jy/dx^j, is a constant and the $(j+1)$th derivative is zero and, further, that any derivative can be approximated by the corresponding ratio of finite differences. (These are described in more detail in Section 6.1 below.) An inspection of such approximations to the derivatives, or of the item differences alone if the x values are equally spaced, will indicate the equation required and from the average value of the items found to be constant an estimate of the coefficient of the highest power of x, b_j, can be obtained. In particular, we have

for a straight line

$$dy/dx = b_1 \doteqdot \Delta y/\Delta x \tag{5.40a}$$

for a quadratic

$$d^2y/dx^2 = 2b_2 \doteqdot \Delta^2 y/\Delta x^2 \tag{5.40b}$$

and for a cubic

$$d^3y/dx^3 = 6b_3 \doteqdot \Delta^3 y/\Delta x^3 \tag{5.40c}$$

Unsmoothed data like the E^0 values of Table 5.2, or of the kind to which the method of averages might be applied, cannot be treated in this way but in such cases the appropriate degree of curvature is generally low and often recognizable from a plot of the points.

As an example of the application of the tabular suitability test and of the method of selected points, consider the values for the specific heat at constant pressure of air at various temperatures in the first two columns of Table 5.7. The first, second, third and

TABLE 5.7. THE SPECIFIC HEAT OF AIR AT CONSTANT PRESSURE

T(°C)	c_p (cal/g)	Δc_p	$\Delta^2 c_p$	$\Delta^3 c_p$	$\Delta^4 c_p$	$c_p - \hat{c}_p$
100	0·2409					−0·0002
		0·0008				
200	0·2417		0·0004			0·0000
		0·0012		−0·0003		
300	0·2429		0·0001		+0·0004	−0·0001
		0·0013		+0·0001		
400	0·2442		0·0002		0·0000	0·0000
		0·0015		+0·0001		
500	0·2457		0·0003		−0·0002	0·0000
		0·0018		−0·0001		
600	0·2475		0·0002		0·0000	0·0000
		0·0020		−0·0001		
700	0·2495		0·0001		+0·0003	−0·0001
		0·0021		+0·0002		
800	0·2516		0·0003		−0·0003	0·0000
		0·0024		−0·0001		
900	0·2540		0·0002			0·0000
		0·0026				
1000	0·2566					0·0000

fourth order differences of the c_p values are given in the adjoining columns. The values of the independent variable, T °C, are equally spaced and the difference between them need not be taken into account for the moment. An inspection of the table reveals that the first order differences are drifting, the second order differences are relatively constant while the remainder fluctuate round zero. It follows that a quadratic equation should represent the data satisfactorily. Furthermore, since the second order differences average about 0·00022, the second derivative is about $0{\cdot}00022/\Delta T^2 = 0{\cdot}00022 \times 10^{-4}$, so the coefficient of T^2 in the fitted equation should be approximately $0{\cdot}00011 \times 10^{-4}$ from eqn. (5.40b). This agrees well with the value of the coefficient as calculated.

In applying the method of selected points to find the quadratic equation, we take three pairs of values well spread over the range of T and substitute them in the general equation of the required form to obtain three simultaneous equations linear in the coefficients. Choosing the second, fifth and last points and substituting leads to

$$a + 200b_1 + 200^2b_2 = 0{\cdot}2417$$

$$a + 600b_1 + 600^2b_2 = 0{\cdot}2475$$

$$a + 1000b_1 + 1000^2b_2 = 0{\cdot}2566$$

Solving these by elimination or by determinants gives

$$c_p = 0{\cdot}2400 + 6{\cdot}25 \times 10^{-6}T + 1{\cdot}03 \times 10^{-8}T^2$$

The differences between the original c_p values and those calculated by this equation are to be found in the last column of Table 5.7. The agreement is very good and no final adjustment is called for. Should the agreement turn out to be unsatisfactory but the differences between the original and calculated values be more or less constant, a correction can be made to a. If the discrepancies drift constantly or curvilinearly with the values of the independent variable, then b_1 and b_2 also require correction but this is only likely to arise from a bad selection of points.

When the values of the independent variable are unequally spaced the differences between them must be taken into account. Consider, for example, the molar heat capacity of ethylene at constant pressure whose values at a few temperatures are given in Table 5.8. In this case the first and subsequent item differences of the molar heats must be divided by the appropriate temperature differences. Thus, the first entries in the two "divided" difference columns are (8·97 – 8·27)/(250 – 200) and (0·0124 – 0·0140)/(275 – 225) and the origin of the other values can be seen by an inspection of the table. Again, from the constancy of the items in the last column, it is apparent that a quadratic equation is sufficient to represent the data.

TABLE 5.8. THE MOLAR HEAT CAPACITY OF ETHYLENE AT CONSTANT PRESSURE

T(°K)	C_p (cal/mole)	Mean T	$\Delta C_p/\Delta T$	Mean T	$\Delta^2 C_p/\Delta T^2$
200	8·27				
		225	0·0140		
250	8·97			250	−0·000032
		275	0·0124		
300	9·59			312·5	−0·000028
		350	0·0103		
400	10·62			387·5	−0·000031
		425	0·0080		
450	11·02			462·5	−0·000031
		500	0·0057		
550	11·59				

The tabular suitability test can be used to see if functions other than polynomials are appropriate. Worthing and Geffner (1943) list the finite differences required in particular cases.

5.4. DEPARTURES FROM THE SIMPLE REGRESSION MODEL

Hitherto, we have considered only those cases in which the incidence of random error on the values of the dependent variable could be regarded as constant and in which the values of the independent variable were exactly known. These assumptions, while perhaps generally applicable, are sometimes plainly not valid and, if adopted, will lead to poor point and interval estimates for the true constants in a regression model. Here we deal briefly with three situations for which the assumptions of the simple linear model of Chapter IV need to be modified and some alteration made in the computational methods described in that chapter.

(1) *The Case of Varying Incidence of Random Error*

It sometimes happens that the error of measurement of values of a variable changes with the level of the values. For instance, in chemical analysis, the error in the amount of a substance found by an analytical method may be proportional to the amount itself. If this is so, we have in place of the model represented by eqns. (4.1 a) and (4.2 a),

$$y = \alpha + \beta x + e \qquad (5.41)$$

where the error, e, is directly proportional to x so that (although it may still be distributed normally about zero) it now has a varying standard deviation,

$$\sigma_e = \sigma'_e x \qquad (5.42)$$

δ'_e being a contant. Consequently, in fitting an equation of the form of (4.1 b) on the basis of a model of this kind, less reliance should be placed on points corresponding to higher values of x and more on those nearer the origin.

If eqn. (5.41) is transformed to

$$(y/x) = \beta + \alpha(1/x) + e/x \qquad (5.43\text{ a})$$

we now have a "dependent" variable, y/x, the errors in whose values are distributed with a constant standard deviation, $\sigma'_e = \sigma_e/x$, and an independent variable, $1/x$, which is still error-free. Equation (5.43 a) can therefore be fitted in the usual way by an equation

$$y' = a' + b'x' \qquad (5.43\text{ b})$$

where $y' = y/x$ and $x' = 1/x$. Thus, we obtain b' as the best estimate of α in eqn. (5.41), a' as the best estimate of β, and a standard error of estimate, s'_e, from which the standard deviations of the y values at any given x value can be found by multiplying by x.

The data of Table 5.9 provide an illustration. The x values are the same as those in Table 4.1 but the y values have been changed to show a degree of scatter rising with the value of x. If we now fit, by the method of Table 4.2, an unconstrained least squares line to the original and to the transformed values we obtain

$$\hat{y} = 4{\cdot}986 + 0{\cdot}88929x, \quad s_e = 4{\cdot}900$$

and

$$\hat{y}' = 0{\cdot}89338 + 4{\cdot}970x, \quad s'_e = 0{\cdot}109$$

or

$$\hat{y} = 4{\cdot}970 + 0{\cdot}89338x, \quad s_e = 0{\cdot}109x, \quad \text{respectively.}$$

In the second case, s_e is 0·98 at $x = 9$ and 6·87 at $x = 63$. It can be seen that there is no great difference between the estimates of the constants of the model but the incorrect standard error of estimate is approximately the average of the values given by the second method over the observed range of the x values.

TABLE 5.9. RESULTS OF TESTS ON AN ANALYTICAL METHOD

Concentration taken, x	Concentration found, y	$y' = y/x$	$x' = 1/x$
9	12·7	1·41111	0·111111
18	23·7	1·31667	0·055556
27	27·2	1·00741	0·037037
36	34·1	0·94723	0·027778
45	50·7	1·12666	0·022222
54	45·7	0·84632	0·018519
63	64·9	1·03016	0·015873

There is a slight inconsistency in the assumption that σ_e is strictly proportional to x when an unconstrained model is adopted but it is unexceptionable if the model be constrained so that the line passes through a given point on the zero ordinate. If this point be the origin, then the model may be represented by

$$y = \beta x + e, \quad \sigma_e = \sigma'_e x \tag{5.44a}$$

and fitted by

$$y' = y/x = b \quad (s'_e \text{ constant}) \tag{5.44b}$$

for which

$$b = \Sigma y'/n \tag{5.45}$$

and

$$s'_e = \sqrt{[(\Sigma y'^2 - (\Sigma y')^2/n)/(n-1)]} = \sqrt{[\Sigma \hat{Y}'^2/(n-1)]} \tag{5.46}$$

(i.e., b is simply the mean of the ratios and s'_e their ordinary estimated standard deviation). If we apply these equations to the data and also, for comparison, (4.25), (4.26) and (4.27) which are derived on the assumption that σ_e is constant, we find

$$\hat{y} = 1{\cdot}09794x, \quad s_e = 0{\cdot}195x$$

and

$$\hat{y} = 1{\cdot}00008x, \quad s_e = 5{\cdot}08, \quad \text{respectively.}$$

Here, the second (incorrect) standard error of estimate is rather lower than the average of $0{\cdot}195x$ over the range of x and the standard error of b by eqn. (4.28), $s_b = 0{\cdot}048$, is considerably below the standard error for the estimate from eqn. (5.45), $s_b = 0{\cdot}195/\sqrt{7} = 0{\cdot}074$. Thus, the differences between the correct values and those obtained by misapplying the equations for the constant error model are considerable although they are partly due here to the pattern of the data.

In this example only the simplest type of standard deviation change has been considered. Brownlee (1960) should be consulted for the general case, the comparison of the constrained and unconstrained weighted regressions and the calculation of a joint confidence region for α and β in the unconstrained case.

(2) *The Case where y is the Difference between Cumulative Values*

The situation sometimes occurs where the values of the dependent variable are obtained as the difference between a total up to one value of the independent variable and that up to the next. Suppose, for example, in a simple chemical kinetic experiment the course of a reaction over time is followed by titration from a burette which is not refilled and suppose the y values required are the differences between successive readings, say 0, $\boldsymbol{a}$, $\boldsymbol{b}$, $\boldsymbol{c}$ ml etc. so that $y_1 = \boldsymbol{a} - 0$, $y_2 = \boldsymbol{b} - \boldsymbol{a}$, $y_3 = \boldsymbol{c} - \boldsymbol{b}$ and so on. Obviously an error in reading $\boldsymbol{a}$ will affect both y_1 and y_2 so that they, and any other pair of adjacent values, are not independent. Mandel (1957) discusses the effects of this lack of independence in connection with a constrained model, which may be written

$$y = \beta x + \Sigma e \tag{5.47}$$

where e is the error of the difference between the successive readings which constitute an individual value of the dependent variable. If the x values are equally spaced at unit intervals from zero, β is estimated by $b = \Sigma y/x_n$, the total of the y values divided by the last x value. If the equations for the constant error constrained model are misapplied in this situation, then b from eqn. (4.25) estimates the β of eqn. (5.47) without bias but eqn. (4.28) grossly underestimates the true standard error of the slope. It is a characteristic of data like these that the points tend to fall on one or other side of the fitted line in long sequences whereas independent y values are, of course, randomly scattered about it. (The expected number of sequences of points is, very approximately, $\sqrt{n}$ in the cumulative case and $n/2$ in the random case.)

(3) *The Case of Errors in Both Variables*

In Section 4.10 (eqns. (4.84a) and (4.84b)) we considered the effect of measuring error in the x and y values on the calculated correlation coefficient and on the regression coefficients of the correlation model. The problems (of what is sometimes called confluence analysis) to which we now briefly refer are those which arise when there is a "functional" relationship between true but unknown values of x and y, say of the form

$$\mu_y = \alpha + \beta\mu_x \quad \text{or} \quad \mu_x = (-\alpha/\beta) + (1/\beta)\,\mu_y \tag{5.48}$$

but the values available to us are themselves of the form

$$x = \mu_x + u, \quad y = \mu_y + v \tag{5.49a, b}$$

where u and v are random errors of zero mean with standard deviations, σ_u and σ_v, respectively. The reader is referred to Acton (1959) and Williams (1959) for general discussions of the situations to which this kind of model applies, and to Halperin (1961) who considers the joint confidence region for α and β and the question of multiple prediction from an estimated relationship.

Although the model represented by eqns. (5.48) and (5.49), or some non-linear extension of it, is perhaps the most appropriate to much physico-chemical data and is certainly more often realistic than the simple regression and correlation models commonly assumed, we rely in practice on the ordinary methods of treatment associated with the latter. The justification is (a) that when the errors in one set of measurements (say those of x) are zero, eqn. (5.48) reduces to eqn. (4.1a) and, if the errors are not zero but relatively small compared with those in the other set, the simple model is still accurate enough for the treatment of the data, and (b) that the same is true under certain conditions of presetting of x to "nominal" values even when the real values are not these exactly. Berkson (1950) established this point and his approach has been modified and extended by Scheffé (1958) who describes other work in the same field.

5.5. Rectification

Much attention has been paid to polynomials because of their general utility but, as has already remarked, they do not take the place of theoretically sound expressions. When such are available, it is the common practice to rectify them and fit the linear expressions so obtained by the method of selected points or by drawing a line through the graphed values by eye. Alternatively, these may follow a pattern for which the theoretical expression is unknown but for which an empirical equation (generally other than a polynomial) is desired. The discovery of a suitable empirical relationship may itself suggest a general law but, more commonly, such an equation is used only to simplify the treatment and handling of results by way of integration, differentiation, interpolation etc., which are discussed in the next chapter. An additional reason for rectification is to simplify the fitting of an equation by least squares when this is done to avoid the subjective element inherent in the approximate methods. If, however, the least squares method is to be applied for the purpose of establishing valid confidence intervals, the practice of rectification leads to problems of weighting as described below.

In the absence of a satisfactory theoretical relationship, the task of discovering a suitable form of empirical equation is a matter of trial and error. First attempts in the two variable case generally involve taking and plotting the logarithms of the values of one or both variables or, rather, of using log-log and semi-log graph paper to effect the same result. The log-log plot will, of course, indicate whether or not y is related to a power of x such as 2 or $\frac{1}{2}$ or -1. In addition to this kind of trial plotting, Lyle (1957) recommends the form

$$y = ax/(b + x)$$

for its wide utility when the data are not markedly curvilinear. He provides a nomograph for assessing its suitability and describes a method of fitting it. Davis (1962) gives a most extensive collection of equations together with examples of their application. In some of the more complicated cases fitting is aided by the use of special tables which Davis provides; the multiple regression procedure can be adapted to others.

TABLE 5.10. SOME TRANSFORMATIONS OF NON-LINEAR EQUATIONS

Non-linear Form	Linear Form	
$y = a + b/x$	$y = a + b(1/x)$	(5.50a, b)
$y = 1/(a + bx)$	$(1/y) = a + bx$	(5.51a, b)
$y = ax/(b + x)$	$(1/y) = (1/a) + (b/a)(1/x)$	(5.52a, b)
$y = x/(a + bx)$	$(1/y) = b + a(1/x)$	(5.53a, b)
	$(x/y) = a + bx$	(5.53c)
$y = a + bx + cx^2$	$(y - a)/x = b + cx$	(5.54a, b)
	$(y - y_0)/(x - x_0) = (b + cx_0) + cx$	(5.54c)
$y = a \exp(bx)$	$\log y = \log a + (b/2{\cdot}303)\, x$	(5.55a, b)
$y = ab^x$	$\log y = \log a + (\log b)\, x$	(5.56a, b)
$y = ax^b$	$\log y = \log a + b(\log x)$	(5.57a, b)
$\left.\begin{array}{l}(y - a)(x - b) = c \\ y = a + c/(x - b)\end{array}\right\}$	$y = a + b(y/x) + (c - ab)(1/x)$	(5.58a, b)

Table 5.10 contains a few examples of curvilinear functions and their linear forms. Some of these correspond to well-known physico-chemical relationships. For example, the adsorption isotherm equations of Langmuir and Freundlich are of the form of eqns. (5.52a) and (5.57a) respectively, y being the specific adsorption and x the equilibrium concentration of the adsorbate in the external phase. Many relationships follow eqn. (5.55a), y being often a velocity or equilibrium constant and x the reciprocal of the absolute temperature, T°K. In these particular cases, linear plots following eqn. (5.55b) yield estimates of the Arrhenius energy of activation and an enthalpy change, respectively, from the slope. When y, as an equilibrium constant, is the solubility of a solid or a gas, Hildebrand (1962) suggests rectification by plotting $\log y$ against $\log x = \log \mathrm{T}$, in which case the slope yields an entropy instead of an enthalpy change. The quadratic form, eqn. (5.54a), for which specific and molar heat examples have already been given, can be rectified by eqn. (5.54b) in the event of a value for a being available (see Section 4.3 above) or by eqn. (5.54c) in which x_0 and y_0 are a pair of

values chosen from the array or picked from a good smooth curve drawn through the points. Examples of eqn. (5.58a) are given in the next section.

While there is no problem in graphing data according to any of the linear forms in Table 5.10 (except for eqn. (5.58b) and thereby getting approximate estimates for the constants of the original non-linear equation, considerable difficulties can arise when least squares estimates and confidence intervals are required for them. Consider eqns. (5.52a) and (5.52b), assuming that y alone is subject to error, this being normally distributed and of constant incidence. Now eqn. (5.43b) has just been used in Section 5.4 as the basis of estimation when the y values are subject to errors directly increasing with x and it was demonstrated there that the wrong procedure leads to incorrect estimates and confidence intervals. Consequently here, in the reverse situation, proper weighting must be used if the constants are to be evaluated by least squares fitting of the linear form. A similar problem arises with most of the linear forms of relationship in Table 5.10. The situation in respect of eqn. (5.58b) is different, however. In this case y remains unchanged on the left-hand side but it also appears in one of the "independent" variables on the right-hand side of this particular equation (one of several possibilities) into which the original has been cast. In consequence, while eqn. (5.58b) does in fact provide fair enough estimates of the true values of the constants, the standard error of estimate found in the usual way (i.e. by the application of eqn. (5.14)) is too low. Of the equations in the table, only (5.50b) is free of both difficulties since in this case the transformation leaves y unchanged on the left-hand side and only involves some manipulation of the supposed error-free x values on the right-hand side.

As an example involving a least squares procedure, consider the fitting of the Arrhenius Equation,

$$\mathrm{k} = \mathrm{A} \,.\, \exp\left(-\mathrm{E_A}/\mathrm{RT}\right) \tag{5.59}$$

which relates the velocity constant, k, of a chemical reaction to the absolute temperature, T°K, by way of the Arrhenius constant, A, and the apparent energy of activation, $\mathrm{E_A}$. Although more direct methods are available, rectification can be adopted and McBride and Villars (1954) and Huyberechts, Halleux and Kruys (1955) use a weighting procedure to fit eqn. (5.55b). Without weighting, this transformation will turn a collection of k distributions of constant standard deviation, non-linear in T, into a collection of log-normal distributions of varying standard deviation, linear in 1/T. Now it appears from (3.60h) that if e is the error in a value of k, $0{\cdot}4343e/\mathrm{k}$ is the error in the corresponding value of log k. This proportionality also applies to the standard deviations of the k and log k distributions at any given value of T, so that if we take $y = \log \mathrm{k}$ as the dependent variable for purposes of fitting, its variance is inversely proportional to k^2 and its values should be weighted inversely according to their variances, i.e. according to k^2.† It should be remembered, however, that in chemical kinetic experimentation k is a quantity derived from times and concentrations and, if these original measurements are normal and of constant variance, its own distributions are not necessarily so characterized. If, for instance, the variance of the values of k derived from the original measurements is proportional to the true k^2 in the first place, then $y = \log \mathrm{k}$ has constant variance and no weighting is called for. All the circumstances should be taken into account in determining the proper weights. In McBride and Villars' example the velocity constants are determined as regression coefficients by a preliminary least squares method and so have standard errors attached. Their paper should be consulted for details of the calculation.

5.6. Miscellaneous Curvilinear Relationships

General methods of fitting equations, particularly according to the principle of least squares, are described by Deming (1943) and Williams (1959) and a shorter account containing several physico-chemical examples is given by Opfell and Sage (1958). We deal briefly with two special cases.

(1) *Hyperbolic Equations*

Consider a model

$$(\mu - \alpha)(x - \beta) = \gamma \quad \text{or} \quad \mu = \alpha + \gamma/(x - \beta) \tag{5.60a}$$

and

$$(\hat{y} - a)(x - b) = c \quad \text{or} \quad \hat{y} = a + c/(x - b) \tag{5.60b}$$

† Strictly speaking this refers to the true, and therefore unknown, value of k and an iterative method is required to determine the best weights. Williams (1959) describes the procedure.

as an equation to be found from an available set of values, those of y being subject to error but distributed about μ, their true mean value at given x.†

Equation (5.60 b) can be transformed in a number of ways. One such is eqn. (5.58 b), here rewritten

$$\hat{y} = a + b(y/x) + (c - ab)(1/x), \qquad (5.61\text{ a})$$

which can be fitted as the multiple regression

$$\hat{y} = a + b_1x_1 + b_2x_2 \qquad (5.61\text{ b})$$

so that the required estimates, a, b and c, are given by a, b_1 and $b_2 + ab_1$.

As has already been remarked, eqn. (5.14) will give an artificially low value for s_e because y appears in the first "independent" variable, x_1, but if the $\hat{y}$ values are found by putting the calculated a, b and c back into eqn. (5.60 b), then s_e can be evaluated directly as $\sqrt{[\Sigma(y - \hat{y})^2/(n - 3)]}$ and an acceptable value obtained. A further improvement can be achieved if the experimental values and the constants b and c are used to adjust the value of a by averaging $[y - c/(x - b)]$. This corrects for rounding errors incurred in fitting the multiple regression. The method appears to work well and, with the directly found s_e, provides quite reasonable confidence intervals for the parameters by way of eqns. (5.21 a), (5.21 b) and (5.22).

A more accurate but rather more complicated procedure involves the rearrangement of eqn. (5.60 b) to

$$\hat{y} = a + c(1/x) + bc(1/x^2 + b/x^3 + b^2/x^4 \ldots) \qquad (5.62)$$

This can be fitted by regressing y on $x_1 = 1/x$ and on x_2, the latter a series expressible in terms of x_1. The multiple regression form is unchanged but x_2, which may be written

$$x_2 = x_1^2(1 + b_tx_1 + b_t^2x_1^2 + \ldots) \qquad (5.63)$$

must be evaluated with the aid of a trial value, b_t. It is necessary to iterate until the trial value coincides with the calculated value of b, b_c, given by b_2/b_1.

† Hey and Hey (1960) deal with the case in which both variables are subject to error. See also Bliss and James (1966).

This method has several features which make it less tedious to apply than might appear at first sight: (a) the series for x_2 is easy to evaluate, (b) several of the quantities required for substitution in the normal and multiplier eqns. ((5.17a) etc.) remain the same, only those involving x_2 itself (i.e., ΣX_2^2, ΣYX_2 and ΣX_1X_2) having to be found afresh in each cycle of iteration, and (c) the successive approximations, b_c, oscillate about the best (final) value and the differences between trial and calculated values, $b_t - b_c$, are nearly linearly related to the trial values when the latter are reasonably good. It follows from (c) that the required value of b, that which equals b_t when $b_t - b_c$ is zero, can be found by interpolation after two or three cycles. It can be seen from eqn. (5.63) that b_tx_1 must be small (made so by adjustment if necessary) for the series to converge rapidly and that if $b_t = 0$ the equation to be fitted becomes a quadratic.

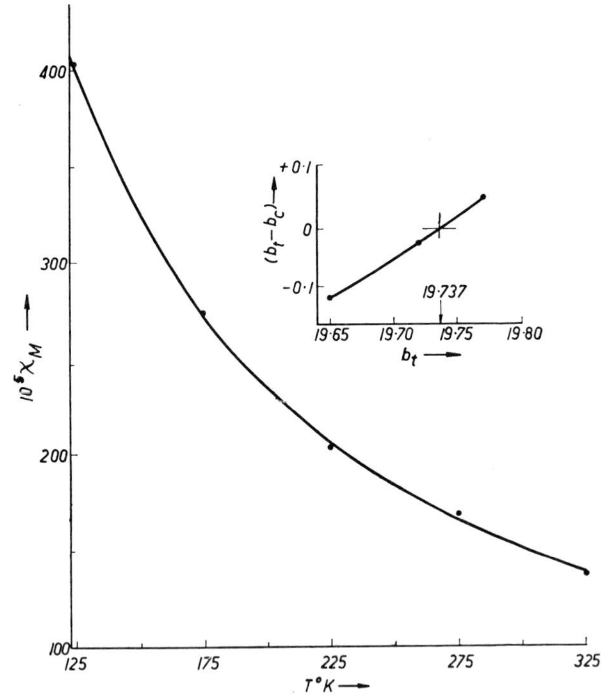

FIG. 5.4. A Hyperbola fitted to the Data of Table 5.11.

The inset graph shows the variation between trial and calculated values of the coefficient b in successive iteration.

A relationship which is in exact accord with eqn. (5.60a) is the Curie–Weiss law,

$$(\chi_M - \alpha)(T - \beta) = \gamma \qquad (5.64)$$

in which χ_M and α are the molar magnetic susceptibility and residual paramagnetism, T and β the absolute temperature and Weiss constant and γ the Curie constant. The latter is related to the number of unpaired electrons in, for example, a complex ion but generally this is unknown so γ has to be estimated along with α and β. In the usual experimental arrangement the temperature is known fairly exactly and may safely be regarded as the independent variable.

Table 5.11(a) contains a set of values, based on some for a hydrated copper formate, which have been artificially scattered so as to show more clearly differences between the results given by the two methods. The values calculated from the adjusted hyperbola obtained by the non-iterative method and those found by the iterative method are given for comparison. The fit of the final iterative curve can be judged from the plot in Fig. 5.4.

In applying the iterative method to these data a trial value of 20 (indicated by the results of the non-iterative method) was first chosen for b_t. With this at T = 125°K, for example, $x_1 = 1/T = 0{\cdot}008$, $b_t x_1 = 0{\cdot}16$ and $x_2 = 0{\cdot}008^2(1 + 0{\cdot}16 + 0{\cdot}16^2 + 0{\cdot}16^3 \ldots + 0{\cdot}16^6) = 76{,}170 \times 10^{-9}$ correct to five significant figures. At T = 325°K only three terms need be taken into the series to achieve the same accuracy. Part (b) of Table 5.11 gives a succession of trial and calculated values, the second, third and fourth of which indicate that b_t and b_c should equalize at 19·737 (see Fig. 5.4). The final trial was made using this value and the standard errors obtained. The slight discrepancy between b_t and b_c in the final fitting is due to y, x_1 and x_2 being taken to only five significant figures. It may be remarked in this connection that, when a three constant equation is fitted to values of two variables, the sum of squares of residuals is rather insensitive to changes in the values of the constants as long as such changes are compensating. Here, for instance, −1·87, 19·64 and 42,675 and −1·68, 19·75 and 42,610 give the same $\Sigma\hat{Y}^2$ and this only slightly above the true minimum. Further, it can be seen from Part (c) of the table that if the points are much scattered the sampling errors of the estimates are relatively great so that good data (rather better than those of Table 5.11) are required for satisfactory results.

TABLE 5.11. HYPERBOLIC VARIATION OF MAGNETIC SUSCEPTIBILITY WITH TEMPERATURE

(a) Data and calculated values

T°K x	$10^5 \chi_M$ (c.g.s.–e.m.u.) y	Non-iterative hyperbola $\hat{y}$	Final iterative hyperbola $\hat{y}$
125	403	403·37	403·13
175	274	272·82	272·77
225	203	205·87	205·91
275	168	165·57	165·24
325	137	137·94	137·90

(b) Trial and calculated values of b in iteration

Trial	b_t	b_c	$b_t - b_c$
First	20·00	19·647	+0·353
Second	19·65	19·768	−0·118
Third	19·77	19·722	+0·048
Fourth	19·72	19·744	−0·024
Final	19·737	19·734	+0·003

(c) Estimates and standard errors

Estimate	Non-iterative hyperbola		Final iterative hyperbola	
	Value	Standard error	Value	Standard error
a	−1·67	14·71	−1·71	14·39
b	19·89	9·39	19·73	9·48
c	42569	5446	42618	5320
s_e	3·035		3·032	

Most of the technical interest in fitting hyperbolic equations appears to be centred on the Antoine equation

$$\log p = A - B/(T + C) \qquad (5.65a)$$

where p is the vapour pressure of a liquid (generally measured in mm), T the temperature (generally in °C) and A, B and C are all adjustable constants. In this case, point or interval estimates of the constants are of little or no theoretical interest. The main requirement is the best possible fitted equation which we write

$$\log \mathrm{p} = a - b/(\mathrm{T} + c) \tag{5.65b}$$

Several approximate methods of fitting this equation have been suggested (see Thomson (1946), Lu (1959) and Miller (1963)) and many least squares treatments described. For the latter purpose, eqn. (5.65b) can be transformed to eqn. (5.58b) but

$$\mathrm{T} \,.\, \log \mathrm{p} = (ac - b) + a\mathrm{T} - c \log \mathrm{p} \tag{5.66}$$

is more convenient in non-iterative computation because the "independent" variables on the right-hand side call for no special treatment (apart from taking the logarithms of the pressures) and the values of the "dependent" variable, $y = \mathrm{T} \,.\, \log \mathrm{p}$, are also those of one of the required cross-products. The form of eqn.(5.65b) appropriate to the iterative method described above is

$$\log \mathrm{p} = a - b(1/\mathrm{T}) + bc(1/\mathrm{T}^2)\,(1 - c/\mathrm{T} + c^2/\mathrm{T}^2 - c^3/\mathrm{T}^3 + \ldots) \tag{5.67}$$

so that $y = \log \mathrm{p}$ can be regressed on $x_1 = 1/\mathrm{T}$ and $x_2 = x_1^2(1 - c_t x_1 + c_t^2 x_1^2 - \ldots)$ until a trial c_t coincides with $c = -b_2/b_1$.

This and similar methods for evaluating the Antoine constants give solutions in which the sum of squares of deviations in log p is minimized and the results so obtained will be biased estimates of the true constants of a model which assumes that the pressure values themselves are normally distributed at fixed and known temperatures. In other words, where this model is appropriate, as in "static" vapour pressure determination, an equation found by minimizing the squared deviations in p will give the better fit to the data. Miller (1963), in an interesting comparison of results, demonstrates that this is so for some typical vapour pressure measurements.

In "dynamic" vapour pressure studies, the pressure is allowed to reach set values and the temperature is then measured. In this situation it would be better to regard temperature as the dependent variable and to fit eqn. (5.65b) in the form

$$\mathrm{T} = -c - b/(\log \mathrm{p} - a) \tag{5.68}$$

The multiple regression approach based on eqns. (5.61) or (5.62) is again appropriate.

An example of purely academic interest and of a rather different though related kind is provided by van der Waals' equation of state (see Hammett, 1952). It differs from the preceeding examples in that it connects four possible variables (temperature, T; pressure, p; volume, V; and amount of gas, n, in moles) and of its three constants one is known exactly. Reducing the variables to three by writing the equation in terms of density, $\varrho = n/V$ (mole/l.), gives

$$(p + \boldsymbol{a}\varrho^2)(1 - \boldsymbol{b}\varrho) = \varrho RT \tag{5.69}$$

where $\boldsymbol{a}$ and $\boldsymbol{b}$ are the "true" values of the intermolecular attractive force constant and the co-volume of the van der Waals model and are to be estimated by a and b respectively from observed data.

No special problem of estimation arises if the pressure of a given amount of gas at constant volume is measured (subject to normal error) at a series of known temperatures. Equation (5.69) can be rewritten

$$p = -\boldsymbol{a}\varrho^2 + [\varrho R/(1 - \boldsymbol{b}\varrho)]\,T \tag{5.70}$$

and estimates of $\boldsymbol{a}$ and $\boldsymbol{b}$ obtained from the intercept and slope of the linear regression of p on T. If, on the other hand, the pressure is measured at constant temperature for varying amounts or volumes of gas (i.e., density becomes the independent variable) we can write

$$p = -\boldsymbol{a}\varrho^2 + RT\varrho(1 + \boldsymbol{b}\varrho + \boldsymbol{b}^2\varrho^2 + \ldots) \tag{5.71}$$

The regression of p on $x_1 = \varrho^2$ and $x_2 = \varrho(1 + b_t\varrho + b_t^2\varrho^2 + \ldots)$ gives the required estimates when the trial value of b, b_t, equals that found from the coefficient of x_2, i.e. $b_c = b_2/RT$. (Note that this regression is constrained to pass through the origin and one of its coefficients is known.)

The values of a and b found by regression analysis lie between those derived from the critical temperature and pressure which are customarily quoted and those based on critical volume which are unwarrantably neglected. For example, for oxygen data at 0 °C and pressures between 25 and 100 atm quoted by Hammett (1952), an iterative procedure along the lines of eqn. (5.71) gives 1·14 and 0·028 approximately for a and b in litre–atmosphere units whereas the critical temperature-critical pressure values are about 1·32 and 0·031 and the values from the critical volume are about 0·82 and 0·025. These discrepancies reflect the inadequacy of the kinetic model on which van der Waals' equation is based and of its derivation from the postulates of that model. Although fitting by least squares gives values differing from those commonly accepted, it is the former which would be required were van der Waals' equation ever used for the accurate representation of pressure–temperature–density data.

(2) *Exponential and Asymptotic Regression*

The exponential regression model

$$\mu = \alpha \,.\, \exp(\beta x) \quad \text{or} \quad y = \alpha \,.\, \exp(\beta x) + e \tag{5.72}$$

may be presumed to underly such relationships as the Arrhenius equation (5.59), when the temperature is known and the reaction velocity constant is subject to normal random error. We discussed the rectification of this kind of relationship in Section 5.6 but its direct fitting is a matter of some difficulty. Brownlee (1960) derives the normal equations,

$$\left.\begin{aligned} \Sigma[y \,.\, \exp(bx)] &= a\Sigma[\exp(2bx)] \\ \Sigma[xy \,.\, \exp(bx)] &= a\Sigma[x \exp(2bx)] \end{aligned}\right\} \tag{5.73}$$

which can be solved by iteration starting from a trial value of b, b_t. Williams (1959) gives an example of the application of a method of this kind for a model of the form

$$\mu = \alpha + \beta \,.\, \exp(\gamma x) \tag{5.74}$$

and shows that few cycles of iteration one required if a good value of the constant in the exponential term be available to start with.

The recent statistical literature offers many methods of fitting these and related regressions, such as the more general form

$$\mu = \alpha + \beta\gamma^x \tag{5.75}$$

Of these, Hartley's (1948) and Patterson's (1956) are the most appropriate for occasional use.

Problems of this kind arise frequently in chemical kinetic studies and several approximate methods of treating data have been proposed in this connection. Consider, for example, a gas reaction obeying the first-order law

$$(\mathrm{p}_\infty - \mathrm{p}) = (\mathrm{p}_\infty - \mathrm{p}_0) \exp(-\mathrm{kt}) \tag{5.76}$$

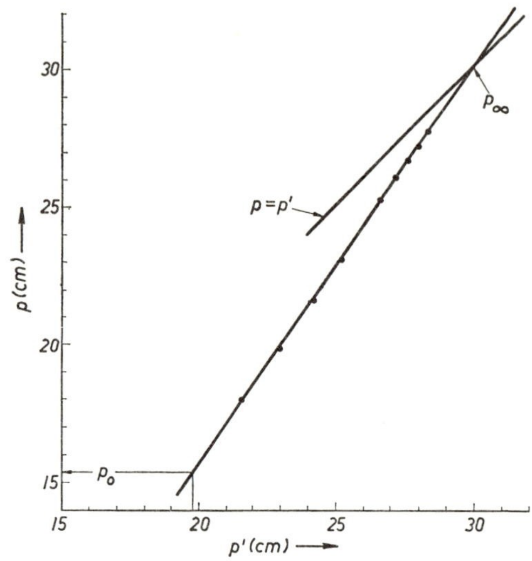

FIG. 5.5. A Plot of the Data of Table 5.12 to determine the Initial and Final Values of Pressure, p.

where p_0 and p_∞ are the initial and final pressures in a reaction vessel, p the observed pressure at time t, and k the velocity constant. It is often difficult or impracticable to measure p_0 and p_∞ but k can be found in various ways even without this knowledge (Guggenheim (1926), Swinbourne (1960)). The method described by Swinbourne is based on a plot of pressure readings, $p_1, p_2 \ldots p_i \ldots p_n$,

TABLE 5.12. PYROLYSIS OF CYCLOHEXYL CHLORIDE AT 384°C

t (min)	p_i (cm)	t + 4 (min)	p_i' (cm)
		4	19·83
2	17·72	6	21·59
4	19·83	8	23·05
6	21·59	10	24·25
8	23·05	12	25·24
10	24·25	14	26·05
12	25·24	16	26·71
14	26·05	18	27·25
16	26·71	20	27·69
18	27·15	22	28·05
20	27·69	24	28·36

made at times, $t_1, t_2 \ldots t_i \ldots t_n$, against those, $p'_1, p'_2 \ldots p'_i \ldots p'_n$, made at later times, $t_1 + \Delta t, t_2 + \Delta t \ldots t_i + \Delta t \ldots t_n + \Delta t$, where Δt is some fixed time interval. Both the first and the second series of measurements obey eqn. (5.76) so we have

for the first series,

$$(p_\infty - p_i) = (p_\infty - p_0) \exp(-kt_i) \qquad (5.77a)$$

and for the second,

$$(p_\infty - p'_i) = (p_\infty - p_0) \exp[-k(t_i + \Delta t)] \qquad (5.77b)$$

Division of one of these equations by the other leads to

$$p_i = p_\infty(1 - \exp(k\Delta t)) + p'_i \,.\, \exp(k\Delta t) \qquad (5.78)$$

so that k can be found from the slope of the straight line through the p_i, p'_i points. In addition, p_∞ is the final value of both series and is given by the intersection of the line through the points and the 45° line, $p_i = p'_i$, while p_0 is the value of p_i corresponding to p'_0 at $t = \Delta t$. These relationships are illustrated in Fig. 5.5 which shows the application of the method to the gas reaction data of Table 5.12. For this example $p_\infty = 30$ cm and $p_0 = 15{\cdot}4$ cm. (For a generalization of the method to other orders, see Swinbourne (1963).)

5.7. Frequency and Distribution Curves

The problems which concern frequency and distribution curves are similar to those which are involved in the fitting of other equations to data, viz. the smoothing or graduation of observed values (here, class or cumulative frequencies), the setting of limits to random variation from the fitted curve and the estimation of the constants of an assumed model. The last brings further problems in its train such as the choice of a suitable function or model and, in the case of heterogeneous data, the choice of two or more such functions and the discovery of a method of resolving the data so as to enable both to be fitted. The reader is referred to Chapters 6 and 7 of Hald's book (1952) for a general discussion covering most of these problems. A few examples of the treatment of frequency data, mainly following Hald's methods, are given here.

(1) *Fitting the Gaussian Function, eqns. (3.11) and (3.12)*

Estimates of the mean and standard deviation of a distribution, found by the methods of Table 2.8 or Table 2.9, can be substituted directly into eqns. (3.11) and (3.12) and the frequency density or cumulative frequency between any chosen limits then calculated. This is done with the aid of the ordinate or area values in Table I of the Appendix.

The use of both ordinates and areas is demonstrated in Table 5.13 where the methods are applied to the data of Table 2.3 using the mean and best estimate of the standard deviation calculated in Table 2.9. Substituting these values in eqn. (3.11) we have

$$f(\%) =$$

$$(100 \times 0{\cdot}01/0{\cdot}0134)\,[(1/\sqrt{(2\pi)})\exp(-(x - 1{\cdot}497)^2/2 \times 0{\cdot}0134^2)]$$

The 100 appears in the first term because we wish to calculate percentages (had we wished to find the actual Gaussian expected frequencies for, say, a χ^2 test, this would have been made 81) and the 0·01 is the class interval width which is introduced because eqn. (3.11) gives the frequency density per unit of the x scale. The term within square brackets is the height of the ordinate of the standard Gaussian frequency function, given in Appendix Table I for the argument $u = |x - \mu|/\sigma$. Consequently, the tabulated value multiplied by the first factor gives the required percentage frequency. In the area method the differences between the tabulated values for the standardized class boundary deviations are multiplied by the total frequency, here 100%.

In the example of Table 5.13 u is approximated by $(x - \bar{x})/s$. There are slight differences between the two sets of results and, in both cases, a slight deficiency which arises from expected values outside the range of the class intervals. It is a characteristic of fitted Gaussian distributions that they indicate frequencies lower than the observed values in the terminal class intervals but otherwise the discrepancies are of the nature of random fluctuations if the model is valid.

Probability paper provides a quick method of fitting normally distributed data. The method is illustrated in Fig. 2.4. Differences between the cumulative percentage frequencies read from the eye-fitted line give class interval frequencies which are adequate for

TABLE 5.13. FITTING THE GAUSSIAN FREQUENCY FUNCTION

(a) Ordinate method					(b) Area method					
Midpoint, x	$x - \bar{x}$	$\|u\|$	Ordinate	f %	Interval boundary, x	$x - \bar{x}$	$\|u\|$	Area	Δ Area	f %
1·460	−0·037	2·76	0·0088	0·66	1·4545	−0·0425	3·172	0·4992	0·0067	0·67
1·470	−0·027	2·01	0·0529	3·95	1·4645	−0·0325	2·425	0·4925	0·0391	3·91
1·480	−0·017	1·27	0·1781	13·29	1·4745	−0·0225	1·679	0·4534	0·1288	12·88
1·490	−0·007	0·52	0·3485	26·01	1·4845	−0·0125	0·933	0·3246	0·2505	25·05
1·500	+0·003	0·22	0·3894	29·06	1·4945	−0·0025	0·187	0·0741 †	0·2864	28·64
1·510	+0·013	0·97	0·2492	18·60	1·5045	+0·0075	0·560	0·2123 †	0·1919	19·19
1·520	+0·023	1·72	0·0909	6·78	1·5145	+0·0175	1·306	0·4042	0·0757	7·57
1·530	+0·033	2·46	0·0194	1·45	1·5245	+0·0275	2·052	0·4799	0·0175	1·75
				99·80	1·5345	+0·0375	2·799	0·4974		99·66

† Areas corresponding to the boundaries 1·4945 and 1·5045 represent differences from the mean ordinate so they are added to obtain the area difference in the second last column.

many purposes. When the cumulative frequency is that actually observed and its values may be regarded as subject to independent errors, the cumulative frequency points can be fitted by a least squares procedure known as *probit analysis*. Finney (1947) should be consulted for details of this method, or Paradine and Rivett (1953) who give concise examples with the necessary tables.

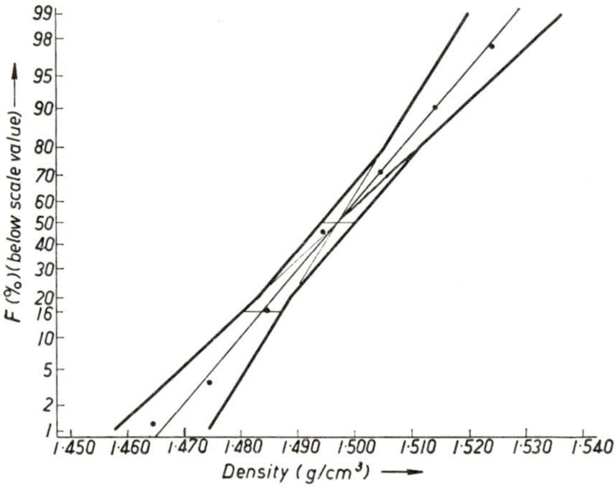

FIG. 5.6. Approximate Limits for the Percentage Cumulative Frequency of a Gaussian Distribution.

The central (parallel) lines represent the effect of the unreliability of $\bar{x}$. The intersecting lines include the effect of the unreliability of s.

Probability paper also provides means of conveniently representing the limits of the true distribution line so that a confidence interval can be read off for the cumulative percentage above or below a particular scale point or, alternatively, for a particular quantile of the distribution. A very approximate method for obtaining such limits, based on one described by Herdan (1948), is illustrated in Fig. 5.6 for the example of Table 5.13. The steps are, taking $\bar{x} =$ 1·497 and $s =$ 0·0134 as before;

(a) draw the cumulative frequency line through the points 50%, 1·497 and 16%, 1·497—0·0134:

(b) find, say, the 95% confidence interval for μ (1·497 $\pm$ 2 $\times$ 0·0134/$\sqrt{81}$ = 1·494 to 1·500 is close enough for the purpose) and draw lines parallel to the first through the points 50%, 1·494, and 50%, 1·500;(c) find the approximate 95% confidence limits for $\mu - \sigma$ (for $\bar{x}$ and s as independent values, $s_{\bar{x}-s} \doteqdot s\sqrt{(1/n + 1/2n)}$

from eqns.
(3.22 b), (3.24 b) and (3.60 a), so $(1{\cdot}497 - 0{\cdot}0134) \pm 2 \times 0{\cdot}0134 \sqrt{(1/81 + 1/162)} = 1{\cdot}480$ to $1{\cdot}487$ is the interval required), plot them on the 16% line and draw lines through these points and the mean point. The outermost parts of the two pairs of lines (one pair parallel, the other intersecting) are the approximate limits required. The centre of the band so obtained represents the effect of uncertainty in the mean and the extremities the effect of uncertainty in the standard deviation of the distribution. The band is certainly unreliable towards the ends of the percentage scale as is the Gaussian model on which it is based but it approximates that derived from more exact calculations.

(2) *Resolving Normal Distributions*

The data of Table 2.5 is represented in histogram form in Fig. 5.7(a). Their bimodal character is clearly shown by the histogram and it is sometimes useful in such a case to resolve or decompose the heterogeneous distribution into hypothetical components, say a pair of normal distributions.

Figure 5.7 illustrates one way of doing this. When the logarithms of the class interval frequencies are plotted against the midpoints (or the frequencies themselves, using semi-log paper), a more or less normal distribution curve is transformed into a parabola which, in the case of fairly smooth data, can be fitted by eye. In the example (see part (b) of the figure) this appears to be the case with the higher frequencies and those above the second mode are unlikely to be affected by the first distribution. Consequently, the branch of the parabola above the second mode can be easily drawn and this, with its image below the mode (shown by a broken line) should represent the second, and most populous, of the two component distributions. Frequencies read from the broken branch of the parabola, italicized in Table 5.14, are presumed to belong to the second component and are subtracted from the total frequencies in the relevant class intervals to obtain values for the first component. The separated frequencies are plotted in cumulative percentage form in Fig. 5.7(c). They appear to follow satisfactorily straight lines between which the curve for the compound distribution as a whole follows a curving path. The Gaussian curves corresponding to the straight lines of part (c) of the figure are reproduced as frequency curves over the histogram in part (a).

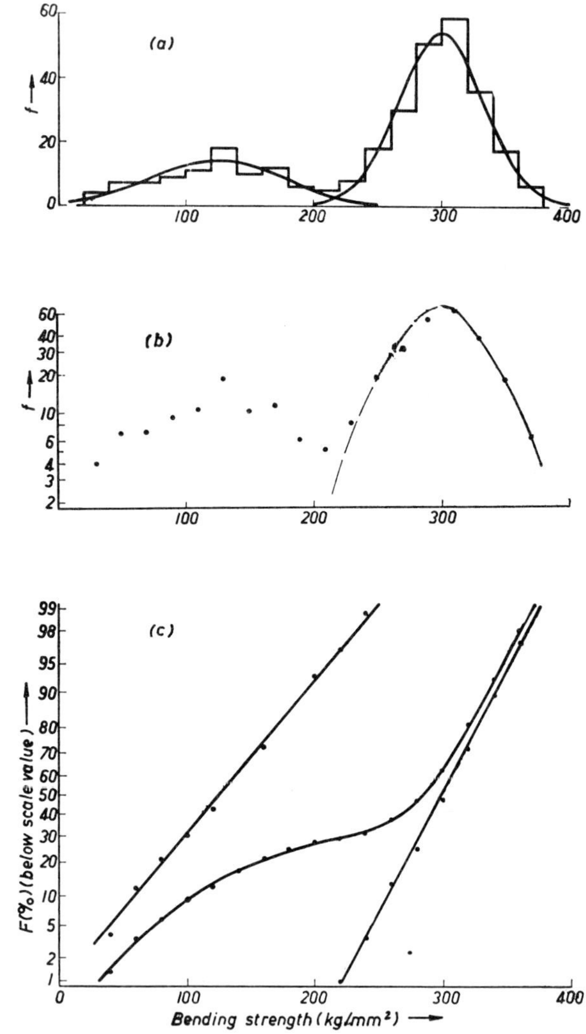

FIG. 5.7. The Resolution of a Heterogeneous Distribution into Two Distributions of Gaussian Form.

The example is one in which the component distributions are rather well separated to start with. However, even if this is not the case, the method described for resolution is still satisfactory so long as one half of one of the component distributions is reasonably clear of the influence of the other. Analytically, the separation can be achieved if the first five moments of the compound distribution are computed (Plummer, 1940) but, because of the unreliability of the estimate of the fifth moment, the method of Bartlett and Smith (1960) which depends on the mode and standard deviation is to be preferred.

TABLE 5.14
RESOLUTION OF A DISTRIBUTION—BENDING STRENGTHS OF GLASS RODS

Bending strength (kg/mm^2)	*f*	*F*%	Upper component				Lower component		
			f calc.	*f* (adjusted)	*f*% (adjusted)	*F*% (adjusted)	*f*	*f*% (adjusted)	*F*% (adjusted)
20—	4	1·3					4	4·44	4·4
40—	7	3·5					7	7·78	12·2
60—	7	5·7					7	7·78	20·0
80—	9	8·6					9	10·00	30·0
100—	11	12·1					11	12·22	42·2
120—	18	17·8					18	20·01	62·2
140—	10	21·0					10	11·11	73·3
160—	12	24·8					12	13·33	86·7
180—	6	26·7					6	6·67	93·3
200—	5	28·3	*2*	2	0·89	0·9	3	3·33	96·7
220—	8	30·9	*6*	6	2·68	3·6	2	2·22	98·9
240—	18	36·6	*17*	17	7·59	11·2	1	1·11	100·0
260—	30	46·2		30	13·39	24·6			
280—	51	62·4		51	22·77	47·3			
300—	59	81·2		59	26·34	73·7			
320—	36	92·7		36	16·07	89·7			
340—	17	98·1		17	7·59	97·3			
360—	6	100·0		6	2·68	100·0			
	314			224	100·00		90	100·00	

(3) *Other Frequency Functions*

Many different equations have been proposed in order to represent the great variety of observed distributions and thus enable observed frequencies to be graduated. There are three main systems of curves: the Pearsonian series based on a relationship for the relative slope of the frequency function itself, the Gram–Charlier series based on derivatives of the Gaussian function, and curves generated by the *method of translation* which are Gaussian for some transformation of the variable. In addition there are numerous miscellaneous equations to be found particularly in the literature relating to particle size distributions.

The Pearsonian curves are typed according to measures of skewness and kurtosis and Pearson and Hartley (1958) and Cowden (1957) provide a $\beta_1 - \beta_2$ chart from which a function appropriate to a particular set of data may be chosen. The fitting of these curves is exhaustively treated by Elderton (1938). The positively skewed distributions commonly encountered in industrial practice are often well described by the Pearsonian Type III and a Gram–Charlier Type A function and by the best known of those obtained by the method of translation, the log-normal distribution. For a comparison of the first two Cowden (1957) should be consulted and Aitchison and Brown (1957) for an extensive description of the last. The log-normal function is much used in graduating particle size distributions and it has the valuable feature that, shape factors apart, it represents the number, linear, surface and volume distributions of a given population of particles with predictable changes in the geometric mean alone. In other words, all these distributions are represented by parallel lines on logarithmic probability paper and the lines are equally spaced. Irani and Callis (1963) give the relationships.†

Some distributions take a particular shape because of truncation or censoring. Hald (1952) describes the fitting of an underlying Gaussian distribution and his method can be extended to the log-normal case.

5.8. A Least Squares Solution of Simultaneous Equations

We conclude with the discussion of a problem which, although rather different from others in this chapter, is solved by a method identical in form with the fitting of a multiple regression. It is the reverse of the multiple regression situation and arises where the

† Espenscheid *et al.* (1964) describe other distributions of the log-normal type used in treating particle size distribution data.

coefficients of a set of equations to be solved are known but their number is more than required for the bare evaluation of the related quantities. For example, the set

$$x + y = 5 \quad x - 2y = 2$$

gives $x = 4$ and $y = 1$, but if in addition we have

$$x + 3y = 9$$

there is no unique solution and only some kind of an average can be obtained. If the equations contained only one unknown, say

$$x = 4, \quad x = 3$$

the average of choice is obviously the arithmetic mean, $\bar{x}$, which is the least squares solution.

In the case of two unknowns, let the equations be

$$a_i x + b_i y = l_i \quad (i = 1, 2 \ldots n) \tag{5.79}$$

and the best solution

$$a_i \bar{x} + b_i \bar{y} = \hat{l}_i \tag{5.80}$$

where $\hat{l}_i$ is the calculated value of the numerical coefficient. It is required to minimize the sum of squares of the deviations

$$\Sigma(l_i - \hat{l}_i)^2 = \Sigma \hat{L}_i^2 = \Sigma(a_i \bar{x} + b_i \bar{y} - l_i)^2 \tag{5.81}$$

After squaring and summing this expression and differentiating with respect to $\bar{x}$ and $\bar{y}$, we obtain the normal equations

$$\left.\begin{aligned} \Sigma a_i^2 \,.\, \bar{x} + \Sigma a_i b_i \,.\, \bar{y} &= \Sigma a_i l_i \\ \Sigma a_i b_i \bar{x} + \Sigma b_i^2 \,.\, \bar{y} &= \Sigma b_i l_i \end{aligned}\right\} \tag{5.82}$$

whose solution gives the averages required. These equations are analogous to those derived in Chapter IV for finding the coefficients of a linear regression line and their obvious extension results in a set analogous to the multiple regression normal equations. Now, for n equations in k unknowns of the form

$$a_i x + b_i y + \cdots k_i w = l_i \tag{5.83}$$

LR

we can write, dropping the subscripts,

$$\left.\begin{aligned}\Sigma a^2 . \bar{x} + \Sigma ab . \bar{y} + \cdots \Sigma ak . \bar{w} &= \Sigma al \\ \Sigma ba . \bar{x} + \Sigma b^2 . \bar{y} + \cdots \Sigma bk . \bar{w} &= \Sigma bl \\ \cdots\cdots\cdots\cdots\cdots\cdots & \\ \Sigma ka . \bar{x} + \Sigma kb . \bar{y} \quad \ldots \Sigma k^2 . \bar{w} &= \Sigma kl\end{aligned}\right\} \tag{5.84}$$

The sum of squares of the residuals can be obtained by substituting the mean values back into the original equations (which provides a useful check, $\Sigma(l_i - \hat{l}_i) = \Sigma\hat{L}_i = 0$) or from the ratio of determinants

$$\Sigma\hat{L}^2 = \begin{bmatrix}\Sigma a^2 & \Sigma ab \ldots & \Sigma ak & \Sigma al \\ \Sigma ba & \Sigma b^2 \ldots & \Sigma bk & \Sigma bl \\ \cdots & \cdots & \cdots & \cdots \\ \Sigma ka & \Sigma kb \ldots & \Sigma k^2 & \Sigma kl \\ \Sigma la & \Sigma lb \ldots & \Sigma lk & \Sigma l^2\end{bmatrix} \div \begin{bmatrix}\Sigma a^2 & \Sigma ab \ldots & \Sigma ak \\ \Sigma ba & \Sigma b^2 \ldots & \Sigma bk \\ \cdots & \cdots & \cdots \\ \Sigma ka & \Sigma kb \ldots & \Sigma k^2\end{bmatrix} = D'/D \tag{5.85}$$

From the sums of squares of the deviations, $\Sigma\hat{L}^2$, a variance can be obtained

$$s^2 = \Sigma\hat{L}^2/(n-k) \quad (f = n - k) \tag{5.86}$$

and from this in turn, the error variances

$$s_{\bar{x}}^2 = s^2 A_a/D; \quad s_{\bar{y}}^2 = s^2 A_b/D; \quad \text{etc.} \quad \text{(5.87a, b, etc.)}$$

in which D is the denominator determinant of eqn. (5.85) and A_a, A_b etc. are the minors of the terms Σa^2, Σb^2 etc. in it. The derivations of eqns. (5.85) and (5.86) and the treatment of various complications such as weighting can be found in Whittaker and Robinson (1944) and in Deming (1943) and a briefer account is given by Topping (1955).

TABLE 5.15

Compound	Gibbs' free energy of complex formation (kcal)
cis-quercitol	$\Delta G = -5{\cdot}26 = \Delta G_B - 2(O_a : H_a) - (O_a : O_a)$
epi-quercitol	$\Delta G = -3{\cdot}36 = \Delta G_B \quad -2(O_1 : O_2)$
scyllo-quercitol	$\Delta G = -0{\cdot}95 = \Delta G_B + 2(O_a : H_a) + (O_a : O_a) - 4(O_1 : O_2)$
cis-inositol	$\Delta G = -8{\cdot}15 = \Delta G_B \quad -3(O_a : O_a)$
epi-inositol	$\Delta G = -5{\cdot}20 = \Delta G_B \quad -(O_a : O_a) - 2(O_1 : O_2)$
myo-inositol	$\Delta G = -1{\cdot}90 = \Delta G_B \quad +(O_a : O_a) - 4(O_1 : O_2)$

TABLE 5.16. LEAST SQUARES SOLUTION OF SIMULTANEOUS EQUATIONS

Equations times a_i	Equations times b_i	Equations times c_i	Equations times d_i
$\bar{x} - 2\bar{y} - \bar{z} = -5{\cdot}26$	$-2\bar{x} + 4\bar{y} + 2\bar{z} = 10{\cdot}52$	$-\bar{x} + 2\bar{y} + \bar{z} = 5{\cdot}26$	
$\bar{x} - 2\bar{w} = -3{\cdot}36$			$-2\bar{x} + 4\bar{w} = 6{\cdot}72$
$\bar{x} + 2\bar{y} + \bar{z} - 4\bar{w} = -0{\cdot}95$	$2\bar{x} + 4\bar{y} + 2\bar{z} - 8\bar{w} = -1{\cdot}90$	$\bar{x} + 2\bar{y} + \bar{z} - 4\bar{w} = -0{\cdot}95$	$-4\bar{x} - 8\bar{y} - 4\bar{z} + 16\bar{w} = 3{\cdot}80$
$\bar{x} - 3\bar{z} = -8{\cdot}15$		$-3\bar{x} + 9\bar{z} = 24{\cdot}45$	
$\bar{x} - \bar{z} - 2\bar{w} = -5{\cdot}20$		$-\bar{x} + \bar{z} + 2\bar{w} = 5{\cdot}20$	$-2\bar{x} + 2\bar{z} + 4\bar{w} = 10{\cdot}40$
$\bar{x} + \bar{z} - 4\bar{w} = -1{\cdot}90$		$\bar{x} + \bar{z} - 4\bar{w} = -1{\cdot}90$	$-4\bar{x} - 4\bar{z} + 16\bar{w} = 7{\cdot}60$
$6\bar{x} - 3\bar{z} - 12\bar{w} = -24{\cdot}82$	$8\bar{y} + 4\bar{z} - 8\bar{w} = 8{\cdot}62$	$-3\bar{x} + 4\bar{y} + 13\bar{z} - 6\bar{w} = 32{\cdot}06$	$-12\bar{x} - 8\bar{y} - 6\bar{z} + 40\bar{w} = 28{\cdot}52$

Normal equations:
eqn. (5.84),

$$
\begin{aligned}
6\bar{x} \qquad - 3\bar{z} - 12\bar{w} &= -24{\cdot}82 \\
8\bar{y} + 4\bar{z} - 8\bar{w} &= 8{\cdot}62 \\
-3\bar{x} + 4\bar{y} + 13\bar{z} - 6\bar{w} &= 32{\cdot}06 \\
-12\bar{x} - 8\bar{y} - 6\bar{z} + 40\bar{w} &= 28{\cdot}52
\end{aligned}
$$

Solution:

$$
\begin{aligned}
\bar{x} &= -2{\cdot}52833 \\
\bar{y} &= 0{\cdot}46167 \\
\bar{z} &= 1{\cdot}89333 \\
\bar{w} &= 0{\cdot}33083
\end{aligned}
$$

Check and evaluation of $\Sigma\hat{L}^2$:

		$\hat{L}$	$\hat{L}^2$
$\bar{x} - 2\bar{y} - \bar{z} + 5{\cdot}26$	$= -2{\cdot}528 - 0{\cdot}923 - 1{\cdot}893$	$= -0{\cdot}085$	$0{\cdot}007225$
$\bar{x} - 2\bar{w} + 3{\cdot}36$	$= -2{\cdot}528 - 0{\cdot}662$	$= +0{\cdot}170$	$0{\cdot}028900$
$\bar{x} + 2\bar{y} + \bar{z} - 4\bar{w} + 0{\cdot}95$	$= -2{\cdot}528 + 0{\cdot}923 + 1{\cdot}893 - 1{\cdot}323$	$= -0{\cdot}085$	$0{\cdot}007225$
$\bar{x} - 3\bar{z} + 8{\cdot}15$	$= -2{\cdot}528 - 5{\cdot}680$	$= -0{\cdot}058$	$0{\cdot}003664$
$\bar{x} - \bar{z} - 2\bar{w} + 5{\cdot}20$	$= -2{\cdot}528 - 1{\cdot}893 - 0{\cdot}662$	$= +0{\cdot}117$	$0{\cdot}013689$
$\bar{x} + \bar{z} - 4\bar{w} + 1{\cdot}90$	$= -2{\cdot}528 + 1{\cdot}893 + 1{\cdot}323$	$= -0{\cdot}058$	$0{\cdot}003664$
		$\Sigma\hat{L} = +0{\cdot}001$†	$\Sigma\hat{L}^2 = 0{\cdot}063767$

(†, rounding error)

TABLE 5.16. *(cont.)*

Determinant and minors: eqn. (5.85),

$$D = \begin{vmatrix} 6 & 0 & -3 & -12 \\ 0 & 8 & 4 & -8 \\ -3 & 4 & 13 & -6 \\ -12 & -8 & -6 & 40 \end{vmatrix} = 27{,}027 \qquad A_a = \begin{vmatrix} 8 & 4 & -8 \\ 4 & 13 & -6 \\ -8 & -6 & 40 \end{vmatrix} = 2656 \qquad A_b = \begin{vmatrix} 6 & -3 & -12 \\ -3 & 13 & -6 \\ -12 & -6 & 40 \end{vmatrix} = 240$$

Similarly: $A_c = 960$, $A_d = 456$

Standard errors and confidence intervals:

eqn. (5.86), $s^2 = 0{\cdot}0638/(6-4) = 0{\cdot}0319$; $s = 0{\cdot}1786$ $(t_{\cdot 1,2} = 2{\cdot}92)$

eqn. (5.87), $s_{\bar{x}} = 0{\cdot}1786\sqrt{(2656/27{,}027)} = 0{\cdot}056$ 90% confidence limits of $\mu_x = -2{\cdot}53 \pm 0{\cdot}16 = \Delta G_B$

$s_{\bar{y}} = 0{\cdot}1786\sqrt{(240/27{,}027)} = 0{\cdot}017$ 90% confidence limits of $\mu_y = 0{\cdot}46 \pm 0{\cdot}05 = (O_a{:}H_a)$

$s_{\bar{z}} = 0{\cdot}1786\sqrt{(960/27{,}027)} = 0{\cdot}034$ 90% confidence limits of $\mu_z = 1.89 \pm 0{\cdot}10 = (O_a{:}O_a)$

$s_{\bar{w}} = 0{\cdot}1786\sqrt{(456/27{,}027)} = 0{\cdot}023$ 90% confidence limits of $\mu_w = 0{\cdot}33 \pm 0{\cdot}07 = (O_1{:}O_2)$

For illustration we take Angyal and McHugh's (1956) data on the interaction energies of axial hydroxyl groups in cyclitols when esterified in boric acid solutions. The measured Gibbs' free energy of complex formation or esterification, ΔG, is equated with the free energy of conformational change, ΔG_B, and such energies of interaction (between axial oxygen atoms, $(O_a:O_a)$; axial oxygen and hydrogen atoms, $(O_a:H_a)$; and oxygen atoms on adjacent carbon atoms, $(O_1:O_2)$) as can arise from the conformation of the compound concerned. The data are given in Table 5.15.

In applying equations (5.84) to these data, we denote the required values of ΔG_B by $\bar{x}$, $(O_a:H_a)$ by $\bar{y}$, $(O_a:O_a)$ by $\bar{z}$ and $(O_1:O_2)$ by $\bar{w}$. To find the sums of squares and products of coefficients it is convenient to set out the equations of Table 5.15, to multiply the coefficients successively by those of $\bar{x}$, $\bar{y}$, $\bar{z}$ and $\bar{w}$, and to add the resulting expressions. The calculations are given in Table 5.16. Since the coefficients of $\bar{x}$ are all unity, the first column in the table contains the original or *conditional* equations themselves. The confidence intervals at the end of the table represent the final results for ΔG_B and the energies of interaction in kcal.

NOTATION OF CHAPTER V

A	Arrhenius constant
A_a, A_b	minors of a determinant
$\boldsymbol{\alpha}$	residual paramagnetism
$\boldsymbol{a}$, $\boldsymbol{b}$, a, b	true and estimated van der Waals constants
A, B, C, a, b, c	true and estimated constants in the Antoine equation
$\alpha, \beta_1, \beta_2, \ldots, a, b_1, b_2 \ldots$	true and estimated pure regression and polynomial coefficients
$\alpha, \beta, \gamma, \ldots, a, b, c, \ldots$	true and estimated regression coefficients (correlation case)
a', b'	regression coefficients with respect to transformed variables
$\boldsymbol{a}, \boldsymbol{b}, \boldsymbol{c}, \ldots$	scale readings (accumulated values of y)
A, B, C, D	coefficients used to solve multiple regression normal equations
A, B, C, D, D'	coefficients in orthogonal regression equations

β	Weiss constant
c (subscript)	cubic; calculated
c_p, C_p	specific heat, molar heat capacity at constant pressure
γ	Curie constant
c_{11}, c_{12} etc.	Gauss multipliers
D, D'	determinants
ΔG	Gibbs' free energy
Δ^j	jth order difference
e_y	error of a y value, the deviation $(y - \mu)$
E_A	apparent or Arrhenius energy of activation
f	degree of freedom; class interval frequency or frequency density
h	class interval width; spacing of x values in orthogonal curvilinear regression
j	highest power of x in a polynomial expression; number of independent variables
k	number of variables in a multiple regression; number of equations
k	velocity constant or reaction rate coefficient
K	a quantity used in evaluating $s_{\hat{y}}$
χ_M	molar magnetic susceptibility
l, $\hat{l}$	actual and estimated numerical coefficient of an equation
$\hat{L}$	the deviation, $l - \hat{l}$
l	(subscript) linear
μ, sometimes $\mu_{y \cdot x}$	true mean of y values at given x
$\mu_{y \cdot xz}$ etc.	true mean of y values at given x and z etc.
$\mu_{y \cdot x(z)}$ etc.	true mean of y values for given x values, all at fixed z value etc.
p	gas or vapour pressure
P	level of significance
q	a sum of variable values used in checking; (as subscript) quadratic
r	degree of replication
r_{xy} etc.	ordinary or crude correlation coefficient between x and y etc.
$r_{xy \cdot z}$ etc.	partial correlation coefficient between x and y, z constant etc.

$r_{y \cdot xz}$ etc.	multiple correlation coefficient between y and x plus z etc.
R	universal gas constant
ϱ	density
σ_a, s_a etc.	true and estimated standard error of constant a etc.
σ_e, s_e	true and estimated standard error of estimate (regression case)
σ'_e, s'_e	true and estimated standard error of estimate with respect to a transformed model
$\sigma_{y \cdot xz}$, $s_{y \cdot xz}$ etc.	true and estimated standard errors of estimate (correlation case)
S_1, S_2 etc.	successive accumulations of y values
t (subscript)	trial in iterative calculation
t	time
T	total
T	temperature
u, v	random errors of x and y values
V	volume
w	range; a variable in multiple correlation
x	a variable in multiple correlation; independent variable in regression
$\mathfrak{x}$	coded value of x, particularly in orthogonal regression
x', y'	transformed values of x and y
x_0, y_0	values of x and y chosen for rectification of a parabola
X, Y etc.	deviations of x, y values etc. from their arithmetic means
ξ_1, ξ_2 etc.	functions of x in orthogonal polynomials
y	a variable in multiple correlation; dependent variable in regression
$\hat{y}$	value of y predicted by an equation
$\hat{y}_l$, $\hat{y}_q$, $\hat{y}_c$	values of y predicted by linear, quadratic and cubic equations
$\bar{\bar{y}}$	a grand mean
$\hat{Y}$	the deviation, $y - \hat{y}$, being the estimate of e_y
z	a variable in multiple correlation; transformed value of y, $y - \alpha$

CHAPTER VI

SMOOTHING, INTERPOLATION, DIFFERENTIATION AND INTEGRATION

IN THE two preceding chapters it has been shown that associated data can be treated by least squares or simpler but equivalent methods to determine the nature and the degree of the relationship between the variables. Because the methods described there lead to the reduction of data to equation form, ancillary operations such as interpolation, differentiation and integration can be carried out by ordinary analytic procedures which normally present little difficulty. When these operations are, however, the main end of data treatment it is often useful to be able to carry them out without going through the intervening stage of equation fitting. In this chapter we review some of the available methods but since they depend in general on the absence of random error in disturbing degree we may require, as a preliminary tool, a means of removing residual scatter. We turn first to this problem.

6.1. THE SMOOTHING OF DATA

Smoothing (the terms *graduation* and *adjustment* are also used in this connection) is the process described above as a procedure for removing the effects of random error so as to replace the scattered values of a dependent variable by others which fall on a smooth curve. Further, since the observations are ordinarily a sample of many possible such values, the smoothed value is the best estimate, obtained in a particular way, of a presumed true value at the same point on the scale of the independent variable. The operation, therefore, does not differ in intent from that of fitting a least squares line or curve although its immediate purpose may simply be the provision of acceptable values for tabulation or subsequent mathematical operations.

(1) *Direct Methods*

The most obvious method of smoothing is to draw a free-hand curve through the graphed points in such a way as to minimize the deviations of the points from the smooth values. Some discussion of simple graphical methods is contained in Chapter I and the principle of smoothing suggested here might be described as that of the "first moment fitting" of Section 1.8.

If the data are so scattered that even approximate curve drawing is difficult, a first attempt to refine the crude representation obtained by plotting the points (and perhaps joining them by line segments) consists in dividing them into groups through each of which a straight line may be drawn. If these lines are fitted together the representation takes the form of a small number of connected segments between which there are discontinuous changes of slope. However, these changes are neither so frequent nor so abrupt as those between segments linking the points themselves and, where the data are few and scattered, this simple type of representation and the resultant smoothing may on occasion be adequate.

The other direct method of obtaining smooth values from raw data, the application of the fitting methods of Chapter IV and more particularly Chapter V, implies the principle of least squares or "second moment fitting" as the criterion of smoothness. Since the procedure requires the operator to choose the form of the equation to be fitted a difficulty arises in that the accuracy of the final values depends to some extent on the wisdom of this choice. Thus, apart from avoiding excessive computational work (one object of the present account), it is desirable to be able to call on methods which do not require decisions as to model to be made but at the same time, more or less automatically, result in the best possible representation of the data with minimization of the degree of scatter.

(2) *Difference Methods*

More generally useful than the direct graphical methods are those in which graphical smoothing is applied in the first instance to first or second order differences derived from the original values.

These quantities were introduced for the purposes of the tabular suitability test in Section 5.3 above. Explicitly: let x be the independent and y the dependent variable and suppose a table be constructed with the values of x ascending (Table 6.5 below is a general example). From the successive values so arranged we can find the *first order differences*, Δx and Δy, thus

$$\Delta y_0 = y_1 - y_0, \; \Delta y_1 = y_2 - y_1,$$

and in general

$$\Delta y_i = y_{i+1} - y_i \tag{6.1a}$$

and, similarly,

$$\Delta x_i = x_{i+1} - x_i \tag{6.1b}$$

where the subscripts 0, 1, 2 etc. refer to values in order of increasing x. (In what follows x_0 and y_0 may be reference values and not necessarily the first pair of values in a set. The notation is commonly used and we adopt it for the purposes of this chapter.) Further, from the first order differences, *second order differences* can be obtained. For example,

$$\Delta^2 y_0 = \Delta y_1 - \Delta y_0, \; \Delta^2 y_1 = \Delta y_2 - \Delta y_1,$$

and in general

$$\Delta^2 y_i = \Delta y_{i+1} - \Delta y_i \tag{6.2}$$

and so on.

From differences such as these we can obtain *divided differences* of the form $\Delta y/\Delta x$ which estimate approximately the corresponding derivatives. Thus, $\Delta y_0/\Delta x_0$ is an average value of dy/dx over the range $x_0 < x < x_1$, $\Delta y_1/\Delta x_1$ over the range $x_1 < x < x_2$ and so on. If the divided differences are plotted against the mid-points of the x intervals, i.e. $\Delta y_0/\Delta x_0$ at $(x_0 + x_1)/2$, $\Delta y_1/\Delta x_1$ at $(x_1 + x_2)/2$ etc., the points will give a graph in which the scatter is somewhat worse than that in the original x versus y graph. (In fact it can be shown that the variance of first order differences is twice, of second order differences six times and of third order differences twenty times the error variance of the original y values.)

In spite of this increased scatter difference graphs can generally be smoothed. One way is by division of the points into groups, drawing line segments of best fit for each group and then joining all such lines together. Since the divided differences represent the derivatives the resultant line or set of lines represents a plot of dy/dx versus x and, since this is continuous but with broken changes of slope, it implies a curve for y versus x in the form of a series of parabolic arcs joined together and with the slopes continuous over the whole range of the independent variable. Alternatively, and more usually if the scatter in the differential graph be not unduly great, it is reasonable to attempt to draw a smooth curve through the points either free-hand or with the aid of a French curve or other instrument. In this case, nothing is implied as to the variation of higher order divided differences (or as to the form of the curve through the original values) except that it should be regular, this being the criterion of smoothness adopted.

Of several possible ways of proceeding, the steps suggested here are (a) find and plot the first order divided differences, (b) draw through the points a curve or set of lines as described above in such a way that the values read from it average or total the same as the raw divided differences (i.e. a first moment fit involving some trial and error), (c) multiply the new divided differences, $\Delta y/\Delta x$(est), by Δx to obtain the corresponding values of Δy, Δy(est), (d) apply the latter to any arbitrarily chosen original value, say y_0, to obtain a trial set of smoothed values, and (e) adjust these by an amount equal to the difference between their mean and the mean of the

TABLE 6.1

SMOOTHING UNEQUALLY SPACED VALUES—LATENT HEATS OF VAPORIZATION OF CO_2

x (T°C)	Δx	$\bar{x}$	y (l, cal/g)	Δy	$\Delta y/\Delta x$	$\Delta y/\Delta x$ (est)	Δy (est)	$\hat{y}'$	$\hat{y}$	$\hat{Y}$
16·0			40·0					40·0	39·8	
	2·0	17·0		2·0	1·0	1·05	2·1			+0·2
18·0			38·0					37·9	37·7	
	3·0	19·5		4·2	1·4	1·20	3·6			+0·3
21·0			33·8					34·3	34·1	
	2·0	22·0		2·8	1·4	1·45	2·9			−0·3
23·0			31·0					31·4	31·2	
	1·0	23·5		2·0	2·0	1·70	1·7			−0·2
24·0			29·0					29·7	29·5	
	2·0	25·0		3·8	1·9	2·00	4·0			−0·5
26·0			25·2					25·7	25·5	
	1·5	26·75		3·2	2·13	2·51	3·8			−0·3
27·5			22·0					21·9	21·7	
	0·5	27·75		1·7	3·4	3·00	1·5			+0·3
28·0			20·3					20·4	20·2	
	1·0	28·5		3·5	3·5	3·42	3·4			+0·1
29·0			16·8					17·0	16·8	
	0·5	29·25		1·8	3·6	4·00	2·0			0·0
29·5			15·0					15·0	14·8	
										−0·1
Total				25·0	20·33	20·33	25·0			−0·1
Mean			27·11					27·33		

Note. According to eqn. (6.1a) the first order differences in the ys and the divided differences should be negative. They appear in the table without sign since their direction of application is obvious.

original set of values. In this way the residual differences between the smoothed and raw values are forcibly adjusted to zero, the final values of y being disposed about the same mean and within nearly the same limits as the raw values to minimize the overall risk of systematic error. Long runs of positive and negative residuals suggest (computational errors apart) over-smoothing in the face of a real change in the form of relationship between y and x or that the errors in the y values contain some component which is additional to ordinary random error.

The various steps in the method described in the last paragraph are applied in Table 6.1 and illustrated in Fig. 6.1 with respect to

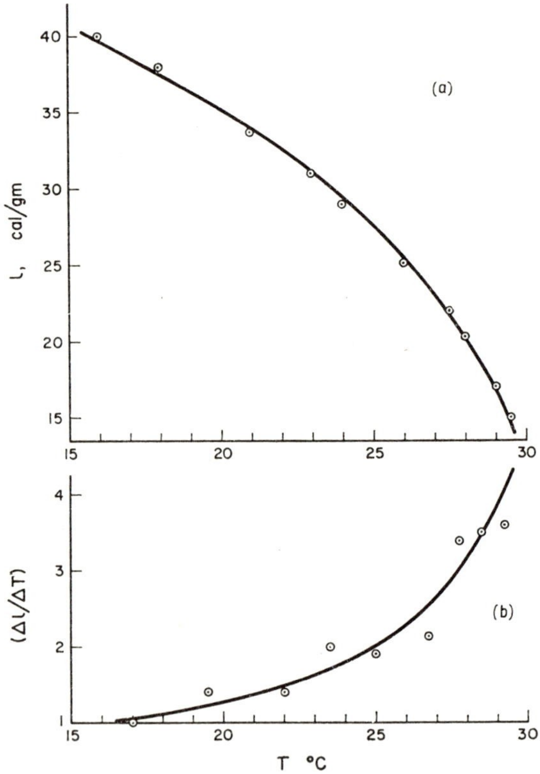

FIG. 6.1. Smoothing by Divided Differences. The Latent Heat of CO_2.

The upper graph shows the experimental points together with a curve drawn through the smoothed values. The lower graph shows the divided differences plotted against the intermediate temperatures. (It is of some assistance to know that l approaches zero asymptotically at about 31°C.)

some data for the latent heat of carbon dioxide at temperatures close to the critical temperature. In this case a somewhat hazardous curve has been drawn through the divided difference points as shown in Fig. 6.1 (b) but the effect of errors in the fitting is reduced in the final result. The estimated item differences, Δy(est), found from the smooth values read from the curve, $\Delta y/\Delta x$(est), have been applied to $y_0 = 40{\cdot}0$ cal/g to obtain the trial set, $\hat{y}'$. Since the latter average 0·2 units higher than the original values, this is the correction applied to yield the final results, $\hat{y}$.

Smoothing methods are generally described only for equally spaced values of the argument and if the original data are themselves reasonably smooth but not equally spaced this situation can sometimes be produced by relatively simple interpolation. In either event, the method described above can be simplified by working directly with the item differences, divided differences being no longer required. The steps now called for are (a) find and plot the Δy values, (b) smooth them by fitting a suitable curve or set of lines so that the new values, Δy(est), have the same average or total as the original, (c) correct an arbitrarily chosen value of y to obtain the trial set, $\hat{y}'$, and (d) adjust the latter by the discrepancy between $\bar{y}$ and $\bar{\hat{y}}'$. If the curve obtained in step (b) is truly smooth (having been obtained, say, by fitting a polynomial) the higher order differences will also be smooth but this is not to be expected of an ordinarily constructed curve and the procedure may be repeated to smooth the second order differences, correct the first order differences and finally correct the original values.

This procedure has been carried out on the data of Table 6.2. These data, given in the first two columns of the table, consist of the converging wavelengths, $\lambda = y$, of lines taken in order, n $= x$, in the absorption spectrum of iodine. The values are already so smooth that no direct graphical curve fitting can produce any sensible improvement and, indeed, any further operations will lead to an additional significant figure. However, the first order differences, Δy in the third column of the table, show some irregularities and smoothed values read from an approximately fitted curve are given in the fourth column. These have been applied as before to $y_0 = 5291$ Å to give a trial set, $\hat{y}'$, and the latter corrected to $\hat{y}$ by the discrepancy between the means, 0·5 Å. Because of the comparative roughness of the adjustment of the first order differences, the second order differences, $\Delta^2 y$ given in the eighth column of the table, still show some irregularities. They have been smoothed by the same graphical method and used to readjust the first order differences. The new values, Δy(est. 2), have been used to derive the finally adjusted set, $\hat{y}_2$, in the same way as before. The sums of the residuals between the smoothed and the original values, designated $\hat{Y}$ and $\hat{Y}_2$, differ from zero by a small rounding error.

TABLE 6.2. SMOOTHING EQUALLY SPACED OBSERVATIONS—ABSORPTION SPECTRUM OF IODINE VAPOUR

x (n)	y (λ, Å)	Δy	Δy (est. 1)	$\hat{y}'$	$\hat{y}$	$\hat{Y}$	$\Delta^2 y$	Δy (est. 2)	$\hat{y}_2$	$\hat{Y}_2$
1	5291			5291·0	5290·3	−0·7			5290·1	−0·9
		18	17·6					17·3		
2	5273			5273·4	5272·7	−0·3	1·0		5272·8	−0·2
		16	16·6					16·5		
3	5257			5256·8	5256·1	−0·9	0·7		5256·3	−0·7
		16	15·9					15·8		
4	5241			5240·9	5240·2	−0·8	0·9		5240·5	−0·5
		16	15·0					15·1		
5	5225			5225·9	5225·2	+0·2	0·7		5225·4	+0·4
		14	14·3					14·4		
6	5211			5211·6	5210·9	−0·1	0·5		5211·0	0·0
		15	13·8					13·8		
7	5196			5197·8	5197·1	+1·1	0·7		5197·2	+1·2
		12	13·1					13·2		
8	5184			5184·7	5184·0	0·0	0·6		5184·0	0·0
		12	12·5					12·6		
9	5172			5172·2	5171·5	−0·5	0·5		5171·4	−0·6
		12	12·0					12·0		
10	5160			5160·2	5159·5	−0·5	0·5		5159·4	−0·6
		13	11·5					11·5		
11	5147			5148·7	5148·0	+1·0	0·5		5147·9	+0·9
		11	11·0					11·0		
12	5136			5137·7	5137·0	+1·0	0·5		5137·9	+1·9
		10	10·5					10·5		
13	5126			5127·2	5126·5	+0·5	0·5		5126·4	+0·4
		10	10·0					10·0		
14	5116			5117·2	5116·5	+0·5	0·4		5116·4	+0·4
		9	9·6					9·6		
15	5107			5107·6	5106·9	−0·1	0·5		5106·8	−0·2
		9	9·1					9·2		
16	5098			5098·5	5097·8	−0·2	0·3		5097·6	−0·4
		9	8·8					8·8		
17	5089			5089·7	5089·0	0·0	0·4		5088·8	−0·2
		9	8·4					8·4		
18	5080			5081·3	5080·6	+0·6	0·4		5080·4	+0·4
		8	8·0					8·1		
19	5072			5073·3	5072·6	+0·6	0·2		5072·3	+0·3
		7	7·8					7·8		
20	5065			5065·5	6064·8	−0·2	0·2		5064·5	−0·5
		7	7·6					7·5		
21	5058			5057·9	5057·2	−0·8			5057·0	−1·0
Total						+0·4				+0·1
Mean	5157·3			5157·8	5157·3				5157·3	

Notes. Differences are given without sign. Both series are negative according to the convention of eqn. (6·1) and (6·2). $\hat{y}$ values are derived from the approximately smoothed first order differences, Δy (est. 1). $\hat{y}_2$ and Δy (est. 2) take into account the second order differences but the intervening adjustments are not given in the table.

Another technique, perhaps more satisfactory from a theoretical point of view, involves graphical integration of an approximation to the differential curve for the data. The required approximation is obtained by erecting on the x axis a series of rectangles having $\Delta y/\Delta x$ as height and the appropriate x interval as base. To avoid lengthy calculations, the rectangles can be grouped and a sloping line drawn from the first to the last boundary in the first group in such a way that the trapezoidal area under the sloping line is equal to the sum of the areas of the rectangles in the group. This condition may be assured by equality in area of the triangles above and below the sloping line, a condition which can be judged fairly accurately by eye. A second trapezoid is now constructed by drawing through the tops of the second group of rectangles another sloping line, this being joined at the boundary to the first line and adjusted in slope so that the algebraic sum of the triangular areas cut off from the corners of the rectangles is zero as before. The process is continued until all the groups of rectangles have been similarly treated.

Ideally, the area under the differential curve from the initial value up to value x is

$$\int_{x_0}^{x} (\mathrm{d}y/\mathrm{d}x)\, \mathrm{d}x = y - y_0 \tag{6.3}$$

A degree of approximation is, of course, introduced when divided differences are used in place of the true derivatives and smoothed by linked straight lines and a small bias may be introduced by progressive under- or over-correction in judging the equality of the triangular areas cut off. Let the bias be equivalent to a multiplier a, then

$$a\int_{x_0}^{x} (\mathrm{d}y/\mathrm{d}x)_{\mathrm{est}}\, \mathrm{d}x = y - y_0 \tag{6.4}$$

where $(\mathrm{d}y/\mathrm{d}x)_{\mathrm{est}}$ are the approximations to $\mathrm{d}y/\mathrm{d}x$, i.e. the smoothed values of the divided differences. The value of a can be estimated by summing the differences, $y - y_0$, over all the experimental values and dividing this sum by the accumulations of integrals up to the various values of x, viz.

$$a = \Sigma\,(y - y_0)/\Sigma\left(\int_{x_0}^{x} (\mathrm{d}y/\mathrm{d}x)_{\mathrm{est}}\, \mathrm{d}x\right) \tag{6.5}$$

This can be used as a correcting factor to adjust the areas under the differential curve from which the final smoothed values are obtained.

To return to the data for the molecular absorption spectrum of iodine, we have $\Delta n = \Delta x$ constant and unity so $\Delta\lambda = \Delta y/\Delta x$ exactly in this case. The values are given again in the third column of Table 6.3 and shown as the heights of the rectangles in Fig. 6.2. For convenience, the data have been divided into three groups at n = 7 and n = 14 and the successively fitted sloping lines are marked AB, BC and CD in the figure. The areas beneath these lines and between the ordinates erected at the various values of x are given in the fourth column of the table and their successive sums, i.e. the areas equivalent to integration of the smoothed item differences up to particular values of x, in the fifth. The sum of the

TABLE 6.3. SMOOTHING BY THE PROGRESSIVE AREA METHOD—WAVELENGTHS, IODINE ABSORPTION SPECTRUM

n	λ	$\Delta\lambda$	Successive areas	Progressive areas	Corrected progressive areas
1	5291				
2	5273	18	17·8	17·8	17·6
3	5257	16	17·1	34·9	34·6
4	5241	16	16·3	51·2	50·7
5	5225	16	15·5	66·7	66·0
6	5211	14	14·7	81·4	80·6
7	5196	15	13·9	95·3	94·4
8	5184	12	13·2	108·5	107·4
9	5172	12	12·6	121·1	119·9
10	5160	12	12·0	133·1	131·8
11	5147	13	11·5	144·6	143·2
12	5136	11	10·9	155·5	154·0
13	5126	10	10·3	165·8	164·2
14	5116	10	9·8	175·6	173·9
15	5107	9	9·3	184·9	183·1
16	5098	9	9·0	195·9	193·6
17	5089	9	8·7	204·6	202·6
18	5080	9	8·3	212·9	210·8
19	5072	8	8·0	220·9	218·7
20	5065	7	7·7	228·6	226·3
21	5058	7	7·3	235·9	233·6
	108,304		235·9	2835·2	2807·0

progressive areas is 2835·2 and this differs from $\Sigma(y - y_0) = 2807$ (conveniently found as the sum of the y values less twenty-one times y_0) by a factor of $a = 2807/2835{\cdot}2 = 0{\cdot}990$ by eqn. (6.5). The smoothed differences, corrected by this factor, are given in the last column of Table 6.3. The smoothed values of the variable can now be constructed as before (they are not given in the table).

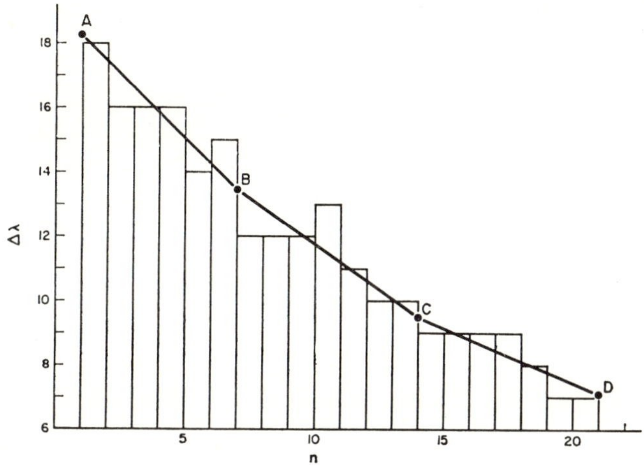

FIG. 6.2. Smoothing by the Progressive Area Method. The Absorption Spectrum of I_2.

(3) *The Least Squares Method for Equally Spaced Values*

A quite different principle for smoothing to the foregoing is to assume that errors in a group of neighbouring points tend to cancel and that their trend can be adequately approximated by a polynomial of low degree. If this is so, we can use as the smoothed value one found by the application of the principle of least squares to the value to be smoothed and its neighbours.

The lowest degrees of polynomial are the zeroth and first so one might use the *moving average* or "moving line". (Since the least squares line passes through the mean point of a set of values, these two are identical if the x values are equally spaced.) In this case the moving average is

$$\hat{y}_0 = (y_{-j} + \cdots y_{-1} + y_0 + y_1 + \cdots y_j)/(2j + 1) \qquad (6.6)$$

where y_0 is now the value to be smoothed and y_{-j} and y_j are the jth above and below it in a table of values equally spaced according to the independent variable so that the total number of values taken into account is $2j + 1$.

Alternatively, one might fit a quadratic or cubic relationship over the $2j + 1$ values and take the fitted value at x_0 as the smoothed value required. For example, if the quadratic form is used, it can be shown that for $j = 2$ (i.e. the *five-point* value),

$$\hat{y}_0 = [17y_0 + 12(y_{-1} + y_1) - 3(y_{-2} + y_2)]/35 \qquad (6.7\,\text{a})$$

For convenience in evaluation, this equation can be recast in terms of the smoothing correction to y_0 and the differences between y_0 and its neighbours, $\Delta y_1 = y_1 - y_0$ etc. The equation now becomes

$$\Delta\hat{y}_0 = [12(\Delta y_{-1} + \Delta y_1) - 3(\Delta y_{-2} + \Delta y_2)]/35 \qquad (6.7\,\text{b})$$

TABLE 6.4. SMOOTHING EQUALLY SPACED VALUES—WAVELENGTHS, IODINE ABSORPTION SPECTRUM

x (n)	y (λ, Å)	$\hat{y}$ (1)	$\hat{Y}$ (1)	$\hat{y}$ (2)	$\hat{Y}$ (2)	$\hat{y}$ (3)	$\hat{Y}$ (3)
1	5291			5291·0	0·0	5290·9	+0·1
2	5273			5273·1	−0·1	5273·3	−0·3
3	5257	5257·4	−0·4	5256·8	+0·2	5256·7	+0·3
4	5241	5241·4	−0·4	5240·8	+0·2	5240·8	+0·2
5	5225	5226·0	−1·0	5225·4	−0·4	5225·3	−0·3
6	5211	5211·4	−0·4	5210·3	+0·7	5210·5	+0·5
7	5196	5197·6	−1·6	5196·6	−0·6	5196·7	−0·7
8	5184	5184·6	−0·6	5183·7	+0·3	5184·0	0·0
9	5172	5171·8	+0·2	5171·8	+0·2	5171·6	+0·4
10	5160	5159·8	+0·2	5159·7	+0·3	5159·5	+0·5
11	5147	5148·2	−1·2	5147·3	−0·3	5147·6	−0·6
12	5136	5137·0	−1·0	5136·0	0·0	5136·2	−0·2
13	5126	5126·4	−0·4	5125·8	+0·2	5125·7	+0·3
14	5116	5116·6	−0·6	5116·2	−0·2	5116·2	−0·2
15	5107	5107·2	−0·2	5106·9	+0·1	5107·0	0·0
16	5098	5098·0	0·0	5098·0	0·0	5097·8	+0·2
17	5089	5089·2	−0·2	5088·9	+0·1	5088·9	+0·1
18	5080	5080·8	−0·8	5080·1	−0·1	5080·2	−0·2
19	5072	5072·8	−0·8	5072·1	−0·1	5072·1	−0·1
20	5065			5064·9	+0·1	5064·7	+0·3
21	5058			5058·1	−0·1	5058·1	−0·1
Total			−9·2		+0·5		+0·2

($\hat{y}$(1) = 5-point moving average; $\hat{y}$(2) and $\hat{y}$(3) = 5- and 7-point polynomials)

For $j = 3$ (i.e. for the *seven-point* value),

$$\hat{y}_0 = [7y_0 + 6(y_{-1} + y_1) + 3(y_{-2} + y_2) - 2(y_{-3} + y_3)]/21 \quad (6.8)$$

Worthing and Geffner (1943) derive the eqns. (6.7) and Whittaker and Robinson (1944) give a general derivation and many additional formulae.

A disadvantage of equations of the form of (6.7) and (6.8) is that they do not cover the first and the last j values of the whole series. However, Guest (1961) gives a method of supplementary polynomial fitting whereby these can be obtained if need be, although estimates of the terminal values are not very reliable. For $j = 2$,

$$\hat{y}_{-2} = (69y_{-2} + 4y_{-1} - 6y_0 + 4y_1 - y_2)/70 \quad (6.9\text{a})$$

and

$$\hat{y}_{-1} = (4y_{-2} + 54y_{-1} + 24y_0 - 16y_1 + 4y_2)/70 \quad (6.9\text{b})$$

where y_{-2} is the initial or final value of the whole set, y_{-1} is the second or second last and y_0 is the third or third last, i.e. the most extreme values covered by eqns. (6.7a) or (6.7b). For $j = 3$ the equations for the first, second and third values from either end of the series are

$$\hat{y}_{-3} = (39y_{-3} + 8y_{-2} - 4y_{-1} - 4y_0 + y_1 + 4y_2 - 2y_3)/42 \quad (6.10\text{a})$$

$$\hat{y}_{-2} = (8y_{-3} + 19y_{-2} + 16y_{-1} + 6y_0 - 4y_1 - 7y_2 + 4y_3)/42 \quad (6.10\text{b})$$

$$\hat{y}_{-1} = (-4y_{-3} + 16y_{-2} + 19y_{-1} + 12y_0 + 2y_1 - 4y_2 + y_3)/42 \quad (6.10\text{c})$$

where y_0 is now the most extreme value covered by eqn. (6.8).

As an example of the application of these equations consider the initial wavelength of Table 6.4. Using the five-point polynomial and smoothing by eqn. (6.9a) we have

$$\hat{y}_{-2} = (69 \times 5291 + 4 \times 5273 - 6 \times 5257 + 4 \times 5241 - 5225)/70$$
$$= 5290{\cdot}97$$

Five-point moving averages and five- and seven-point polynomial values for the wavelengths of the iodine spectrum are given in Table 6.4. In the cases of the polynomials the terminal values are filled in using eqns. (6.9) and (6.10) but similar end corrections cannot be devised satisfactorily to supplement moving averages when the data are markedly curvilinear.

The least squares method has less smoothing effect than those previously described unless many points are taken into account but suitable variants of the method are relatively easy to apply with the aid of a small calculator, they involve no subjective judgements and make no demands on principle beyond the unexceptionable ones mentioned above. It is worthy of remark in this connection that the residuals between smoothed and original values are of less significance than is the case in the comparison of equations fitted to all the data. Small residuals, however, often imply inadequate smoothing while large residuals need not necessarily be an indication of over-smoothing. Their examination is an aid in detecting systematic error such as is shown by the moving averages in the third column of Table 6.4.

(4) *Other Methods*

Many other methods of smoothing have been devised. The reader should consult Whittaker and Robinson (1944) for a very elaborate and theoretically satisfactory procedure based on considerations of probability and Yule and Kendall (1950) for methods of treating frequency distribution data.

6.2. Interpolation and Extrapolation

If the values of a dependent variable, y, are known for specified values of the independent variable, x, say at $x_0, x_1, x_2, \ldots, x_n$, the process of finding the appropriate y value at some unlisted value of x within the range x_0 to x_n is called *interpolation* and that of finding a value of y for one of x outside the range is called *extrapolation*. *Inverse interpolation* is the process of finding a value of x corresponding to a given value of y. Many methods are available for interpolation but there can be no completely satisfactory method for extrapolation in the absence of exact knowledge of the relationship between x and y.

Interpolation and smoothing are obviously related processes and the methods for each bear a close similarity.

(1) *Direct Methods*

For rapid and repeated interpolation, either direct or inverse, over the whole range of the independent variable a graph or an equation is most suitable.

Graphing is the most obvious and direct method and consists simply in plotting the tabulated data and drawing a smooth curve through the points by eye with regard to the usual precautions. The desired value of y can then be read off at the appropriate value of x or vice versa. The method is very rapid but its accuracy depends on the ability of the worker to draw a fair curve through the points at his disposal and unfortunately his ability to do this is diminished when the points are scattered enough to justify this procedure in favour of better but more tedious methods. Extrapolation can only be carried out graphically with any degree of confidence when the data at the relevant end of the recorded range approach linearity or if rectification is feasible.

Undoubtedly better than graphing is to fit an empirical equation (or one of a form theoretically indicated should such be available) to the data by least squares and to interpolate by substitution as required.

(2) *Proportional Parts and Three-point Interpolation*

Less time consuming than fitting an equation to the whole of a set of data is to adopt an empirical expression, say a polynomial of low degree, for the points embracing the position of interpolation. The simplest relationship which can be assumed is a linear one between adjacent values so that, in fact, the desired value is obtained by proportional parts.

If x_0, y_0 and x_1, y_1 are adjacent pairs of values between which it is desired to interpolate, then it can be shown that

$$\hat{y} = y_0 + (\Delta y_0/\Delta x_0)(x - x_0) = y_0 + k\Delta y_0 \qquad (6.11\text{a})$$

or, alternatively,

$$\hat{x} = x_0 + (\Delta x_0/\Delta y_0)(y - y_0) = x_0 + l\Delta x_0 \qquad (6.11\text{b})$$

where $\Delta x_0 = x_1 - x_0$ and $\Delta y_0 = y_1 - y_0$ as before and k and l are the relative positions of interpolation:

$$k = (x - x_0)/\Delta x_0; \quad l = (y - y_0)/\Delta y_0 \qquad (6.11\text{c, d})$$

Even when the relationship is curvilinear, eqn. (6.11) might give the required accuracy if the reference points be closely spaced but it is generally more realistic to assume a quadratic relationship between the neighbouring points, say x_0, y_0, x_1, y_1 and x_2, y_2. The required equation now has the form

$$\hat{y} = y_0 + b(x - x_0) + c(x - x_0)^2 \qquad (6.12)$$

If the point x_0, y_0 be taken as an arbitrary origin, then b and c can be found by solving

$$(y_1 - y_0) = b(x_1 - x_0) + c(x_1 - x_0)^2 \qquad (6.13a)$$

$$(y_2 - y_0) = b(x_2 - x_0) + c(x_2 - x_0)^2 \qquad (6.13b)$$

If Δx is constant, these expressions simplify to

$$b = (2y_1 - \tfrac{1}{2}y_2 - 1\tfrac{1}{2}y_0)/\Delta x \qquad (6.14a)$$

and

$$c = (\tfrac{1}{2}y_2 - y_1 + \tfrac{1}{2}y_0)/\Delta x^2 \qquad (6.14b)$$

For inverse interpolation similar equations can be written down with the positions of y and x reversed.

Consider the application of these expressions to find the vapour pressure of water at $x = \mathrm{T} = 75{\cdot}6\,°\mathrm{C}$ from the data of Table 6.6. Using the nearest two and the nearest three pairs of values respectively we have, from eqn. (6.11a),

$$\mathrm{p} = \hat{y} = 277{\cdot}2 + (24{\cdot}2/2)\,(75{\cdot}6 - 74) = 296{\cdot}56 \text{ mm}$$

and, from eqns. (6.14a) and (6.14b), $b = 11.675$ and $c = 0{\cdot}2125$ so, from eqn. (6.12),

$$\mathrm{p} = \hat{y} = 277{\cdot}2 + 11{\cdot}675 \times 1{\cdot}6 + 0{\cdot}2125 \times 2{\cdot}56 = 296{\cdot}42 \text{ mm}$$

Linear interpolation based on the known theoretical relationship, i.e. using the reciprocals of the temperatures in °K and four place logarithms of the vapour pressures, gives 296·4 mm as the required value.

(3) *The Gregory–Newton Method*

This method, first proposed by Gregory in 1670, is quick and accurate but it is limited in application to smooth equally spaced data and is less suitable than the methods above for repetitive work.

Consider a set of data in which the independent variable takes the regular values x_0, $x_1 = x_0 + \Delta x$, $x_2 = x_0 + 2\Delta x$ etc., the corresponding values of the dependent variable being y_0, y_1, y_2 etc. Now we can set up a table in the form of Table 6.5 in which Δy, $\Delta^2 y$ etc. are the first, second and higher order differences and the number of first order differences is n, one less than the number of points.

TABLE 6.5. SUCCESSIVE ITEM DIFFERENCES

x	y	Δy	$\Delta^2 y$	$\Delta^3 y$
x_0	y_0			
		Δy_0		
$x_0 + \Delta x$	y_1		$\Delta^2 y_0$	
		Δy_1		$\Delta^3 y_0$
$x_0 + 2\Delta x$	y_2		$\Delta^2 y_1$	
		Δy_2		$\Delta^3 y_1$
$x_0 + 3\Delta x$	y_3		$\Delta^2 y_2$	
		Δy_3		
....	..		....	
		...		
$x_0 + i\Delta x$	y_i		....	
		Δy_i		
....	..		$\Delta^2 y_i$	
		...		
....	..		....	
		Δy_{n-1}		
$x_0 + n\Delta x$	y_n			

It can be shown that any value of y, $\hat{y}$, is related to a reference value, say y_0, by the expression

$$\hat{y} = y_0 + k\Delta y_0 + [k(k-1)/2!]\,\Delta^2 y_0 + [k(k-1)(k-2)/3!]\,\Delta^3 y_0 \ldots \tag{6.15}$$

where k is the point of interpolation on the x scale in terms of spacing units from x_0 and is given by eqn. (6.11c). The equation is valid for all *positive* values of k whether fractional or integral.

Consider again interpolation in the data of Table 6.6 to find the vapour pressure at 75·6°C. Taking $x_0 = 74$ and $y_0 = 277{\cdot}2$ as the reference values we have, from eqn. (6.11c), $k = (75{\cdot}6 - 74)/2 = 0{\cdot}8$, so

$$\begin{aligned} p = \hat{y} &= 277{\cdot}2 + 0{\cdot}8 \times 24{\cdot}2 + (0{\cdot}8 \times -0{\cdot}2/2)\,1{\cdot}7 \\ &\qquad + (0{\cdot}8 \times -0{\cdot}2 \times -1{\cdot}2/6)\,0{\cdot}2 \\ &= 277{\cdot}2 + 19{\cdot}36 - 0{\cdot}136 + 0{\cdot}0064 = 296{\cdot}43 \text{ mm} \end{aligned}$$

when the third order difference is taken into eqn. (6.15). (The fourth order difference is obviously of negligible effect.) Had the substitution been taken only as far as the second order difference the result of the three-point interpolation, 296·42 mm, would have been obtained since the second order difference is the last in the quadratic relationship. Note that had p = 301·4 mm at 76°C been chosen as the reference point, $x - x_0$ would have taken the value −0·4 and to keep k a positive number it would have been necessary to reverse the order of the values, i.e. to tabulate in order of decreasing values of x so as to make Δx also negative. It should also be noted that there is no need for the reference points to be as close as possible to the position of interpolation; 72°C and 254·6 mm would have served as well as those chosen in Table 6.6. They lead to 296·41 mm for p at 75·6°C.

TABLE 6.6. INTERPOLATION—VAPOUR PRESSURE OF WATER

x (T °C)	y (p, mm)	Δy	$\Delta^2 y$	$\Delta^3 y$	$\Delta^4 y$
70	233·6				
72	254·6				
74	*277·2*				
		24·2			
76	301·4		*1·7*		
		25·9		*0·2*	
78	327·3		1·9		*0·1*
		27·8		0·3	
80	355·1		2·2		
		30·0			
82	385·1				

The Gregory–Newton expression can be used for inverse interpolation but only by successive approximation. Suppose we require the temperature at which the vapour pressure is 300 mm. Now $\Delta T/\Delta p$ in the relevant interval is 2/24·2 = 0·083 deg/mm so T = 76 − 1·4 × 0·083 ≑ 75·9° is a good first approximation to the value at 300 mm. Substituting this into eqn. (6.15), using the first four terms, we obtain a pressure of 300·15 mm. T is therefore still too high by about 0·15 × 0·083 = 0·012°. Further interpolation at 75·9 − 0·012 = 75·888° does not alter the third decimal and 75·89° might be taken as a reasonable value, considering likely rounding errors in the original data.

(4) *The Newton Method*

This method, of which the Gregory–Newton method is a special case, is based on divided differences and is applicable to unequally spaced data. The nature and interrelationship of the various orders of divided differences are shown in Table 6.7. In this table, for simplicity, the symbol Δ is used for a divided difference, the superscript indicating order above the first and the subscript the position of the difference in the set.

TABLE 6.7. SUCCESSIVE DIVIDED DIFFERENCES

x	y	Δ	Δ^2	Δ^3
x_0	y_0			
		$(y_1-y_0)/(x_1-x_0)=\Delta_0$		
x_1	y_1		$(\Delta_1-\Delta_0)/(x_2-x_0)=\Delta_0^2$	
		$(y_2-y_1)/(x_2-x_1)=\Delta_1$		$(\Delta_1^2-\Delta_0^2)/x_3-x_0)=\Delta_0^3$
x_2	y_2		$(\Delta_2-\Delta_1)/(x_3-x_1)=\Delta_1^2$	
		$(y_3-y_2)/(x_3-x_2)=\Delta_2$		
x_3	y_3			
...	...			

In terms of these differences, Newton's expression for the interpolated value is

$$\hat{y} = y_0 + (x - x_0)\Delta_0 + (x - x_0)(x - x_1)\Delta_0^2$$

$$+ (x - x_0)(x - x_1)(x - x_2)\Delta_0^3 + \dots \qquad (6.16)$$

where x is the point of interpolation and y_0 is a suitable reference value.

Consider again interpolation to find the vapour pressure of water at 75·6 °C but now with the available data unequally spaced as in Table 6.8. Applying eqn. (6.16) as far as the third order divided

TABLE 6.8. INTERPOLATION—VAPOUR PRESSURE OF WATER

x (T °C)	y (p, mm)	Δx	Δy	Δ	Δ^2	Δ^3
74	277·2					
		1	11·9	11·9		
75	289·1				1·3/6 = 0·217	
		5	66·0	13·2		0·04/8 = 0·005
80	355·1				1·8/7 = 0·257	
		2	30·0	15·0		
82	385·1					

difference as shown in the table, we obtain

$$\begin{aligned} p = \hat{y} &= 277{\cdot}2 + (75{\cdot}6 - 74)\,11{\cdot}9 + (75{\cdot}6 - 74)(75{\cdot}6 - 75)\,0{\cdot}217 \\ &\quad + (75{\cdot}6 - 74)(75{\cdot}6 - 75)(75{\cdot}6 - 80)\,0{\cdot}005 \\ &= 277{\cdot}2 + 19{\cdot}04 + 0{\cdot}208 - 0{\cdot}021 = 296{\cdot}43 \text{ mm} \end{aligned}$$

It is obvious that the fourth order difference term can only affect the third place of decimals.

(5) *Interpolation with Two Independent Variables*

It is sometimes necessary to interpolate in a table in which the values of the dependent variable are given for two arguments whose own values may or may not be equally spaced. Table 6.9 illustrates the situation in a general way and Table 6.10 contains some rounded values of the Joule–Thomson coefficient for carbon dioxide at a few temperatures and pressures.

TABLE 6.9

$x \backslash y$	y_0	y_1	y_2
x_0	z_{00}	z_{01}	z_{02}
x_1	z_{10}	z_{11}	...
x_2	z_{20}	...	...

Approximate values can be obtained by the method of proportional parts described above. In the two argument case we have

$$\hat{z} = z_{00} + k\Delta_x z + l\Delta_y z + kl\Delta^2_{xy} z \tag{6.17}$$

where k and l are the points of interpolation with respect to interval on the x and y scales given by eqns. (6.11c) and (6.11d), $\Delta_x z$ and $\Delta_y z$ are the differences between the z values in column and row,

TABLE 6.10. JOULE–THOMSON COEFFICIENTS FOR CO_2, μ (deg/atm)

T (°K)	p (atm)		
	10	40	60
300	1·084	1·018	0·968
325	0·908	0·802	0·723
350	0·785	0·678	0·602

$z_{10} - z_{00}$ and $z_{01} - z_{00}$, and $\Delta^2_{xy}z$ is the difference with respect to both arguments, $z_{00} + z_{11} - z_{01} - z_{10}$.

If we wish to find the Joule–Thomson coefficient for CO_2 at 320 °K and 30 atm by eqn. (6.17), we have

$$\begin{aligned}\mu = \hat{z} &= 1{\cdot}084 + (20/25)(0{\cdot}908 - 1{\cdot}084) \\ &\qquad + (20/30)(1{\cdot}018 - 1{\cdot}084) \text{ etc.} \\ &= 1{\cdot}084 - 0{\cdot}141 - 0{\cdot}044 - 0{\cdot}021 \\ &= 0{\cdot}878 \text{ deg/atm.}\end{aligned}$$

This result could also have been obtained by interpolating in stages, i.e. by finding the 30 atm values at 300° and 325° using eqns. (6.11 a), 1·040 and 0·837 respectively, and then interpolating between these two at 320° or, alternatively, by finding the 320° values at 10 and 40 atm and interpolating at 30 atm., it being immaterial which independent variable is taken first.

A better result, equivalent to fitting a curved (second order) surface over the points of Table 6.9, is given by a variant of the Newton expression, eqn. (6.16), which now takes the form (Whittaker and Robinson, 1944)

$$\begin{aligned}\hat{z} = z_{00} &+ [z_{00}/(x_0 - x_1) + z_{10}/(x_1 - x_0)](x - x_0) \\ &+ [z_{00}/(y_0 - y_1) + z_{01}/(y_1 - y_0)](y - y_0) \\ &+ [z_{00}/(x_0 - x_1)(x_0 - x_2) + z_{10}/(x_1 - x_0)(x_1 - x_2) \\ &\qquad + z_{20}/(x_2 - x_0)(x_2 - x_1)](x - x_0)(x - x_1) \\ &+ [z_{00}/(y_0 - y_1)(y_0 - y_2) + z_{01}/(y_1 - y_0)(y_1 - y_2) \\ &\qquad + z_{02}/(y_2 - y_0)(y_2 - y_1)](y - y_0)(y - y_1) \\ &+ [z_{00}/(x_0 - x_1)(y_0 - y_1) + z_{01}/(x_0 - x_1)(y_1 - y_0) \\ &+ z_{10}/(x_1 - x_0)(y_0 - y_1) + z_{11}/(x_1 - x_0)(y_1 - y_0)] \times \\ &\qquad \times (x - x_0)(y - y_0) \qquad (6.18)\end{aligned}$$

(Whittaker and Robinson also quote a variant of the Gregory–Newton expression, eqn. (6.15), for use when the x and y values are equally spaced.) Interpolating again in the data of Table 6.10, eqn. (6.18) gives 0·875 deg/atm for the Joule–Thomson coefficient of the gas at 320 °K and 30 atm. The first three terms of eqns. (6.18) and of (6.17) are the same.

(6) *Other Methods and Applications*

There are many other methods for interpolation (several having special application to mathematical and statistical tables) which are to be found in the references already cited. Special methods for treating frequency distribution data, e.g. for deriving class interval frequencies from mid-point values and for estimating quantiles, are given by Yule and Kendall (1950).

6.3. Graphical and Tabular Differentiation

Differentiation or the evaluation of the derivative of a function from available data is a common enough physico-chemical problem of whose type the determination of the rate of change of e.m.f. of an electrochemical cell with temperature to determine the entropy change of the cell reaction ($\Delta S = zF(\partial E/\partial T)_p$, z being the electronic transfer and F the Faraday) and the determination of partial specific volumes by the intercept method are sufficient examples. In many instances differentiation at a single point is all that is required. The first example is of this kind, the problem being generally to find dE/dT at some given temperature from e.m.f.-temperature values measured at constant pressure. In other cases the operation must be carried out over the whole range of available data as in the second example where the specific volume-composition curve for a liquid mixture is so treated.

(1) *Direct Methods*

The simplest direct method is the graphical procedure whereby the points are plotted, a fair curve drawn through them with the usual precautions, the tangent found at the point of interest and its slope determined.

The main difficulty lies in the location of the tangent. Roughly in order of accuracy, this can be done by (a) rotating a straight edge round the point so that the eye can gauge the equality of the angles between the edge and the curve on either side of the point, (b) using a front-surface mirror to find the normal to the tangent by placing the mirror edge at the point and rotating it until the curve up to the mirror and its reflection appear as one, continuous and smooth (the tangent slope being found as the reciprocal of the normal slope or by graphical means) or (c) by construction of chords in the vicinity of the tangent which approach the true tangent in slope. This last method is illustrated in Fig. 6.3. Chords are drawn through pairs of points, located with compasses so as to be equidistant from the point of interest, and extended to a point of intersection. A line drawn through the point of intersection and the point of interest on the curve approximates the tangent at the latter. Obviously, if the curvature is known to be uniform, the slope of the tangent can taken as that of the chord itself.

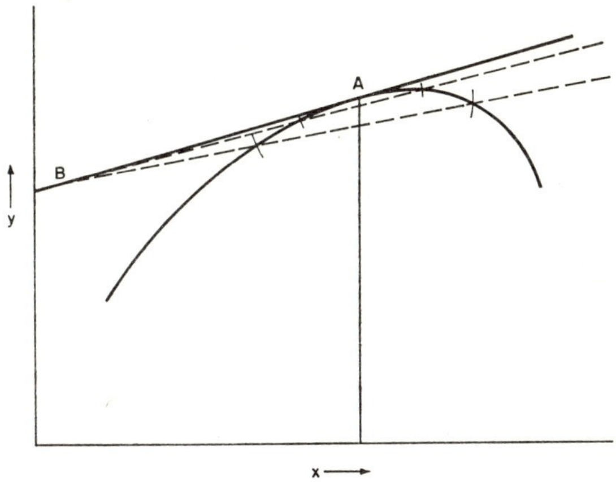

FIG. 6.3. Graphical Differentiation—The Chord and Tangent Method.
A is the point at which the tangent is required. *B* is the point of intersection of two close chords.

Careful plotting and smooth curve drawing is, of course, most necessary in graphical work. The tangent slope can be found most precisely when the tangent itself lies at 45° to the x axis and in the case of single determinations the scales can be chosen with this in view. The use of a clear plastic ruler with the index line scribed on its underside is also an aid to accuracy if the simple procedure, (a) above, is used.

If the form of the function underlying the data is known, or if it can be approximated by a polynomial of suitable degree, least squares curve fitting and subsequent differentiation is to be preferred for repetitive work.

(2) *Difference Methods*

For equally spaced values the Gregory–Newton expression, eqn. (6.15), can be differentiated to give

$$dy/dx = (1/\Delta x)[\Delta y_0 + (2k - 1)\Delta^2 y_0/2 + (3k^2 - 6k + 2)\Delta^3 y_0/6 + (4k^3 - 18k^2 + 22k - 6)\Delta^4 y_0/24 + \ldots] \quad (6.19)$$

where dy/dx is the estimate required, k is the point of differentiation on the x scale according to eqn. (6.11c) and Δy_0, $\Delta^2 y_0$ etc. are the successive item differences in relation to a chosen reference point, x_0, y_0, as shown in Table 6.5. By further differentiation an expression for the second derivative can be obtained, this being

$$d^2y/dx^2 = (1/\Delta x^2)[\Delta^2 y_0 + (k - 1)\Delta^3 y_0 + (6k^2 - 18k + 11)\Delta^4 y_0/12 \ldots] \quad (6.20)$$

$\Delta y_0/\Delta x$ and $\Delta^2 y_0/\Delta x^2$, divided differences in the range covering the position of differentiation, are seen to be the first approximations to the derivatives in each case.

When the values of the independent variable are unequally spaced modifications of the Newton expression, eqn. (6.16), can be used. For the first two derivatives in terms of the divided differences as shown in Table 6.7 we have

$$dy/dx = \Delta_0 + (m_0 + m_1)\Delta_0^2 + (m_0 m_1 + m_0 m_2 + m_1 m_2)\Delta_0^3 + \ldots \quad (6.21)$$

$$d^2y/dx^2 = 2[\Delta_0^2 + (m_0 + m_1 + m_2)\Delta_0^3 + (m_0 m_1 + m_0 m_2 + m_0 m_3 + m_1 m_2 + m_1 m_3 + m_2 m_3)\Delta_0^4 + \ldots] \quad (6.22)$$

where $m_0 = x - x_0$, $m_1 = x - x_1$ etc.

We illustrate the application of eqns. (6.19) and (6.21) using again the data of Tables 6.6 and 6.8. Wishing now to find the derivative, dp/dT, at 75·6°C taking 74°, 277·2 mm for reference as before, we have in the first case $k = 0{\cdot}8$ and

$$\begin{aligned} dp/dT &= dy/dx \\ &= \tfrac{1}{2}[24{\cdot}2 + (1{\cdot}6 - 1)\,1{\cdot}7/2 + (1{\cdot}92 - 4{\cdot}8 + 2)\,0{\cdot}2/6 \ldots] \\ &= 12{\cdot}1 + 0{\cdot}255 - 0{\cdot}015 \\ &= 12{\cdot}34 \text{ mm/°C} \end{aligned}$$

In the second case we have $m_0 = 75{\cdot}6 - 74 = 1{\cdot}6$, $m_1 = 75{\cdot}6 – 75 = 0{\cdot}6$, $m_2 = 75{\cdot}6 – 80 = -4{\cdot}4$ etc. So, substituting in eqn. (6.21),

$$\begin{aligned} \mathrm{d}p/\mathrm{d}T &= \mathrm{d}y/\mathrm{d}x \\ &= 11{\cdot}9 + 2{\cdot}2 \times 0{\cdot}217 + (0{\cdot}96 - 7{\cdot}04 - 2{\cdot}64)\, 0{\cdot}005 \\ &= 11{\cdot}9 + 0{\cdot}477 - 0{\cdot}044 \\ &= 12{\cdot}33 \text{ mm/°C.} \end{aligned}$$

(The slight difference between the two results is mainly due to an inconsistency between the two sets of original values introduced by rounding.)

(3) *A Semi-Graphical Method*

A difficulty which may arise in the practical application of the item and divided difference methods is that basic data are often not sufficiently well smoothed or free of random errors. Good smoothing is necessary to them since the effect of random fluctuation rapidly increases with the order of difference. It may, therefore, be preferable on occasion to adopt a semi-graphical procedure in which the normal tabulation method is used to find the dominant first order differences followed by a graphical method to find the corrections arising from the higher order differences. In this way the fluctuations deriving from the original values are removed without going through all the steps of a smoothing procedure.

The method is best described by way of an example. Consider the data for the density of water at various temperatures in Table 6.11 and suppose the coefficient of density increase is required at each of the tabulated temperatures. The steps of the method are: (a) draw up a table of first order divided differences, (b) plot them as height on their intervals on the independent variable scale as base so as to obtain a series of rectangles, (c) draw a smooth curve through the tops of the rectangles in such a way that the area beneath it is equal to the sum of the areas of the rectangles, and (d) read off the required values of the derivative from the curve. The curve in step (c) can be positioned by inspection of the triangular or nearly triangular areas cut off from the tops of the rectangles by it, the procedure being, in fact, virtually the same as that for smoothing described in sub-section 6.1(2) above. It is immaterial to the method whether or not the original values be equally spaced.

TABLE 6.11
DIFFERENTIATION—VARIATION OF THE DENSITY OF WATER WITH TEMPERATURE

T (°C)	ϱ (g/cm³)	$\Delta\varrho$	$\bar{T}$	$\Delta\varrho/\Delta T$	$\Delta\varrho/\Delta T$ (from graph)
0	0·99987				+0·000065
		+0·00013	2	+0·000032	
4	1·00000				0·000000
		−0·00012	6	−0·000030	
8	0·99988				−0·000060
		−0·00036	10	−0·000090	
12	0·99952				−0·000115
		−0·00055	14	−0·000138	
16	0·99897				−0·000162
		−0·00074	18	−0·000185	
20	0·99823				−0·000205
		−0·00091	22	−0·000228	
24	0·99732				−0·000245
		−0·00106	26	−0·000265	
28	0·99626				−0.000285
		−0·00121	28	−0·000308	
32	0·99505				−0·000320

Figure 6.4 illustrates the procedure described. Estimates of the derivatives, read from the curve, are given in the last column of Table 6.11. This process, step (d) of the suggested procedure, is simple graphical interpolation between the mean temperatures of the intervals.

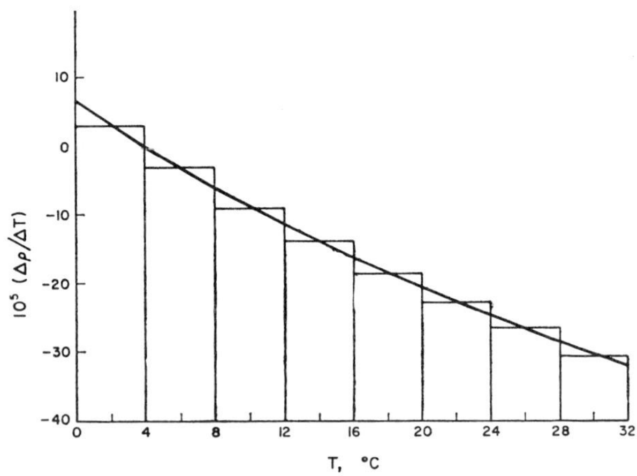

FIG. 6.4. Semi-graphical Differentiation. The Rate of Change of the Density of Water with Temperature.

6.4. Graphical and Tabular Integration

The problem of integrating experimental data is another which arises frequently in both pure and applied research and many techniques are available for carrying out the operation. A few of the more conventional and, perhaps, more generally convenient are described here.

(1) *The Graphical Method*

Basically, this consists of drawing the curve between the limits over which it is desired to integrate and measuring the area between curve and axis. The second step is usually the most difficult part of the procedure. One method is to count the squares beneath the curve on the graph paper on which it is drawn and this is accurate enough if the paper scale divisions are sufficiently small and time is not a consideration. On the other hand, by taking large enough divisions a quick approximation can also be obtained in this way. A second simple procedure is to cut out the area corresponding to the integral and weigh the piece on an analytical balance. The area is obtained by dividing the weight of the piece by the weight per unit area of the paper determined in a separate weighing. A better and perhaps generally quicker method than either of these is to find the average ordinate over the range of integration. It depends, however, on an accurate graphical construction based on a good curve. Worthing and Geffner (1943) should be consulted for details. Lastly, the area can be determined by mechanical devices such as the planimeter.

(2) *The Equation Method*

When an equation of amenable form can be fitted to the data, the definite integral can be found by the ordinary processes of integral calculus. The integration step is particularly simple when a polynomial is appropriate but the fitting process is much more time-consuming than any graphical procedure.

(3) *The Trapezoidal Rule and Simpson's Rules*

Undoubtedly the most convenient methods are those which may be used directly on the tabulated values and do not necessarily involve an intervening graphical or curve fitting step. The trapezoidal method is one such. In this, the region to be estimated is, in effect, divided into a series of smaller trapezia whose areas can be found from the product of width and average height and summed to obtain the whole. Thus, the first trapezium has as base $\Delta x_0 = x_1 - x_0$ and the average height of the bounding ordinates is $(y_0 + y_1)/2$ and so on, so the rule can be expressed by

$$\int_{x_0}^{x_n} y \, . \, dx = [\Delta x_0(y_0 + y_1)/2 + \Delta x_1(y_1 + y_2)/2 + \dots \Delta x_{n-1}(y_{n-1} + y_n)/2] \quad (6.23)$$

where $x_0, y_0 \dots x_n, y_n$ are the successive pairs of values, those of x being in order of magnitude. If the values are equally spaced, i.e. Δx is constant, eqn. (6.23) becomes

$$\int_{x_0}^{x_n} y \, . \, dx = \Delta x[(y_0 + y_n)/2 + y_1 + y_2 + y_3 + \dots y_{n-1}] \quad (6.24)$$

The assumption underlying the trapezoidal rule is that the curve for the data can be adequately represented by a series of linear segments between the given points, an approximation which may hold well enough for integration when the values are closely spaced with respect to the range of interest. In principle, more accurate expressions are obtainable by assuming that the curve for the data can be represented by curves of polynomials of the second, third and higher degrees over groups of neighbouring points. The rules known as Simpson's *one-third* and *three-eighth* rules, for example, are based on quadratic and cubic arcs respectively. They may be written

$$\int_{x_0}^{x_n} y \, . \, dx = (\Delta x/3)\,[(y_0 + y_n) + 4(y_1 + y_3 + y_5 + \dots) + 2(y_2 + y_4 + y_6 + \dots)] \quad (6.25)$$

and

$$\int_{x_0}^{x_n} y \, . \, dx = (3\Delta x/8)\,[(y_0 + y_n) + 2(y_3 + y_6 + y_9 + \dots) + 3(y_1 + y_2 + y_4 + y_5 + y_7 + y_8 + \dots)] \quad (6.26)$$

Eqn. (6.25) is applicable over an even number of intervals (an odd number of ordinates) and eqn. (6.26) over a number of intervals divisible by three. Δx must, of course, be constant.

It is generally assumed that eqn. (6.26) is the most and eqn. (6.24) the least accurate of these expressions but Jenson and Jeffreys (1963) show that the trapezoidal rule gives the best result when the data are not very smooth. Whittaker and Robinson (1944) quote theoretical corrections in terms of higher order differences. The correction to the trapezoidal rule, eqn. (6.24), another expression due to Gregory, is

$$-(\Delta x/12)(\Delta y_{n-1} - \Delta y_0) - (\Delta x/24)(\Delta^2 y_{n-2} + \Delta^2 y_0) - (19\Delta x/720)(\Delta^3 y_{n-3} - \Delta^3 y_0)$$

the item differences being those of Table 6.5.

Arbitrary curves based mainly on equally spaced data can often be most speedily integrated by using Simpson's *one-third* rule over an even number of intervals covering most of the range and the trapezoidal rule over the remainder. If any special function is known to apply over part of the range of the independent variable, it should of course be used. This is the situation in the following example.

The standard entropy, S^0, of an element can be found by integrating the quotient of atomic heat capacity and temperature from

TABLE 6.12
ATOMIC HEAT CAPACITY OF ALUMINIUM

T (°K)	C_p (cal/g-atom)	C_p/T
15	0·022	0·00147
30	0·203	0·00677
50	0·913	0·02005
70	1·851	0·02645
90	2·729	0·03032
110	3·451	0·03138
130	3·989	0·03075
150	4·427	0·02950
170	4·776	0·02810
190	5·045	0·02655
210	5·251	0·02485
230	5·418	0·02355
250	5·557	0·02225
270	5·677	0·02105
290	5·778	0·01991
298·2	5·817	0·01950

0 °K to 298·2 °K or 25 °C. Some data for aluminium are given in Table 6.12. These being equally spaced at 20° intervals except for the values at the beginning and end of the range, the required interval can be split so

$$S^0 = \int_0^{298\cdot2} (C_p/T)\,dT$$

$$= \int_0^{15} (C_p/T)\,dT + \int_{15}^{50} (C_p/T)\,dT$$

$$+ \int_{50}^{290} (C_p/T)\,dT + \int_{290}^{298\cdot2} (C_p/T)\,dT$$

$$= \text{Area}(1) + \text{Area}(2) + \text{Area}(3) + \text{Area}(4)$$

and evaluated as follows.

Area (1) by using the Debye expression for the low temperature variation of C_p, $C_p = kT^3$ where $k = 6\cdot51 \times 10^{-6}$ by substitution of 15° and 0·022 cal/g-atom:

$$\int_0^{15} (6\cdot51 \times 10^{-6})\,T^2\,dT \qquad = 0\cdot007 \text{ e.u.}$$

Area (2) by eqn. (6.23):

$$15(0\cdot00147 + 0\cdot00677)/2 + 20(0\cdot00677 + 0\cdot02005)/2 \quad = 0\cdot330 \text{ e.u.}$$

Area (3) by eqn. (6.25):

$$(20/3)\left\{\begin{Bmatrix}0\cdot02005\\+0\cdot01991\end{Bmatrix} + 4\begin{Bmatrix}0\cdot02645\\+0\cdot03138\\+0\cdot02950\\+0\cdot02655\\+0\cdot02355\\+0\cdot02105\end{Bmatrix} + 2\begin{Bmatrix}0\cdot03032\\+0\cdot03075\\+0\cdot02180\\+0\cdot02485\\+0\cdot02225\end{Bmatrix}\right\}$$

$$= 6\cdot310 \text{ e.u.}$$

Area (4) by eqn. (6.23):

$$8\cdot2(0\cdot01991 + 0\cdot01950)/2 \qquad = 0\cdot159 \text{ e.u.}$$

$$\text{Total} = 6\cdot806 \text{ e.u.}$$

Thus, the required standard entropy is 6·806 e.u. (entropy units or cal/g-atom.).

We are tempted to conclude this discussion with the remarks of Yule and Kendall concerning smoothing and interpolation (but which also apply to differentiation and integration at least in so far as they depend on the former or are carried out by similar methods). They describe these processes as being "as much artistic as scientific. No absolute rules can be laid down, judgment must be used, and it is the experienced craftsman who is likely to get the best results with the least labour".

Notation of Chapter VI

a	correcting factor
a, b, c	coefficients of a quadratic equation
$\mathrm{C_p}$	heat capacity at constant pressure
Δ, Δ^2 etc.	first, second etc. order divided differences
$\Delta x, \Delta y, \Delta^2 x$ etc.	first, second etc. order item differences
E	e.m.f.
F	the Faraday
j	number of values before and after y_0 used in smoothing
k	point of interpolation within an interval on the scale of x
k	a constant
l	point of interpolation within an interval on the scale of y
l	latent heat of vaporization
λ	wavelength
m_0, m_1 etc.	point of differentiation with respect to x_0, x_1 etc.
μ	Joule–Thomson coefficient
n	number of differences between values in a set
n	order of spectral line
p	pressure
ϱ	density
S	entropy
T	temperature
x, y	variables, y generally being dependent
x_0, y_0	initial or reference values in a set

$y_0, \hat{y}_0, \Delta\hat{y}_0$	raw and smoothed values at the point of smoothing and the difference between them
$\hat{y}, \hat{y}'$	corrected and uncorrected smoothed values of y
$\hat{Y}$	the difference between unsmoothed and smoothed values of y
z	a variable, dependent on x and y
z	electronic transfer in a cell reaction

CHAPTER VII

DIMENSIONAL ANALYSIS IN PHYSICO-CHEMICAL PROBLEMS

So FAR this volume has been concerned with the handling of the numerical component of measurements. However, a complete expression of the result of a measurement is not only the value obtained (and its precision) but should include the unit in which the measurement has been taken. As will be seen later, the units in terms of which measurements are taken are not, at least in all advanced sciences, independent. The relationship between units is often just as important and just as informative as the relationship between numerical measurements and, what is more important, the necessary relationships between units in physical measurement limit the relations between the measurements themselves. Indeed, a complicated problem in the physical sciences can often be simplified by first eliminating the restraints on the system produced by the nature of the units of measurement.

7.1. UNITS

A measurement in physical science consists, in principle, in the comparison of the unknown quantity with some standard quantity *of the same nature*, the comparison being effected by a method of superposition in which a number of units and subunits are superimposed until the resultant gives the same effect as that of the unknown quantity. Direct methods of superposition can be used, in fact as well as in principle, for measuring many relatively simple physical quantities such as length, volume, force, mass and electrical resistance.

For a large number of physical properties, indeed for the great majority, it is just not possible to superimpose one unit upon another directly and thereby to gain two units. The units of temperature, for example, cannot be superimposed, nor can the units of viscosity. The only method of measurement available in such instances consists in the use of a transducer which translates changes in the property to be measured into movement of a pointer over a scale (see Chapter I). In a mercury-in-glass temperature transducer, changes in temperature cause the surface of a mercury thread to move over a linear scale ruled on a glass tube. This scale is ruled in units of temperature and not in units of length. Alternatively, a platinum resistance temperature transducer converts changes in temperature to electrical resistance changes which are then measured on an electrical resistance bridge calibrated in units of temperature.

The translation of temperature changes to spatial or resistance changes was used initially simply as a matter of convenience. However, progress in methods of measurement has run parallel with progress in the theory relating the not-directly-measurable scale of temperature with a number of other physical properties, and there are now many different ways of obtaining scales of temperature all of which have been proved empirically to give, within the accuracy of measurements using them, identical scales and most of which are based on self-consistent theoretical models.

The methodology of temperature measurement can be applied quite generally to any measurement in Nature. Any phenomenon whatsoever producing a real effect on the physical world can be caused to operate through a transducer to produce a linear change, namely the movement of a pointer on a linear scale. Providing only that the change produced is a monotonic (or single-valued) function of the effect required to be measured, the scale used can be described as a scale for measurement of the effect. The scale however seldom becomes of any use in physical science until it becomes part of a self-consistent scheme using different transducers for the same effect.

Summarizing then, measurements are taken by superposition in line scales or by the movement of pointers over ruled linear or circular scales. Complex properties not directly converted to the movement over ruled scales can be caused to operate through transducers converting changes in the physical properties into changes in the spatial scales. The ruled scales, however, are not necessarily scales of length or angle but scales ruled in terms of the operation of the transducers used. While all types of measurements may in practice be recorded on similarly ruled spatial scales the relationships between the measurements themselves are dominated by the characteristics of the transducers and are not necessarily related to one another. Later, however, we shall see that physical measurements are not all independent of one another and it is the nature of their interdependence which is the thesis of this chapter.

7.2. Systems of Units

A complete statement of a measurement of a physical quantity must give not only the numerical measure and its estimated precision but also the unit in which the measurement has been taken. There are often a number of different units in which the measurement of any particular physical property can be taken. Some of these units are simple legal definitions of multiples or submultiples of one another. Thus, miles, yards, feet and inches are different units for the measurement of length with simple integral ratios between them. The kilometre, metre and centimetre are a similar class of units of length with simple numerical relationship between different members in the class. However, the standard for any one class of unit does not bear any simple integral relationship to the standard in a different class and can usually only be determined by the same type of experimental measurement as that employed in determining the relationship between different physical variables. Nevertheless, a physical quantity which can be measured in any one of the units in a particular class can equally readily be measured in any member of a different class of units, provided only that the second class is appropriate for the same type of measurement. The ensemble of all classes of units suitable for measuring any particular quantity (irrespective of whether it may give a very large number, or a number of the order unity, or a very small number) may be called a *system of units*.

There is, then, one system of units available for every type of quantity which it may be desired to measure. The property which endows a measurement with the capacity to be appropriate to one particular system of units is known as the dimensions of that measurement and, in turn, as the dimensions of the property of which the measurement has been taken.

7.3. DIMENSIONS

The dimensions of a physical property are indicative of the system of units appropriate to measurements of that property. Dingle (1942) and Campbell (1942) indeed have both claimed that the concept of dimensions is more appropriate to the type of measurement than to the property to be measured. All quantities which can be measured in the same way have the same dimensions. Thus heat, work and acoustical, luminous or radiative energy all have the dimensions of energy. The fact that different classes of units may be employed for the different properties does not appear in the expression for their dimensions. The dimensions of a property, then, ignore the values of any numerical constant used in the definition of their units. Also, there are many examples of properties which are physically distinct but which, at least on ordinary conventions, have the same dimensions. Thus action and angular momentum have the same dimensions. Work and torque have the same dimensions. Again, there is a large number of different properties quite distinct from one another which have zero dimensions. Angle, strain, Reynolds number, Mach number, and the fine structure constant are typical examples of such dimensionless quantities.

The classes of physical dimensions are not all independent. Thus, in a description of the kinematics of a single particle, the properties of length, time, velocity and acceleration may be used to describe the system. Of these four properties, however, only two may be taken as independent and there are, therefore, two independent dimensions at the most. In systems more complicated than the kinematics of a single particle there may be more than two independent appropriate dimensions. However, systematic physical theory demands that the units and thus the dimensions of many of the complicated physical properties are numerically related to those of the simple properties; consequently while there may be a very large number of different physical dimensions, there are only a limited number of independent physical dimensions. Indeed the whole fabric of physical theory arises from these limitations on independent dimensions and, therefore, on independent units in physical science. Where, as in systems other than physical ones, there are no dimensional relationships between units then there can be (in spite of accuracy of measurement) no strand of systematic theory like that underlying the physical sciences.

In physical science in general all dimensions are not singly interrelated. There must be more than one independent dimension otherwise the whole system of units could, at least in principle, be related by purely deductive chains of reasoning. In such a case experimental investigation and inductive reasoning would be pointless. Pure mathematics is a subject devoid of any dimensional complexities and as such has no element of experimentation as understood in the physical sciences. On the other hand, although physical science cannot be derived by a purely deductive process, it is profitless to attempt to derive all of the relations of physical science by purely inductive processes. These two facets of science are due to the existence of a finite number of independent physical dimensions.

The dimension of a property will hereafter be written as an algebraic symbol enclosed in a square bracket. For the length, l, we write the dimension of length as [l], and for the time, t, the dimension of time as [t], and for the velocity, v, the dimension of velocity as [v], or alternatively $[lt^{-1}]$. Alternatively, had we selected [v] as an independent dimension, the dimension of time could have been written as $[lv^{-1}]$.

7.4. Dimensional Homogeneity

Any equation in physical theory expresses a condition of balance between two or more factors in a system. As a perfectly general representation of such a system we write

$$Q_1 + Q_2 + Q_3 + \cdots Q_n = 0 \tag{7.1}$$

where the individual terms may be single properties, algebraic products of two or more groups, or differential or integral expressions. However, if the equation is to be valid for all conditions, then each of the individual terms Q_1, Q_2, Q_3 etc., must be capable of being measured in the same units, or in other words, have the same dimensions. If this does not hold, and one of the units employed is changed, the equation can no longer be true. When each term in the equation has the same dimensions it is said to be dimensionally homogeneous. All equations which are true expressions of laws of nature are dimensionally homogeneous and any which is not can only be true for certain fixed values for all the units of measurement employed. Such inhomogeneous expressions may be used under strictly controlled conditions in engineering and technological application but they have no place in basic physical science.

Let us suppose in eqn. (7.1) that there are n different terms and that k different properties are used to define all the terms. Q_2, for example, might be a product of two of the k properties. Since the dimensions of all the properties are the same we have $[Q_1] = [Q_2] = [Q_3]$ and so on giving $n - 1$ simultaneous equations in the total number of k dimensions. In such an expression therefore, there are $k + 1 - n$ independent dimensions and $n - 1$ dependent dimensions. The independent dimensions can of course be selected in a large number of ways. The most general set of independent dimensions, useful within a whole realm of physical science, is known as the fundamental dimensions. Selection of a set of fundamental dimensions is just as arbitrary as the selection of a set of independent dimensions. However, it is often a matter of convenience to have a set of fundamental dimensions corresponding to the set of fundamental standard units.

As an illustration of a derivation of a set of independent dimensions consider a simple dynamical system. The behaviour of a body with initial velocity, v_0, acceleration, a, and displacement, s, in time, t, is subject to the equation,

$$s = v_0t + \tfrac{1}{2}at^2 \tag{7.2a}$$

The dimension of s is [l] and, since the numerical constant does not enter into the dimensional expression, we get for this

$$[l] = [v][t] + [a][t]^2 \tag{7.2b}$$

In a like manner, an equation or equations for any physical system may be expressed as a relation or relations between the dimensions of the properties concerned. The dimensional equation may be obtained by dropping all purely numerical factors and replacing all the physical variables by their dimensions. A differential coefficient of one particular property, x, with respect to another, y, namely dx/dy, has the dimensions $[x][y]^{-1}$; the second differential coefficient d^2x/dy^2 has the dimensions $[x][y]^{-2}$: while the integral $\int x \,.\, dy$ has the dimensions $[x][y]$.

The condition for dimensional homogeneity in eqn. (7.2a) gives

$$[vt] = [v][t] = [l] \tag{7.2c}$$

and

$$[at^2] = [a][t]^2 = [l] \tag{7.2d}$$

as two equations in the four unknown dimensions. Two of these dimensions therefore may be selected in any arbitrary manner. The most convenient pair, in this case length and time, may be selected as independent and fundamental, and in terms of these

$$[v] = [l][t]^{-1} \tag{7.3}$$

$$[a] = [l][t]^{-2} \tag{7.4}$$

As another simple example, consider the definition of force, f, in terms of Newton's second law of motion: force is equal to the rate of change of momentum:

$$f = d(mv)/dt \tag{7.5a}$$

In terms of dimensions,

$$[f] = [m][v]/[t] \tag{7.5b}$$

The total number of terms, n, is two and the total number of dimensions is four. Consequently for this system there are $4 + 1 - 2 = 3$ independent dimensions. For most dynamical systems there are, likewise, three convenient independent and fundamental dimensions which are derived from conditions for the balance of force (or torque).

Equation (7.5a) is the simplest of the balance of force equations. If any further terms are added to it, for example to take into account elastic and damping forces, such forces being proportional to s and ds/dt respectively, then we must add one property for each extra term. Accordingly the number

$$k + 1 - n$$

remains three. However, in systems more complicated than those of dynamics (such as thermal, electrical, chemical and cosmological systems), the number of fundamental dimensions may be different from three. Many of the equations in these systems involve one or more of the constants of Nature; the gas constant, Planck's constant, the Rydberg constant, the velocity of light, the universal constant of gravitation etc. These and other like quantities when entered into equations in dimensions can either be regarded as dimensional terms, contributing to the total value of k in the system or as pure numbers accidentally set to irrational values by our unhappy choice of fundamental standards. Thus if we regard the velocity of light, c, as a pure number, then length and time have the same dimensions and one second of time is equivalent to $3{\cdot}0 \times 10^{10}$ cm.

Generally speaking, while attempts to demote constants of Nature to those of dimensionless rank reduce the required number of fundamental dimensions and, at least in appearance, simplify the theories of physical science, they correspondingly reduce the power and usefulness of dimensional methods. It is often preferable to accept five to six fundamental dimensions in certain complex physico-chemical systems rather than to reduce the complexity by arbitrary selection of one or more of the constants of Nature as purely numerical quantities for there is often no unambiguous way in which this selection can be effected.

7.5. Systems of Dimensions

While the selection of fundamental dimensions is arbitrary, the variables of space and time have some prior claim to preference because they enter into practically all physical systems and into many chemical ones. The dimensions of length and time, [l] and [t], are accordingly two almost universal choices for fundamental dimensions.

In dynamical systems three fundamental dimensions are indicated. The best-known system uses mass, [m], as the third fundamental dimension. However, in the gravitational or technical systems, force, [f], is often taken as an alternative third fundamental dimension and in physico-chemical systems and others involving energy transformations, the energy, [E], may be used instead. Comparative dimensions of the more common dynamical properties in these three systems are given in Table 7.1. The three systems are, of course, alternative and interchangeable.

In thermal systems the equation expressing complete equilibrium has three terms less than the number of physical variables. In consequence there are four fundamental dimensions in such systems. The dimensions used may be based on any of the three common dynamical systems with one additional fundamental dimension of a purely thermal nature. This extra fundamental dimension is usually taken either as that of temperature, [T], or entropy, [S]. The former is the more usual. A list of typical thermal proper-

TABLE 7.1. DIMENSIONS OF SOME COMMON DYNAMICAL PROPERTIES

Property	Physical system	Technical system	Energetic system
Length	$[l]$	$[l]$	$[l]$
Time	$[t]$	$[t]$	$[t]$
Velocity	$[l]\,[t]^{-1}$	$[l]\,[t]^{-1}$	$[l]\,[t]^{-1}$
Frequency	$[t]^{-1}$	$[t]^{-1}$	$[t]^{-1}$
Mass	$[m]$	$[f]\,[t]^2\,[l]^{-1}$	$[E]\,[t]^2\,[l]^{-2}$
Momentum	$[m]\,[l]\,[t]^{-1}$	$[f]\,[t]$	$[E]\,[l]^{-1}\,[t]$
Angular momentum	$[m]\,[l]^2\,[t]^{-1}$	$[f]\,[l]\,[t]$	$[E]\,[t]$
Force	$[m]\,[l]\,[t]^{-2}$	$[f]$	$[E]\,[l]^{-1}$
Pressure	$[m]\,[l]^{-1}\,[t]^{-2}$	$[f]\,[l]^{-2}$	$[E]\,[l]^{-3}$
Surface tension	$[m]\,[t]^{-2}$	$[f]\,[l]^{-1}$	$[E]\,[l]^{-2}$
Power	$[m]\,[l]^2\,[t]^{-3}$	$[f]\,[l]\,[t]^{-1}$	$[E]\,[t]^{-1}$
Energy	$[m]\,[l]^2\,[t]^{-2}$	$[f]\,[l]$	$[E]$
Action	$[m]\,[l]^2\,[t]^{-1}$	$[f]\,[l]\,[t]$	$[E]\,[t]$
Energy flux	$[m]\,[t]^{-3}$	$[f]\,[l]^{-1}\,[t]^{-1}$	$[E]\,[l]^{-2}\,[t]^{-1}$
Moment of inertia	$[m]\,[l]^2$	$[f]\,[l]\,[t]^2$	$[E]\,[t]^2$
Modulus of elasticity	$[m]\,[l]^{-1}\,[t]^{-2}$	$[f]\,[l]^{-2}$	$[E]\,[l]^{-3}$
Viscosity	$[m]\,[l]^{-1}\,[t]^{-1}$	$[f]\,[l]^{-2}\,[t]$	$[E]\,[l]^{-3}\,[t]$
Density	$[m]\,[l]^{-3}$	$[f]\,[l]^{-4}\,[t]^2$	$[E]\,[l]^{-5}\,[t]^2$
Diffusivity	$[l]^2\,[t]^{-1}$	$[l]^2\,[t]^{-1}$	$[l]^2\,[t]^{-1}$

ties together with the corresponding dimension in a number of thermal systems is given in Table 7.2.

The last two properties listed in Table 7.2 deserve special comment. They are examples of constants of Nature of the type mentioned above. The gas constant, R, occurs in the expression for the perfect gas law,

$$pV = RT \tag{7.6}$$

The Stefan–Boltzmann constant, σ, occurs in the expression for the amount of radiation, E, leaving unit area, A, of a black body in unit time at temperature T, namely

$$d(E/A)/dt = \sigma T^4 \tag{7.7}$$

In Table 7.2, R and σ have been treated as dimensional constants of Nature. On the other hand, selection of either R or σ as purely numerical quantities would involve expressing [T], or [S] in terms of the dynamical fundamental dimensions, while the selection of both R and σ as numerical quantities would lead to inconsistent relationships. One of the two acts of selecting R and σ as numerical

TABLE 7.2. DIMENSIONS OF SOME COMMON THERMAL PROPERTIES

Property	Thermo-physical systems	Thermo-technical systems	Thermo-energetic systems	Entropo-energetic systems
Fundamentals	[l] [t] [m] [T]	[l] [t] [f] [T]	[l] [t] [E] [T]	[l] [t] [E] [S]
Heat	$[m]\ [l]^2\ [t]^{-2}$	[f] [l]	[E]	[E]
Temperature	[T]	[T]	[T]	$[E]\ [S]^{-1}$
Entropy	$[m]\ [l]^2\ [t]^{-2}\ [T]^{-1}$	$[f]\ [l]\ [T]^{-1}$	$[E]\ [T]^{-1}$	[S]
Coefficient of expansion	$[T]^{-1}$	$[T]^{-1}$	$[T]^{-1}$	$[E]^{-1}\ [S]$
Heat capacity	$[l]^2\ [t]^{-2}\ [T]^{-1}$	$[l]^2\ [t]^{-2}\ [T]^{-1}$	$[l]^2\ [t]^{-2}\ [T]^{-1}$	$[l]^2\ [t]^2\ [E]^{-1}\ [S]$
Thermal conductivity	$[m]\ [l]\ [t]^{-3}\ [T]^{-1}$	$[f]\ [t]^{-1}\ [T]^{-1}$	$[l]^{-1}\ [t]^{-1}\ [E]\ [T]^{-1}$	$[l]^{-1}\ [t]^{-1}\ [S]$
Heat transfer coefficient	$[m]\ [t]^{-3}\ [T]^{-1}$	$[f]\ [l]^{-1}\ [t]^{-1}\ [T]^{-1}$	$[l]^{-2}\ [t]^{-1}\ [E]\ [T]^{-1}$	$[l]^{-2}\ [t]^{-1}\ [S]$
Convection modulus	$[l]^{-3}\ [T]^{-1}$	$[l]^{-3}\ [T]^{-1}$	$[l]^{-3}\ [T]^{-1}$	$[l]^{-3}\ [E]\ [S]$
Gas constant	$[l]^2\ [t]^{-2}\ [T]^{-1}$	$[l]^2\ [t]^{-2}\ [T]^{-1}$	$[l]^2\ [t]^{-2}\ [T]^{-1}$	$[l]^2\ [t]^{-2}\ [E]^{-1}\ [S]$
Stefan–Boltzmann constant	$[m]\ [t]^{-3}\ [T]^{-4}$	$[f]\ [l]^{-1}\ [t]^{-1}\ [T]^{-4}$	$[l]^{-2}\ [t]^{-1}\ [E]\ [T]^{-4}$	$[l]^{-2}\ [t]^{-1}\ [E]^{-3}\ [S]^4$

TABLE 7.3. DIMENSIONS OF ELECTRICAL AND MAGNETIC QUANTITIES

System	Electromagnetic	Electrostatic	Electro-physical	Electro-technical	"Practical"
Properties	$[l]\ [t]\ [m]\ [\mu]$	$[l]\ [t]\ [m]\ [\varepsilon]$	$[l]\ [t]\ [m]\ [Q]$	$[l]\ [t]\ [f]\ [Q]$	$[l]\ [t]\ [R]\ [I]$
Energy	$[m]\ [l]^2\ [t]^{-2}$	$[m]\ [l]^2\ [t]^{-2}$	$[m]\ [l]^2\ [t]^{-2}$	$[f]\ [l]$	$[I]^2\ [R]\ [t]$
Power	$[m]\ [l]^2\ [t]^{-3}$	$[m]\ [l]^2\ [t]^{-3}$	$[m]\ [l]^2\ [t]^{-3}$	$[f]\ [l]\ [t]^{-1}$	$[I]^2\ [R]$
Electric charge	$[m]^{1/2}\ [l]^{1/2}\ [\mu]^{-1/2}$	$[m]^{1/2}\ [l]^{3/2}\ [t]^{-1}\ [\varepsilon]^{1/2}$	$[Q]$	$[Q]$	$[I]\ [t]$
Electric displacement	$[m]^{1/2}\ [l]^{-3/2}\ [\mu]^{-1/2}$	$[m]^{1/2}\ [l]^{-1/2}\ [t]^{-1}\ [\varepsilon]^{1/2}$	$[Q]\ [l]^{-2}$	$[Q]\ [l]^{-2}$	$[I]\ [l]^{-2}\ [t]$
Electric field intensity	$[m]^{1/2}\ [l]^{1/2}\ [t]^{-2}\ [\mu]^{1/2}$	$]m]^{1/2}\ [l]^{-1/2}\ [t]^{-1}\ [\varepsilon]^{-1/2}$	$[m]\ [Q]^{-1}\ [l]\ [t]^{-2}$	$[f]\ [Q]^{-1}$	$[I]\ [R]\ [l]^{-1}$
Capacitance	$[l]^{-1}\ [t]^2\ [\mu]^{-1}$	$[l]\ [\varepsilon]$	$[m]^{-1}\ [Q]^2\ [l]^{-2}\ [t]^2$	$[f]^{-1}\ [Q]^2\ [l]^{-1}$	$[R]^{-1}\ [t]$
Current	$[m]^{1/2}[l]^{1/2}\ [t]^{-1}\ [\mu]^{-1/2}$	$[m]^{1/2}\ [l]^{3/2}\ [t]^{-2}\ [\varepsilon]^{1/2}$	$[Q]\ [t]^{-1}$	$[Q]\ [t]^{-1}$	$[I]$
Voltage	$[m]^{1/2}[l]^{3/2}\ [t]^{-2}\ [\mu]^{1/2}$	$[m]^{1/2}\ [l]^{1/2}\ [t]^{-1}\ [\varepsilon]^{-1/2}$	$[m]\ [Q]^{-1}\ [l]^2\ [t]^{-2}$	$[f]\ [Q]^{-1}\ [l]$	$[I]\ [R]$
Resistance	$[l]\ [t]^{-1}\ [\mu]$	$[l]^{-1}\ [t]\ [\varepsilon]^{-1}$	$[m]\ [Q]^{-2}\ [l]^2\ [t]^{-1}$	$[f]\ [Q]^{-2}\ [l]\ [t]$	$[R]$
Magnetic flux	$[m]^{1/2}\ [l]^{3/2}\ [t]^{-1}\ [\mu]^{1/2}$	$[m]^{1/2}\ [l]^{1/2}\ [\varepsilon]^{-1/2}$	$[m]\ [Q]^{-1}\ [l]^2\ [t]^{-1}$	$[f]\ [Q]^{-1}\ [l]\ [t]$	$[I]\ [R]\ [t]$
Magnetic induction	$[m]^{1/2}\ [l]^{-1/2}\ [t]^{-1}\ [\mu]^{1/2}$	$[m]^{1/2}\ [l]^{-3/2}\ [\varepsilon]^{-1/2}$	$[m]\ [Q]^{-1}\ [t]^{-1}$	$[f]\ [Q]^{-1}\ [l]^{-1}\ [t]$	$[I]\ [R]\ [l]^{-2}\ [t]$
Magnetomotive force	$[m]^{1/2}\ [l]^{1/2}[t]^{-1}\ [\mu]^{-1/2}$	$[m]^{1/2}\ [l]^{3/2}\ [t]^{-2}\ [\varepsilon]^{1/2}$	$[Q]\ [t]^{-1}$	$[Q]\ [t]^{-1}$	$[I]$
Reluctance	$[l]^{-1}\ [\mu]^{-1}$	$[l]\ [t]^{-2}\ [\varepsilon]$	$[m]^{-1}\ [Q]^2\ [l]^{-2}$	$[f]^{-1}\ [Q]^2\ [l]^{-1}\ [t]^{-2}$	$[R]^{-1}\ [t]^{-1}$
Inductance	$[l]\ [\mu]$	$[l]^{-1}\ [t]^2\ [\varepsilon]^{-1}$	$[m]\ [Q]^{-2}\ [l]^2$	$[f]\ [Q]^{-2}\ [l]\ [t]^2$	$[R]\ [t]$
Permittivity	$[l]^{-2}\ [t]^2\ [\mu]^{-1}$	$[\varepsilon]$	$[m]^{-1}\ [Q]^2\ [l]^{-3}\ [t]^2$	$[f]^{-1}\ [Q]^2\ [l]^{-2}$	$[R]^{-1}\ [l]^{-1}\ [t]$
Permeability	$[\mu]$	$[l]^{-2}\ [t]^2\ [\varepsilon]^{-1}$	$[m]\ [Q]^{-2}\ [l]$	$[f]\ [Q]^{-2}\ [t]^2$	$[R]\ [l]^{-1}\ [t]$

quantities is clearly false and we prefer to take both as equally false and so make the dimension of temperature independent of the dynamical dimensions.

A similar state of affairs arises in connection with the dimensions of the electrical and magnetic quantities. It is possible to express the dimensions of all electromagnetic quantities in terms of the dynamical dimensions either from the definition of the electrostatic charge (by taking the dielectric permittivity as a numerical quantity) or, alternatively, from the definition of the magnetic pole (taking the magnetic permeability as a numerical quantity). The two systems, however, are dimensionally inconsistent and we prefer to regard both as false and introduce in electromagnetic quantities one more fundamental dimension. In the electrostatic system this is the permittivity, ε, in the electromagnetic the permeability, μ. Systems based on either ε or μ as a fundamental dimension have a serious disadvantage in that a number of important derived electrical properties have dimensions with half integer exponents. Other systems based on the electric charge, Q, or the electric current, I, as fundamental dimensions give simpler expressions for most of the derived dimensions. The systems based on Q are of more fundamental appeal from basic electrical theory but the systems based on I correspond more closely to the modern system of fundamental units. The dimensions of common electrical and magnetic quantities using some of the possible systems are given in Table 7.3.

Thermionic, thermo-electric and thermo-magnetic properties, in which both thermal and electric phenomena are combined, require five fundamental dimensions. The dimensions of some such properties with [l], [t], [m], [T] and [Q] taken as fundamental are given in Table 7.4.

TABLE 7.4
DIMENSIONS OF THERMO-ELECTRIC AND THERMO-MAGNETIC PROPERTIES

Property	Dimensions
Thermionic work function	$[l]^2\,[t]^{-2}$
Constant of thermionic emission	$[t]^{-1}\,[T]^{-2}\,[Q]\,[l]^{-2}$
Peltier coefficient	$[l]^2\,[t]^{-2}\,[m]\,[Q]^{-1}$
Thermo-electric power	$[l]^2\,[t]^{-2}\,[m]\,[T]^{-1}\,[Q]^{-1}$
Thomson coefficient	$[l]^2\,[t]^{-2}\,[m]\,[T]^{-1}\,[Q]^{-1}$
Nernst coefficient	$[l]^3\,[t]^{-1}\,[m]\,[T]^{-1}\,[Q]^{-2}$

7.6. Some Discussion of the Length Dimension

Measurements of spatial quantities are unique in that a single number seldom suffices to locate a point in space whereas a single number will locate a point in the scales of time, temperature or electric charge. In space three numbers are required to locate a point and this type of multiplicity extends to many of the properties which are derived from the dimensions of space (or length) combined with other dimensions. The quantities which possess this type of multiplicity are referred to as vector or tensor quantities.

We have already seen that each term in any particular equation in physical science has the same physical dimensions. In addition, if the equation applies to a space of more than one geometrical dimension, then every term must have the same tensor nature or must be homogeneous in geometrical dimensions. There is indeed no really clear-cut differentiation between geometrical and physical dimensions. In a general physical system there is however no preferential direction in space and the x, y and z co-ordinates are interchangeable in a way in which the T, t and m measurements are not, so that there is some justification for regarding the difference between x, y and z as of a lesser order than the difference between the fundamental physical dimensions. However there are a number of physical conditions in which all the directions in space are not necessarily equivalent. For example, in anisotropic crystalline matter, Nature herself appears to use different linear scales along the different crystal axes in the sense that simple geometric relations between the different crystal faces are obtained only when distances measured along each axis are referred to the lattice parameter for that axis. Again in many engineering systems concerned with cylindrical bodies, such as flow in pipes, there is evidence that measurements along the line of flow are not equivalent to those across the line. In dimensional analyses of such systems the down-wind and cross-wind directions are treated as if dimensionally non-equivalent.

Further, in simple dynamical systems many pairs of properties, of which one alone has been derived from angular quantities, have the same physical dimensions without necessarily being of the same nature. Work and torque, and angular momentum and action are two such pairs. To avoid ambiguity it has been suggested by Stubbings (1948) that angle, $[\alpha]$, become an auxiliary fundamental dimension which would give the following for the above properties.

Work,

$$[m][l][t]^{-2}$$

Torque,

$$[m][l]^2\,[t]^{-2}\,[\alpha]^{-1}$$

Action,

$$[m][l]^2\,[t]^{-1}$$

Angular momentum,

$$[m][l]^2\,[t]^{-1}\,[\alpha]^{-1}$$

which then are dimensionally distinct.

Angle, which in principle is measured by the quotient of one length measurement by a perpendicular length measurement (arc by radius) can only be regarded as a quantity of zero dimensions if the two perpendicular directions in space are regarded as completely equivalent. The convention of taking angle as an auxiliary dimension is thus equivalent to the assumption that at least two of the geometrical dimensions are physically distinct. Also it may be claimed that there are no fast rules regarding $[x]$, $[y]$ and $[z]$ as being distinct, or all or some equivalent to [l]. The matter is entirely one of convenience. Williams (1892) early suggested that the three linear dimensions should be treated separately and this suggestion has been more recently discussed by Huntley (1952).

7.7. Dimensionless Ratios or Numerics

Special interest in the theory of dimensions is attached to quantities of zero dimensions. These are the combinations in which all the exponent indices of the fundamental dimensions vanish. Such quantities accordingly have numerical values which are completely independent of the units in terms of which the quantities are measured, provided only that a consistent set of standards is adopted for these units. Thus, the relationship between dimensionless quantities is one which we can expect to express a measure of fundamental reality independent of any artificial correction arising from our standards of measurement. Indeed, in engineering practice correlation between dimensionless quantities often succeeds in coordinating wide ranges of measurements on very different systems in a unified presentation not possible by any other means. Dimensionless quantities are also sometimes called *numerics*.

A set of numerics can always be derived from any equation of equilibrium such as eqn. (7.1). Let us divide all the terms in that equation by the first, thus getting

$$Q_2/Q_1 + Q_3/Q_1 \cdots Q_n/Q_1 = -1 \qquad (7.8\,a)$$

which is an equation in $n - 1$ dimensionless quantities. We could alternatively have divided every term in eqn. (7.1) by the second term obtaining

$$Q_1/Q_2 + Q_3/Q_2 + \cdots Q_n/Q_2 = -1 \qquad (7.8\,b)$$

which is an equivalent dimensionless equation in $n - 1$ dimensionless terms in which $n - 2$ are different from those in eqn. (7.8 a). In all, a large number, $n(n - 1)/2$, dimensionless quantities can be derived from an expression of the type of eqn. (7.1) but of these only $n - 1$ are independent combinations of the experimental parameters.

The dimensionless quantities which arise in this manner are of course ratios of physical measurements of the same nature. Thus, had eqn. (7.1) expressed a balance of a number of types of forces, then the dimensionless ratios in eqns. (7.8 a) and (7.8 b) would each be the ratio of pairs of forces.

A well-known example occurs in a flowing fluid stream. Here, the conditions for equilibrium or the balance of the forces per unit volume give for a chemically reacting system in one geometrical dimension (the x direction) the following expression:

$$\varrho(\partial v/\partial t) + \varrho v(dv/dx) + (dp/dx) - \eta(d^2v/dx^2) - \varrho vk\Delta V/V = 0 \qquad (7.9)$$

where ϱ is the density, v the velocity, p the pressure, η the viscosity, k the reaction rate and $\Delta V/V$ the fractional volume change (or rate of volume change per unit volume) resulting from chemical reaction which, in this system, becomes a single variable. By taking the ratios of pairs of properties in the equation we arrive at a number of dimensionless quantities some of which have been given specific names. Thus, from the first and the second terms we obtain a quantity known as the Stroudhal number having the (cancelling) dimensions of $[v][t]/[x]$; from the second and the third terms the

Newton number, $[p]/[\varrho][v]^2$; and from the second and the fourth the Reynolds number, $[\varrho][v][x]/[\eta]$. The ratio of the first to the fifth terms has the dimensions of $[k][t][\Delta V/V]$ which in turn is the product of two simple dimensionless quantities, one the product of time and reaction rate and the other the fractional change in volume.

The number of independent dimensionless ratios which can be obtained from any particular expression can be obtained directly from what is variously known as the Riabouchinski, Buckingham or Π theorem. This theorem states that if there are n dimensions (or n different physical measurements marked, say, from $Q_1, Q_2, \ldots, Q_n$) required to specify the state of equilibrium of any particular system and if these dimensions can all be expressed in terms of k fundamental dimensions then there are $n - k - 1$ different ways in which the various measurements may be combined to form dimensionless products of the type

$$\Pi_1 = Q_1^{\ a_1} . Q_2^{\ b_1} \ldots Q_n^{\ n_1}$$

$$\Pi_2 = Q_1^{\ a_2} . Q_2^{\ b_2} \ldots Q_n^{\ n_2} \text{ etc.}$$

(where the exponents are integral numbers—positive, negative or zero).

By replacing the experimental dimensional quantities, Q (n in number) by the dimensionless products, Π ($n - k - 1$ in number) we reduce an expression of order n to the simpler order $n - k - 1$ by taking into account the fact that the expression must be dimensionally homogeneous. Accordingly, an increase in the value of k will reduce the number of independent dimensionless products and simplify the resultant dimensionless product expression. However, the act of increasing the number of fundamental dimensions involves, as we have seen earlier, the promotion of a numerical quantity to the rank of a dimensional constant of Nature. If this constant of Nature occurs in the system, the number of Qs is increased by unity so that the value of $n - k - 1$, and the complexity of the derived dimensionless expressions, is unaltered. Indiscriminate proliferation of fundamental dimensions adds nothing to the power of dimensional analysis.

As an example of an application of the Π theorem we shall consider the flow of a fluid in a pipe. The experimental factors which may be expected to control the flow rate, with their associated dimensions in [m], [l] and [t] are

d, the diameter of the pipe,	[l]
i, the loss of hydraulic head per unit length of pipe,	$[m]\,[l]^{-2}\,[t]^{-2}$
v, the velocity,	$[l]\,[t]^{-1}$
η, the viscosity of the fluid,	$[m]\,[l]^{-1}\,[t]^{-1}$
ϱ, the density of the fluid,	$[m]\,[l]^{-3}$

With no prior knowledge from experimental data or from considerations of dimensional homogeneity we might expect the condition for equilibrium to be some arbitrary expression in five independent variables, namely

$$\varphi(\mathrm{d}, \mathrm{i}, \mathrm{v}, \eta, \varrho) = 0 \tag{7.10}$$

On applying the Π theorem with $n = 5$ and $k = 3$ there is only one independent dimensionless product, Π_1, and we may put

$$\Pi_2 = \varphi(\Pi_1)$$

where Π_2 is any other dimensionless product. The values of Π_1 and Π_2 may be selected from the array of possible dimensionless products

$$\mathrm{dv}\varrho/\eta,\ \mathrm{id}/\varrho\mathrm{v}^2,\ \eta\mathrm{v}/\mathrm{id}^2,\ \varrho^2\mathrm{v}^3/\mathrm{i}\eta, \text{ and } \varrho\mathrm{id}^3/\eta^2$$

The equation for flow, for example, could be expressed in the form

$$(\mathrm{id}/\varrho\mathrm{v}^2) = \varphi(\mathrm{dv}\varrho/\eta) \tag{7.11}$$

which means that a single series of measurements can be used to determine the form of the function $\varphi(\mathrm{dv}\varrho/\eta)$. Equation (7.11) will thus give all the physical information which could possibly be derived from the really far more complicated eqn. (7.10).

As mentioned above many of the better known dimensionless quantities have been given special names. These numerics, some of which are listed in Table 7.5, may conveniently be divided into two classes: one which is concerned only with the ratio of physical properties and one which includes process variables such as velocities, times, and distances. Dimensionless ratios are commonly denoted by double letter symbols.

TABLE 7.5. SOME DIMENSIONLESS QUANTITIES

(a) Ratios of physical properties			(b) Process dimensionless variables		
Combination of properties	Name	Symbol	Combination of properties	Name	Symbol
$c_p\eta/k$	Prandtl number	Pr	$vd\varrho/\eta$	Reynolds number	Re
			$p/\varrho v^2$	Newton number	Ne
$\eta/\varrho D$	Schmidt number	Sm	$\varrho dv^2/\gamma$	Weber number	We
			v^2/dg	Froude number	Fr
$k/c_p\varrho D$	Lewis number	Le	hd/k	Nusselt number	Nu
			$dv\varrho c_p/k$	Peclet number	Pe
			$\varrho vc_p/\sigma lT^3$	Radiation or Thring number	Th
			vt/d	Stroudhal number	
			$h/c_p\varrho v$	Stanton number	
			E/RT	Arrhenius group	
			v/c	Mach number	

(Symbols denoting particular properties are defined in the notation appendix at the end of this chapter.)

7.8. APPLICATIONS OF DIMENSIONAL THEORY

The condition for dimensional homogeneity in any expression for a physical law gives rise to certain necessary relationships between the variables. These necessary relationships, while never enough to completely fix numerical solutions for the expression, can often produce a considerable degree of simplification, particularly in systems for which the fundamentals are very incompletely known. Many of the best known applications of dimensional methods arise from these considerations.

Dimensional methods are of considerable value in the checking of formulae. Every term in such a formula must have the same dimensions and if dimensions do not balance a mistake may have occurred in remembering or copying the equation. For example, suppose an expression for the rate of fall, v, of a sphere of radius, r, and density, ϱ_s, in a liquid of density, ϱ, and viscosity, η, is given as

$$v = (2/9)\,(gr/\eta)\,(\varrho_s - \varrho) \qquad (7.12a)$$

The dimensions of the various terms entering the equation are

for $\quad v, \quad [l][t]^{-1}$

$g, \quad [l][t]^{-2}$

$r, \quad [l]$

$\eta, \quad [m][l]^{-1}[t]^{-1}$

and for

$(\varrho_s - \varrho), \quad [m][l]^{-3}$

The dimensions of the left-hand term of eqn. (7.12a) are, therefore, $[l]\,[t]^{-1}$, but those on the right, $[l]\,[t]^{-2}\,[l]\,[m]\,[l]^{-3}/[m]\,[l]^{-1}\,[t]^{-1}$, reduce to $[t]^{-1}$ so eqn. (7.12a) is dimensionally inhomogeneous and is not correct. Furthermore, the analysis indicates that the error is in the power of a property having the dimensions of [l]. In point of fact, the correct expression for the rate of fall of a sphere under Stokes' Law is

$$v = (2/9)\,(gr^2/\eta)\,(\varrho_s - \varrho) \tag{7.12b}$$

which may readily be verified as dimensionally homogeneous. A similar quick check on the dimensional homogeneity of remembered formulae or formulae copied many times from the original (particularly in circularized technical publications) can often eliminate obvious mistakes.

In advanced chemical or physical theory, groups of measured quantities occur in complex transcendental expressions. Thus the groups (E/RT) and $(h\nu/\varkappa T)$ occur in the form of an exponential in many equations relating to activated rate processes or for expressions of thermodynamical equilibrium. Now an exponential and many other similar transcendental expressions can be expanded in ascending (or descending) powers of the argument. Thus

$$\exp(x) = 1 + x + x^2/2! + x^3/3! \ldots x^n/n! \tag{7.13}$$

$$\sin x = x - x^3/3! + x^5/5! - \ldots \tag{7.14}$$

and

$$\sinh x = x + x^3/3! + x^5/5! + \ldots \tag{7.15}$$

Any term, x, which, like those above, is the argument of a transcendental function must then appear as a series of different powers of x in the expanded form. If, however, x has any dimensional quality, the sums of the ascending (or descending) powers of x cannot be dimensionally homogeneous. Accordingly, all complex terms which occur as the argument of any transcendental term in a physical expression must be dimensionless as, for example, the group $(h\nu/\varkappa T)$ in the commonly encountered term, $\exp(h\nu/\varkappa T)$. Similarly, the expression for the charge density, σ, in relation to the electrophoretic or zeta potential, ζ, in a medium containing ions of common charge, ez, of concentration C, is

$$\sigma = 2\sqrt{(DN\varkappa TC/2000\,\pi)}\,\sinh(\zeta\, ez/2\varkappa T) \tag{7.16}$$

Consequently, $(\zeta ez/2\varkappa T)$ is a dimensionless quantity.

An exception to the general rule that arguments in transcendental equations or terms must be dimensionless may occur in the case of the logarithm. Thus, the Clausius–Clapeyron equation can be written

$$d(\ln p)/dT = L/RT^2 \tag{7.17a}$$

where L is the latent heat of vaporization and the other symbols have the meanings given in this chapter. Now the pressure, p, is not a dimensionless quantity and a logarithm can be rewritten as

$$\ln x = \int dx/x$$

and no matter what the dimensions of x, $\ln x$ and accordingly $\log x$ or the logarithm to any base is a dimensionless quantity.

Another application of dimensional theory is concerned with the reduction of the number of physical variables in complex problems using the principles of the Π theorem. If it is known or suspected that a certain dimensional variable Q_1 is dependent on a number of dimensional variables Q_2, Q_3 etc., including any necessary constants of Nature, then the value of Q_1 can only be proportional to certain groups of power products of the independent variables, of such a nature as to have the same dimensions as that of Q_1. On the basis of dimensional homogeneity, Q_1 can only be given by expressions of the type

$$Q_1 = KQ_2^aQ_3^bQ_4^c \ldots \tag{7.18}$$

where K is a numerical constant and a, b, c, etc., are power exponents which may be positive, negative or zero. There may be only one possible solution for the exponents, in which case Q_1 must be a unique function of the independent variables apart from the numerical constant. If there be more than one possible solution, Q_1 may be a bi- or multi-variable function of the independent variables. If no solution exists, then Q_1 cannot be dependent on the selected properties.

The evaluation of the exponents involves the solution of a series of simultaneous equations in number equal to the number of fundamental dimensions in the system concerned. For example, suppose without detailed analysis we wish to know the form of the relationship between the rate at which molecules impinge on a surface, i.e. the number, N, of moles striking unit area in unit time and such independent variables as the pressure, p, the molecular weight, M, and the temperature, T. In all such gas kinetic systems the dimensional constant of Nature, R, the gas constant, or its molecular equivalent, $\varkappa$, the Boltzmann constant, is involved. An arbitrary power relationship thus takes the form

$$N = Kp^aR^bT^cM^d \tag{7.19a}$$

Let [m], [l], [t] and [T] be taken as fundamental dimensions. Then, the equivalent dimensional equation is

$$[l]^{-2}\,[t]^{-1} = [m]^a\,[l]^{-a}\,[t]^{-2a}\,[m]^b\,[l]^{2b}\,[t]^{-2b}\,[T]^{-b}\,[T]^c\,[m]^d \tag{7.19b}$$

The condition for dimensional homogeneity requires that the exponents in each fundamental dimension should separately balance. Accordingly,

$$\begin{aligned} \text{for} \qquad [m],&\quad a+b+d=0\\ [l],&\quad -a+2b=-2\\ [t],&\quad -2a-2b=-1\\ [T],&\quad -b+c=0 \end{aligned} \tag{7.20}$$

These four equations may be satisfied only if $a = 1$ and $b = c = d = -\frac{1}{2}$. Accordingly, the relationship between N and the independent variables must take the form

$$N = K(p/\sqrt{(MRT)} \tag{7.21}$$

In the kinetic theory of ideal gases the numerical constant takes the value $1/\sqrt{2\pi}$.

Any similar application to the reduction of physical data involves the solution of a set of simultaneous equations in the indices. The number of such equations is equal to the number of fundamental dimensions, and the number of variable indices is equal to the number of independent dimensional variables and constants of Nature relevant to the problem. If the number of equations (i.e. the number of fundamental dimensions) is less than the number of variables an infinite number of solutions can be obtained. Such an example occurs in the pipe flow problem considered above. In such a case one or more of the exponents becomes arbitrary and dimensional analysis leaves the problem in terms of an arbitrary function of one or more groups of properties which, on the requirement of dimensional homogeneity, must be dimensionless groups.

A unique solution, such as that given above, is obtainable if the number of variables is equal to the number of equations. However, in certain instances the equations in indices are found to be inconsistent. Such equations will always be inconsistent if the determinant of the coefficients of exponents in the set of simultaneous equations is equal to zero. For example, it has been suggested that a relationship might be found between the property of thermal conductivity, **k**, of dimensions $[m]\,[l]\,[t]^{-3}\,[T]^{-1}$ and other thermal properties such as the compressibility per unit mass, $\boldsymbol{\chi}$, of dimensions $[m]^{-2}\,[l]^{4}\,[t]^{2}$, the thermal expansion, $\boldsymbol{\beta}$, of dimensions $[m]^{-1}\,[l]^{3}\,[T]^{-1}$, the specific heat, c_p, of dimensions $[l]^{2}\,[t]^{-2}\,[T]^{-1}$ and the temperature T of dimensions [T]. Writing

$$\mathbf{k} = \chi^a \beta^b c_p^c T^d \tag{7.22}$$

we obtain for the set of simultaneous equations for the indices

$$\begin{aligned} -2a - b \qquad\qquad &= 1 \\ 4a + 3b + 2c \qquad &= 1 \\ 2a \qquad - 2c \qquad &= 3 \\ - b - c + d &= -1 \end{aligned} \tag{7.23}$$

The determinant in the coefficients of the indices, namely

$$\begin{bmatrix} -2 & -1 & 0 & 0 \\ 4 & 3 & 2 & 0 \\ 2 & 0 & -2 & 0 \\ 0 & -1 & -1 & 1 \end{bmatrix}$$

may readily be seen to vanish. Accordingly the set of eqns. (7.23) are inconsistent and no expression of the type eqn. (7.22) can be found.

The case where the number of equations in the indices is greater than the number of variables could also lead to an impossible solution. However, such cases do not arise for, as we have shown above, the total number of fundamental dimensions appropriate to any particular physical system cannot be greater than the number of independent dimensional quantities defining the system.

In the physico-chemical field, dimensional analysis has been used extensively in the development of surface properties, starting from the early treatise by Bashforth and Adams (1883). The shapes taken up by fluid surfaces (bubbles or drops) under the joint influence of surface tension and gravitational forces were shown by these authors to depend upon a dimensionless ratio of these two factors, namely

$$\beta = gr^2 \Delta\varrho/\gamma \tag{7.24}$$

where r is the radius of curvature of the surface, g is the acceleration due to gravity at the point concerned, γ is the surface tension, and $\Delta\varrho$ the density difference across the interface. The tables given by Bashforth and Adams express β in terms of dimensionless geometrical ratios concerned with the properties of the surface: the ratios of radii of droplets to radii of tubes from which they form, the ratio of heights to diameters of free drops, the ratio of volumes of isolated droplets to the volumes of spheres on the same radii. These tables have been used to derive experimental corrections for many of the standard methods of measuring surface tension such as the drop weight, bubble pressure and drop height methods.

Dimensional methods have also been used in the reduction of some of the more complicated expressions in chemical kinetics. Consider a system in which a single molecular species is undergoing first-order decomposition. If the concentration of the reactant is C units at any time t, then the rate of the decomposition, $-dC/dt$ is given by the equation

$$-dC/dt = k_1 C \tag{7.25}$$

signifying that the rate of decrease of the concentration is proportional to the number of decomposing molecules present. For a decomposition following an *n*th order decomposition,

$$-dC/dt = k_n C^n \tag{7.26}$$

Equations (7.25) and (7.26) can be integrated, subject to the boundary condition that C takes the value C_0 (the initial concentration) at $t = 0$, to give

$$k_1 t = \ln (C_0/C) \tag{7.27}$$

and

$$k_n t = [1/(n - 1)]\,[1/C^{n-1} - 1/C_0^{n-1}] \tag{7.28}$$

These expressions can be written in a somewhat simpler form by using the dimensionless quantities

$$\alpha = C/C_0,$$

the relative concentration at

$$\tau = k_0 C_0^{n-1} t,$$

the relative time. In terms of these quantities the course of all reactions follows the law

$$\ln \alpha = -\tau,$$

for the first order, and

$$\alpha^{(1-n)} - 1 = (n - 1)\,\tau$$

for any other order. Curves due to Powell (1953) showing the relation between α and τ for the first four integral orders are given in Fig. 7.1. The curves are unique for order of the reaction.

While the use of dimensionless quantities in the examples above permits a uniform presentation for all reactions of the same order the degree of simplification in the application of these principles is not particularly striking. However, in more complicated systems the achievable degree of simplification becomes more pronounced. Thus, in a complex first order reaction in which component **A** first changes to **B** with a rate constant k_1 and **B** subsequently changes to **C** with a rate constant k_2, the value of the concentrations of **A**, **B** and **C** in a time t becomes

$$A = A_0 \exp(-k_1 t) \tag{7.29a}$$

$$B = [A_0 k_1/(k_2 - k_1)]\,[\exp(-k_1 t) - \exp(-k_2 t)] \tag{7.29b}$$

$$C = A_0\{1 - [k_2 \exp(-k_1 t) - k_1 \exp(-k_2 t)]/(k_1 - k_2)\} \tag{7.29c}$$

where A_0 is the initial concentration of **A** and A, B and C are the actual concentrations of **A**, **B** and **C**. By using dimensionless ratios

$$\alpha = A/A_0, \quad \beta = B/A_0, \quad \gamma = C/A_0, \quad \tau = k_1 t, \quad \varkappa = k_2/k_1 \qquad (7.30\,\text{a, b, c, d, e})$$

the above expressions reduce to:

$$\alpha = \exp(-\tau) \qquad (7.31\,\text{a})$$

$$\beta = [\exp(-\tau) - \exp(-\varkappa\tau)]/(\varkappa - 1) \qquad (7.31\,\text{b})$$

$$\gamma = 1 + [\varkappa \exp(-\tau) - \exp(-\tau)]/(1 - \varkappa) \qquad (7.31\,\text{c})$$

Similar applications of dimensionless quantities to reduce complexities in progressive or competing processes have been described by Swain (1944), French (1950), French and McIntyre (1951) and others.

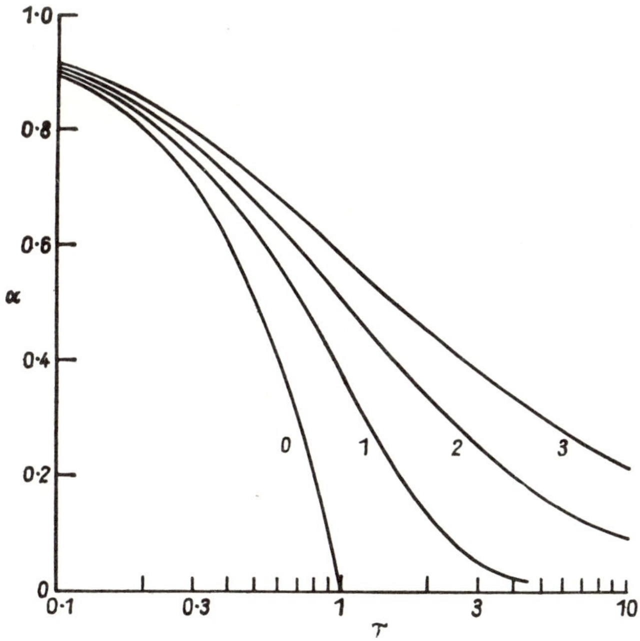

FIG. 7.1. Fraction remaining, α, as a Function of Time Parameter τ, for Zero, First-, Second-, and Third-order Reactions according to R. E. Powell (1953).

(Reproduced by permission of J. Wiley & Sons, New York.)

By and large, however, the principles of dimensional analysis have not played as full a part in physico-chemical investigations as they deserve and have not been paid due regard in many of the definitions of the "constants" of physical chemistry. For example the velocity constants, k, for reactions of order n, namely those of eqn. (7.26) have different physical dimensions for each different order of reaction and may in certain cases involve not only fractional but non-rational values of n. A similar situation arises in the integralist approach to the rheology of semi-fluid material. The Nutting–Scott Blair equation for stiffness, ψ, reads

$$\psi = S^{\beta}t^{k}/\sigma \tag{7.32}$$

where σ is the strain, S the stress, t the time and β and k constants for any particular semi-fluid. The dimensions of the property, ψ, thus depend both on β and k and may involve non-rational indices.

7.9. The Principle of Similarity

Earlier in this chapter we have seen that an equation of equilibrium (static or dynamic) can always be reduced to an expression between dimensionless quantities and is in the simplest possible form when it is thus expressed. By their very nature the numerical value of the dimensionless quantities occurring in such an expression are independent of the system of units or the size of the units, provided only that measurements are carried out in a consistent manner. It follows accordingly that such dimensionless expressions will be valid no matter what the scale of the measurement or the scale on which the operations are carried out. Two different physical systems of different geometrical size may then be regarded as behaving in a precisely similar manner if all of the dimensionless ratios defined from the equation of equilibrium have the same values for the two different systems. The two systems are then said to be dimensionally *similar* or to show a *condition of similarity*. The smaller of the two may be described as a *model* for the other which is generally called the *prototype*.

Different classes of similarity may be realized depending on the nature of the equilibrium equation used to describe the system. Systems exhibit geometrical similarity if the ratios of all corresponding geometrical parameters are the same on both model and prototype. Systems exhibit dynamical similarity if the ratios of all pairs of forces on one system are the same as the corresponding ratios on the other. Conditions defined from a balance of forces equation thus give rise to dynamical similarity. Conditions defined from a balance of thermal forces may similarly give rise to thermal similarity, and those from a balance of chemical rate changes to a condition of chemical similarity.

The example of the isothermal flow in a pipe considered in Section 7.7 above was finally expressible in the dimensionless form:

$$(\mathrm{id}/\varrho \mathrm{v}^2) = \varphi(\mathrm{dv}\varrho/\eta) \tag{7.11}$$

Two such systems in pipes of different diameter and with different fluids (i.e. different ϱ and η) will *normally* give rise to the same values of $(\mathrm{id}/\varrho \mathrm{v}^2)$ if the $(\mathrm{dv}\varrho/\eta)$ takes the same value irrespective of whether or not each of four dimensional properties d, v, ϱ and η have been changed. It is required only that the combination $(\mathrm{dv}\varrho/\eta)$, or the Reynolds number, be unchanged. The precautional word "normal" was used in this statement because under certain conditions such extreme values of one or more of the dimensional properties may be required to satisfy the conditions of similarity that the basic equation from which the dimensionless equation was derived be no longer valid. For example, the argument leading to eqn. (7.11) assumed that a certain limited number of properties control the frictional forces. Under extreme conditions other factors must be taken into account. Thus, for supersonic flow ($\mathrm{v} > \mathrm{c}$, where c is velocity of sound) another dimensionless quantity the Mach number, v/c, becomes at least as important as, and finally more important than, the Reynolds number. We then must amend eqn. (7.11) to read

$$(\mathrm{id}/\varrho \mathrm{v}^2) = \varphi(\mathrm{dv}\varrho/\eta, \mathrm{v}/\mathrm{c}) \tag{7.33}$$

in which $(\mathrm{id}/\varrho \mathrm{v}^2)$ is now a bivariate function. Extreme values of v in the model might be demanded if the model diameter is small and viscosity simultaneously large.

At another extreme range of conditions, namely when d is so small as to be comparable with the thickness of the electrical double layer on the solid-liquid interface, sets of new factors including the zeta potential and the ionic strength begin to control the frictional resistance and eqn. (7.11) must be modified in other ways. The influence of these modifications appears in the operation of filtration. (See Ruth (1935).) Again, eqn. (7.11) must be modified and extended if, as a result of the development of frictional heat in the pipe, isothermal conditions are not maintained.

Summarizing therefore, we may say that an expression of balance in a natural system with a limited number of physical variables is an approximation only but may well be valid, with reasonable accuracy, over a limited range of the variables concerned. However, at extreme values new factors become important. It is essential therefore that similarity principles should be used only when it can be established that the new physical conditions still fall within the range of validity of the equation of equilibrium.

A further complication may arise if the conditions of equilibrium are controlled by more than one factor. Let us again consider the example of a liquid flowing in a hydraulic gradient, but now suppose that the forces concerned are not only inertial and frictional factors but also include others due to the presence of a pressure gradient, grad p, acceleration due to unsteady state flow and the influence of surface forces operating on a surface of varied curvature having a hydraulic radius, r. Now we must take for the condition of balancing forces per unit volume

$$\varrho \Delta^2 i = -\varrho(\partial v/\partial t) - \varrho(v\Delta v) - \text{grad}\, p + \gamma\, \text{grad}\,(1/r) + \eta \Delta^2 v \qquad (7.34a)$$

On dividing every term in this expression by the third term we get

$$\Delta^2 i/v\Delta v = -1 - (\partial v/\partial t)/v\Delta v - p/\varrho v^2 - \gamma/\varrho r v^2 + (\eta/\varrho)\,(\Delta^2 v/v\Delta v) \qquad (7.34b)$$

Further, on taking all the spatial quantities as equivalent and expressing the dimensionless ratio on the left-hand side as an arbitrary function of the others, it can be shown that

$$(id/\varrho v^2) = \varphi(p/\varrho v^2,\ vt/d,\ \varrho d v^2/\gamma,\ \varrho d v/\eta) \qquad (7.35)$$

in which the frictional number $(id/\varrho v^2)$ is now a quadravariant function of the Newton number, the Stroudhal number, the Weber number and the Reynolds number. These four quantities must take the same values in the model and prototype if complete similarity is to be attained.

If the variables of the model are denoted by m and of the prototype by p then

$$p_m/\varrho_m v_m^2 = p_p/\varrho_p v_p^2 \tag{7.36a}$$

$$d_m/v_m t_m = d_p/v_p t_p \tag{7.36b}$$

$$\varrho_m d_m v_m^2/\gamma_m = \varrho_p d_p v_p^2/\gamma_p \tag{7.36c}$$

$$\varrho_m d_m v_m/\eta_m = \varrho_p d_p v_p/\eta_p \tag{7.36d}$$

If the geometrical scale ratio of model to prototype, namely d_p/d_m is equal to δ and if the same fluid is used in both cases, then condition (7.36d) requires that v_m/v_p should be δ, while condition (7.36c) requires that v_m/v_p should be $v\delta$, condition (7.36b) requires that v_m/v_p should be $1/\delta$ and condition (7.36a) requires that v_m/v_p should be 1. These four conditions are mutually inconsistent. It therefore follows that a complete condition of similarity between two systems can only be obtained if each one of the physical properties (viscosity, density, surface tension, etc.) of the systems also be changed. Occasionally the required properties for dimensional similarity may be unattainable.

A similar situation can occur if the complete expressions for true equilibrium require more than one type of balance. An expression of all conditions of balance in a chemically reacting system may involve consideration of three types of balance equations, namely a material balance, an energy balance, and a momentum balance. Damköhler (1936) first obtained conditions of similarity for such triple-fold equations of balance and again he was led to the conclusion that several of the required conditions were unattainable.

7.10. Partial Similarity

When the complete conditions for similarity are inconsistent, it sometimes happens that one alone of the various factors controlling the state of equilibrium of the system is completely or almost completely dominant so that the influence of the other factors may then amount only to minor corrections. In chemical kinetics, for example, we may speak of the rate controlling factor. When a distinctive rate controlling factor exists we may proceed to set up conditions of similarity by ignoring the contributions due to other factors. States of similarity established under these conditions are referred to as conditions of partial similarity. We may also establish conditions of partial similarity where two or anything less than the full number of independent dimensionless quantities as defined by the Π theorem are taken into account. Partial similarity is of value and applications of its principles valid when and only when one or more of the controlling factors are distinctly dominant, at least over a limited region of all the variable scales. Of course, for different ranges of the variables, different control variables may be dominant. Thus, the upthrust exerted by a moving fluid or an aerofoil depends on the Reynolds number and on the Mach number. The first requires that v be proportional to the reciprocal of a linear dimension and the second that v be independent of the same (see Table 7.5). If one fluid only (air) is to be used these conditions are inconsistent. However, at subsonic speeds the contribution due to the Reynolds number is dominant and a condition of partial similarity may be established in terms of this factor. At supersonic speeds the contribution due to the Mach number is dominant and completely different conditions for partial similarity may be established.

An important consideration relevant to the conditions of partial similarity as concerned with chemical systems was raised by Laupichler (1938) who studied the role of different factors controlling the rate of the water gas reaction in a commercial reactor. The overall rate was the sum of the reciprocal of two terms, one due to chemical reaction and the other due to the rate of removal of the products of the reaction (i.e., to the process of mass diffusion). Under special conditions either factor could be the rate controlling factor. For a diffusion-controlled system Johnstone and Thring (1957) refer to a dynamical regime and they refer the chemically controlled process to a chemical regime. In the first case the appropriate model calls for dynamical similarity and the second for chemical similarity.

Special interest in systems showing partial similarity centres around the different geometrical dimensions. Complete similarity requires not only dynamical, chemical and other similarity but also a condition of geometrical similarity in which the model has the same shape as the prototype. This is tantamount to assuming that all the three linear dimensions have a separate physical dimensional significance and must be balanced independently in the attempt to set the model conditions. In many engineering and chemical applications the geometrical shape of vessels concerned does not appreciably affect the performance of the system. Accordingly the model may be geometrically deformed relative to the prototype in order to help balance the non-geometrical conditions for similarity. Thus, in seeking for models of a catalysed reaction, Bosworth (1947a) proposed that the catalyst grains in the smaller model should actually be larger than the catalyst grains in a larger reactor in order to maintain chemical similarity. This argument implies that a workable model has only partial similarity to its prototype with the (generally insignificant) geometrical conditions ignored, but the fact that some such conditions are ignored limits the range of usefulness of the model concept. A useful partial model, for example, could not be made so small that the required catalyst grains were larger than the reactor itself. Thus partial similarity can only be used for a limited change in the scale of operations.

7.11. Chemical Similarity

The problem of scaling-up a chemical reaction is one of the major technical problems of an expanding chemical industry. The traditional approach has been that of cautious approximation. A reaction known to proceed in a controlled and favourable manner in the laboratory is studied on successively larger and larger scale units until finally a factory scale unit can be designed. In many branches of engineering (particularly in mechanical engineering) dimensional methods have been applied systematically to problems of this nature. Models which are dynamically similar to the full scale equipment may be constructed and studied in the laboratory. Until comparatively recently little attention has been given to the possibility of using a similar technique in the chemical technologies, but the problem is intrinsically more complicated in that the state of equilibrium in a chemical reactor involves the balance of many factors and each equation of balance can give rise to a distinct set of dimensionless ratios required to be considered in establishing the complete conditions for similarity.

In the papers of Damköhler (1936), Johnston (1939), and Bosworth (1947b), the equations of balance represent those due to the conservation of mass for each atomic species entering into the reaction, and those due to the conservation of energy and momentum. These equations then give rise to a series of dimensionless quantities. For a perfectly general system of n reacting substances in which both a change in volume and a change in temperature (heat being removed either by conduction, convection or radiation) can occur, there are $5 + 2n$ dimensionless ratios and thus $5 + 2n$ conditions for similarity. Some of these conditions may be difficult to meet. For example, it appears that for combined reaction-kinetic and dynamical similarity it is necessary both that the flow rate and the chemical reaction rate should be faster on the smaller scale reactor. Similarly, in a catalyst reactor it is required that the grain size and the catalyst activity should be greater in the smaller scale reactor. If the principles of chemical similarity are generally to be used it appears either that reaction vessels can be designed for only a fraction of their potential capacity or else that it is necessary to study (parallel to any industrial development) systems of the same order and kinetic nature but which exhibit faster reactions than the ones of industrial importance. This type of work may well form an important research field in any institution devoted to the study, rather than to the production, of the results of chemical technology.

Among the other factors controlling the complete conditions of chemical similarity there may, and frequently will, be conditions for any fixed system which are mutually inconsistant. Under each set of conditions it is often possible, by ignoring the contribution of minor factors in the equations of equilibrium, to obtain usable conditions of partial similarity. Where only one, or a limited small number, of independent factors are dominating the system it is frequently profitable to set up a usable partial model which belongs to a different sort of system altogether. Notable examples occur in the use of electrical analogue machines such as the heat and mass flow analyser at the University of Columbia (Paschkis, 1947). Somewhat more sophisticated applications have been made by studying processes of evaporation using dimensionless ratios which were taken as the analogues of corresponding properties concerned with convective heat transfer (Powell, 1940). By eliminating the effect of one of the chemical kinetic factors, Bosworth (1949) was able to co-ordinate much of the data connected with the chemical attack on a metal surface by corrosive acid mixtures by setting up a partial analogue with thermal systems involving forced and natural convection. The curves found to fit the relationship between dimensionless quantities in the thermal system were shown also to fit the relationship between corresponding dimensionless quantities in the chemical system.

NOTATION OF CHAPTER VII

a	acceleration
A	area
α	angle
a, *b*, *c*, etc.	exponents
A, B, C	concentrations of reactants **A**, **B**, **C**
α, β, γ	dimensionless ratios of concentrations
β	coefficient of thermal expansion; Bashforth and Adams' numeric; a constant
c	velocity of light; velocity of sound
c_p	specific heat at constant pressure
C	concentration; C_0 = initial concentration
γ	surface tension
d	diameter or linear dimension
D	diffusion coefficient
δ	dimensionless ratio of diameters
e	electronic charge
E	energy
ε	permittivity
η	coefficient of viscosity
f	force
g	acceleration due to gravity
h	heat transfer coefficient; Planck's constant
i	loss of hydraulic head
I	current
k	specific reaction rate; a constant
k	number of fundamental dimensions
k	thermal conductivity
$\varkappa$	Boltzmann's constant; dimensionless ratio of specific reaction rates
K	a numerical constant
l	length
L	latent heat
m	mass
m	(as subscript) model

M	molecular weight
μ	permeability
n	number of terms in an equation; order of reaction
N	Avogadro's number; number of moles
ν	frequency
p	pressure
p	(as subscript) prototype
Π	a numeric or dimensionless product. The common (double letter) symbols for well-known numerics are given in Table 7.5
Q	a measured property or general term in an equation; electric charge
r	radius
R	universal gas constant
ϱ	density
s	displacement
S	entropy; stress
σ	charge density; strain; Stefan–Boltzmann constant
ψ	stiffness
t	time
T	temperature
τ	relative time
v	velocity
V	volume
x, y, z	variables; spatial coordinates
χ	compressibility
z	charge on an ion
ζ	zeta potential

APPENDIX

TABLES OF STATISTICAL FUNCTIONS

Sources and References

The tables in this appendix are adapted in the main from the *Biometrika Tables for Statisticians* edited by E. S. Pearson and H. O. Hartley (Pearson and Hartley, 1966) and are reproduced with the permission of the Biometrika Trustees. Those not derived from this source are separately acknowledged.

Tables of statistical functions and those designed as aids in computing are now as readily available as tables of logarithms and only a short working set is included here. For more detailed extensive collections the reader is referred to the *Biometrika Tables*, the books on statistical methods listed in the general bibliography (especially Dixon and Massey, 1957) and to the following:

Fisher, R. A. and Yates, F. (1963) *Statistical Tables for Biological, Agricultural and Medical Research*, 6th Ed., Oliver & Boyd, Edinburgh.

Hald, A. (1952) *Statistical Tables and Formulas*, Wiley, New York.

Lindley, D. V. and Miller, J. C. P. (1953) *Cambridge Elementary Statistical Tables*, Cambridge.

Barlow's Tables of Squares, Cubes etc. of Integers up to 12,500 (1941), 4th Ed., Spon, London.

The last is of great assistance in computation.

Table I. Ordinates and Areas

(a) Ordinates

u	0	1	2	3	4	5	6	7	8	9
0·0	·39894	·39892	·39886	·39876	·39862	·39844	·39822	·39797	·39767	·39733
0·1	·39695	·39654	·39608	·39559	·39505	·39448	·39387	·39322	·39253	·39181
0·2	·39104	·39024	·38940	·38853	·38762	·38667	·38568	·38466	·38361	·38251
0·3	·38139	·38023	·37903	·37780	·37654	·37524	·37391	·37255	·37115	·36973
0·4	·36827	·36678	·36526	·36371	·36213	·36053	·35889	·35723	·35553	·35381
0·5	·35207	·35029	·34849	·34667	·34482	·34294	·34105	·33912	·33718	·33521
0·6	·33322	·33121	·32918	·32713	·32506	·32297	·32086	·31874	·31659	·31443
0·7	·31225	·31006	·30785	·30563	·30339	·30114	·29887	·29659	·29430	·29200
0·8	·28969	·28737	·28504	·28269	·28034	·27798	·27562	·27324	·27086	·26848
0·9	·26609	·26369	·26129	·25888	·25647	·25406	·25164	·24923	·24681	·24439
1·0	·24197	·23955	·23713	·23471	·23230	·22988	·22747	·22506	·22265	·22025
1·1	·21785	·21546	·21307	·21069	·20831	·20594	·20357	·20121	·19886	·19652
1·2	·19419	·19186	·18954	·18724	·18494	·18265	·18037	·17810	·17585	·17360
1·3	·17137	·16915	·16694	·16474	·16256	·16038	·15822	·15608	·15395	·15183
1·4	·14973	·14764	·14556	·14350	·14146	·13943	·13742	·13542	·13344	·13147
1·5	·12952	·12758	·12566	·12376	·12188	·12001	·11816	·11632	·11450	·11270
1·6	·11092	·10915	·10741	·10567	·10396	·10226	·10059	·09893	·09728	·09566
1·7	·09405	·09246	·09089	·08933	·08780	·08628	·08478	·08329	·08183	·08038
1·8	·07895	·07754	·07614	·07477	·07341	·07206	·07074	·06943	·06814	·06687
1·9	·06562	·06438	·06316	·06195	·06077	·05959	·05844	·05730	·05618	·05508
2·0	·05399	·05292	·05186	·05082	·04980	·04879	·04780	·04682	·04586	·04491
2·1	·04398	·04307	·04217	·04128	·04041	·03955	·03871	·03788	·03706	·03626
2·2	·03547	·03470	·03394	·03319	·03246	·03174	·03103	·03034	·02965	·02898
2·3	·02833	·02768	·02705	·02643	·02582	·02522	·02463	·02406	·02349	·02294
2·4	·02239	·02186	·02134	·02083	·02033	·01984	·01936	·01888	·01842	·01797
2·5	·01753	·01709	·01667	·01625	·01585	·01545	·01506	·01468	·01431	·01394
2·6	·01358	·01323	·01289	·01256	·01223	·01191	·01160	·01130	·01100	·01071
2·7	·01042	·01014	·00987	·00961	·00935	·00909	·00885	·00861	·00837	·00814
2·8	·00792	·00770	·00748	·00727	·00707	·00687	·00668	·00649	·00631	·00613
2·9	·00595	·00578	·00562	·00545	·00530	·00514	·00499	·00485	·00470	·00457
3·0	·00443	·00430	·00417	·00405	·00393	·00381	·00370	·00358	·00348	·00337
3·1	·00327	·00317	·00307	·00298	·00288	·00279	·00271	·00262	·00254	·00246
3·2	·00238	·00231	·00224	·00216	·00210	·00203	·00196	·00190	·00184	·00178
3·3	·00172	·00167	·00161	·00156	·00151	·00146	·00141	·00136	·00132	·00127
3·4	·00123	·00119	·00115	·00111	·00107	·00104	·00100	·00097	·00094	·00090
3·5	·00087	·00084	·00081	·00079	·00076	·00073	·00071	·00068	·00066	·00063
3·6	·00061	·00059	·00057	·00055	·00053	·00051	·00049	·00047	·00046	·00044
3·7	·00042	·00041	·00039	·00038	·00037	·00035	·00034	·00033	·00031	·00030
3·8	·00029	·00028	·00027	·00026	·00025	·00024	·00023	·00022	·00021	·00021
3·9	·00020	·00019	·00018	·00018	·00017	·00016	·00016	·00015	·00014	·00014
4·0	·00013	·00013	·00012	·00012	·00011	·00011	·00011	·00010	·00010	·00009
4·1	·00009	·00009	·00008	·00008	·00008	·00007	·00007	·00007	·00006	·00006
4·2	·00006	·00006	·00005	·00005	·00005	·00005	·00005	·00004	·00004	·00004
4·3	·00004	·00004	·00004	·00003	·00003	·00003	·00003	·00003	·00003	·00003
4·4	·00002	·00002	·00002	·00002	·00002	·00002	·00002	·00002	·00002	·00002

4·36 ·00003; 4·45 ·00002; 4·60 ·00001

Table (a) gives the height of the ordinate of the standardized curve at $u = (x - \mu)/\sigma$, the area under the curve being unity.

OF THE GAUSSIAN CURVE

(b) Areas

u	0	1	2	3	4	5	6	7	8	9
0·0	·00000	·00399	·00798	·01197	·01595	·01994	·02392	·02790	·03188	·03586
0·1	·03983	·04380	·04776	·05172	·05567	·05962	·06356	·06749	·07142	·07535
0·2	·07926	·08317	·08706	·09095	·09483	·09871	·10257	·10642	·11026	·11409
0·3	·11791	·12172	·12552	·12930	·13307	·13683	·14058	·14431	·14803	·15173
0·4	·15542	·15910	·16276	·16640	·17003	·17364	·17724	·18082	·18439	·18793
0·5	·19146	·19497	·19847	·20194	·20540	·20884	·21226	·21566	·21904	·22240
0·6	·22575	·22907	·23237	·23565	·23891	·24215	·24537	·24857	·25175	·25490
0·7	·25804	·26115	·26424	·26730	·27035	·27337	·27637	·27935	·28230	·28524
0·8	·28814	·29103	·29389	·29673	·29955	·30234	·30511	·30785	·31057	·31327
0·9	·31594	·31859	·32121	·32381	·32639	·32894	·33147	·33398	·33646	·33891
1·0	·34134	·34375	·34614	·34850	·35083	·35314	·35543	·35769	·35993	·36214
1·1	·36433	·36650	·36864	·37076	·37286	·37493	·37698	·37900	·38100	·38298
1·2	·38493	·38686	·38877	·39065	·39251	·39435	·39617	·39796	·39973	·40147
1·3	·40320	·40490	·40658	·40824	·40988	·41149	·41308	·41466	·41621	·41774
1·4	·41924	·42073	·42220	·42364	·42507	·42647	·42786	·42922	·43056	·43189
1·5	·43319	·43448	·43574	·43699	·43822	·43943	·44062	·44179	·44295	·44408
1·6	·44520	·44630	·44738	·44845	·44950	·45053	·45154	·45254	·45352	·45449
1·7	·45543	·45637	·45728	·45818	·45907	·45994	·46080	·46164	·46246	·46327
1·8	·46407	·46485	·46562	·46638	·46712	·46784	·46856	·46926	·46995	·47062
1·9	·47128	·47193	·47257	·47320	·47381	·47441	·47500	·47558	·47615	·47670
2·0	·47725	·47778	·47831	·47882	·47932	·47982	·48030	·48077	·48124	·48169
2·1	·48214	·48257	·48300	·48341	·48382	·48422	·48461	·48500	·48537	·48574
2·2	·48610	·48645	·48679	·48713	·48745	·48778	·48809	·48840	·48870	·48899
2·3	·48928	·48956	·48983	·49010	·49036	·49061	·49086	·49111	·49134	·49158
2·4	·49180	·49202	·49224	·49245	·49266	·49286	·49305	·49324	·49343	·49361
2·5	·49379	·49396	·49413	·49430	·49446	·49461	·49477	·49492	·49506	·49520
2·6	·49534	·49547	·49560	·49573	·49585	·49598	·49609	·49621	·49632	·49643
2·7	·49653	·49664	·49674	·49683	·49693	·49702	·49711	·49720	·49728	·49736
2·8	·49744	·49752	·49760	·49767	·49774	·49781	·49788	·49795	·49801	·49807
2·9	·49813	·49819	·49825	·49831	·49836	·49841	·49846	·49851	·49856	·49861
3·0	·49865	·49869	·49874	·49878	·49882	·49886	·49889	·49893	·49896	·49900
3·1	·49903	·49906	·49910	·49913	·49916	·49918	·49921	·49924	·49926	·49929
3·2	·49931	·49934	·49936	·49938	·49940	·49942	·49944	·49946	·49948	·49950
3·3	·49952	·49953	·49955	·49957	·49958	·49960	·49961	·49962	·49964	·49965
3·4	·49966	·49968	·49969	·49970	·49971	·49972	·49973	·49974	·49975	·49976
3·5	·49977	·49978	·49978	·49979	·49980	·49981	·49981	·49982	·49983	·49983
3·6	·49984	·49985	·49985	·49986	·49986	·49987	·49987	·49988	·49988	·49989
3·7	·49989	·49990	·49990	·49990	·49991	·49991	·49992	·49992	·49992	·49992
3·8	·49993	·49993	·49993	·49994	·49994	·49994	·49994	·49995	·49995	·49995
3·9	·49995	·49995	·49996	·49996	·49996	·49996	·49996	·49996	·49997	·49997

4·01 ·49997; **4·11** ·49998; **4·26** ·49999

Table (b) gives the area between the mid-ordinate and *u* taken positive.

TABLE II. PERCENTAGE POINTS

	t			
f	P = 0·1	P = 0·05	P = 0·01	P = 0·1
1	6·31	12·71	63·66	2·71
2	2·92	4·30	9·92	4·61
3	2·35	3·18	5·84	6·25
4	2·13	2·78	4·60	7·78
5	2·02	2·57	4·03	9·24
6	1·94	2·45	3·71	10·64
7	1·89	2·36	3·50	12·02
8	1·86	2·31	3·36	13·36
9	1·83	2·26	3·25	14·68
10	1·81	2·23	3·17	15·99
11	1·80	2·20	3·11	17·28
12	1·78	2·18	3·05	18·55
13	1·77	2·16	3·01	19·81
14	1·76	2·14	2·98	21·06
15	1·75	2·13	2·95	22·31
16	1·75	2·12	2·92	23·54
17	1·74	2·11	2·90	24·77
18	1·73	2·10	2·88	25·99
19	1·73	2·09	2·86	27·20
20	1·72	2·09	2·85	28·41
21	1·72	2·08	2·83	29·62
22	1·72	2·07	2·82	30·81
23	1·71	2·07	2·81	32·01
24	1·71	2·06	2·80	33·20
25	1·71	2·06	2·79	34·38
26	1·71	2·06	2·78	35·56
27	1·70	2·05	2·77	36·74
28	1·70	2·05	2·76	37·92
29	1·70	2·05	2·76	39·09
30	1·70	2·04	2·75	40·26
40	1·68	2·02	2·70	51·8
50	1·68	2·01	2·68	63·2
60	1·67	2·00	2·66	74·4
80	1·66	1·99	2·64	96·6
100	1·66	1·98	2·63	118·5

P is the two-sided significance level for t and r but one-sided for χ^2.

OF THE DISTRIBUTIONS OF t, χ^2 AND r (FOR $\varrho = 0$)

χ^2		r		
P = 0·05	P = 0·01	P = 0·1	P = 0·05	P = 0·01
3·84	6·63	0·988	0·997	1·000
5·99	9·21	0·900	0·950	0·990
7·81	11·34	0·805	0·878	0·959
9·49	13·28	0·729	0·811	0·917
11·07	15·09	0·669	0·754	0·875
12·59	16·81	0·621	0·707	0·834
14·07	18·48	0·582	0·666	0·798
15·51	20·09	0·549	0·632	0·765
16·92	21·67	0·521	0·602	0·735
18·31	23·21	0·497	0·576	0·708
19·68	24·72	0·476	0·553	0·684
21·03	26·22	0·457	0·532	0·661
22·36	27·69	0·441	0·514	0·641
23·68	29·14	0·426	0·497	0·623
25·00	30·58	0·412	0·482	0·606
26·30	32·00	0·400	0·468	0·590
27·59	33·41	0·389	0·456	0·575
28·87	34·81	0·378	0·444	0·561
30·14	36·19	0·369	0·433	0·549
31·41	37·57	0·360	0·423	0·537
32·67	38·93	0·352	0·413	0·526
33·92	40·29	0·344	0·404	0·515
35·17	41·64	0·337	0·396	0·505
36·42	42·98	0·330	0·388	0·496
37·65	44·31	0·323	0·381	0·487
38·89	45·64	0·317	0·374	0·478
40·11	46·96	0·311	0·367	0·470
41·34	48·28	0·306	0·361	0·463
42·56	49·59	0·301	0·355	0·456
43·77	50·89	0·296	0·349	0·449
55·8	63·7	0·257	0·304	0·393
67·5	76·2	0·231	0·273	0·354
79·1	88·4	0·211	0·250	0·325
101·9	112·3	0·183	0·217	0·283
124·3	135·8	0·164	0·195	0·254

TABLE III. PERCENTAGE POINTS

f_2 \ f_1	Upper value, P = 0·05,			
	1	2	3	4
1	161 648	200 800	216 864	225 900
2	18·5 38·5	19·0 39·0	19·2 39·2	19·2 39·2
3	10·1 17·4	9·55 16·0	9·28 15·4	9·12 15·1
4	7·71 12·2	6·94 10·6	6·59 9·98	6·39 9·60
5	6·61 10·0	5·79 8·43	5·41 7·76	5·19 7·39
6	5·99 8·81	5·14 7·26	4·76 6·60	4·53 6·23
7	5·59 8·07	4·74 6·54	4·35 5·89	4·12 5·52
8	5·32 7·57	4·46 6·06	4·07 5·42	3·84 5·05
9	5·12 7·21	4·26 5·71	3·86 5·08	3·63 4·72
10	4·96 6·94	4·10 5·46	3·71 4·83	3·48 4·47
12	4·75 6·55	3·89 5·10	3·49 4·47	3·26 4·12
15	4·54 6·20	3·68 4·77	3·29 4·15	3·06 3·80
20	4·35 5·87	3·49 4·46	3·10 3·86	2·87 3·51
30	4·17 5·57	3·32 4·18	2·92 3·59	2·69 3·25
40	4·08 5·42	3·23 4·05	2·84 3·46	2·61 3·13
60	4·00 5·29	3·15 3·93	2·76 3·34	2·53 3·01
120	3·92 5·15	3·07 3·80	2·68 3·23	2·45 2·89
∞	3·84 5·02	3·00 3·69	2·60 3·12	2·37 2·79

f_1 corresponds to the greater expected or numerator variance.

OF THE DISTRIBUTION OF F

Lower value, P = 0·025

5	6	9	12	24	∞
230	234	241	244	249	254
922	937	963	977	997	1018
19·3	19·3	19·4	19·4	19·5	19·5
39·3	39·3	39·4	39·4	39·5	39·5
9·01	8·94	8·81	8·74	8·64	8·53
14·9	14·7	14·5	14·3	14·1	13·9
6·26	6·16	6·00	5·91	5·77	5·63
9·36	9·20	8·90	8·75	8·51	8·26
5·05	4·95	4·77	4·68	4·53	4·36
7·15	6·98	6·68	6·52	6·28	6·02
4·39	4·28	4·10	4·00	3·84	3·67
5·99	5·82	5·52	5·37	5·12	4·85
3·97	3·87	3·68	3·57	3·41	3·23
5·29	5·12	4·82	4·67	4·42	4·14
3·69	3·58	3·39	3·28	3·12	2·93
4·82	4·65	4·36	4·20	3·95	3·67
3·48	3·37	3·18	3·07	2·90	2·71
4·48	4·32	4·03	3·87	3·61	3·33
3·33	3·22	3·02	2·91	2·74	2·54
4·24	4·07	3·78	3·62	3·37	3·08
3·11	3·00	2·80	2·69	2·51	2·30
3·89	3·73	3·44	3·28	3·02	2·72
2·90	2·79	2·59	2·48	2·29	2·07
3·58	3·41	3·12	2·96	2·70	2·40
2·71	2·60	2·39	2·28	2·08	1·84
3·29	3·13	2·84	2·68	2·41	2·09
2·53	2·42	2·21	2·09	1·89	1·62
3·03	2·87	2·57	2·41	2·14	1·79
2·45	2·34	2·12	2·00	1·79	1·51
2·90	2·74	2·45	2·29	2·01	1·64
2·37	2·25	2·04	1·92	1·70	1·39
2·79	2·63	2·33	2·17	1·88	1·48
2·29	2·17	1·96	1·83	1·61	1·25
2·67	2·52	2·22	2·05	1·76	1·31
2·21	2·10	1·88	1·75	1·52	1·00
2·57	2·41	2·11	1·95	1·64	1·00

TABLE IV. PERCENTAGE POINTS

k \ c	Upper value, P = 0·05;						
	0·0	0·5	1·0	1·5	2·0	2·5	3·0
3	5·99 9·21	6·47 9·92	6·89 10·47	7·20 10·78	7·38 10·81	7·39 10·50	7·22 9·83
4	7·81 11·34	8·24 11·95	8·63 12·46	8·96 12·86	9·21 13·11	9·38 13·18	9·43 13·03
5	9·49 13·28	9·88 13·81	10·24 14·30	10·57 14·71	10·86 15·03	11·08 15·25	11·24 15·34
6	11·07 15·09	11·43 15·58	11·78 16·03	12·11 16·44	12·40 16·79	12·65 17·07	12·86 17·27
7	12·59 16·81	12·94 17·27	13·27 17·70	13·59 18·10	13·88 18·46	14·15 18·77	14·38 19·02
8	14·07 18·48	14·40 18·91	14·72 19·32	15·03 19·71	15·32 20·07	15·60 20·39	15·84 20·67
9	15·51 20·09	15·83 20·50	16·14 20·90	16·44 21·28	16·73 21·64	17·01 21·97	17·26 22·26
10	16·92 21·67	17·23 22·06	17·54 22·45	17·83 22·82	18·12 23·17	18·39 23·50	18·65 23·80
11	18·31 23·21	18·61 23·59	18·91 23·97	19·20 24·33	19·48 24·67	19·76 25·00	20·02 25·31
12	19·68 24·72	19·97 25·10	20·26 25·46	20·55 25·81	20·83 26·15	21·10 26·48	21·36 26·79
13	21·03 26·22	21·32 26·58	21·60 26·93	21·89 27·28	22·16 27·62	22·43 27·94	22·69 28·25
14	22·36 27·69	22·65 28·04	22·93 28·39	23·21 28·73	23·48 29·06	23·75 29·38	24·01 29·69
15	23·68 29·14	23·97 29·49	24·24 29·83	24·52 30·16	24·79 30·49	25·05 30·80	25·31 30·11

† Merrington and Thompson (1946). The tabulated values are maximum should be consulted for the slightly lower values which obtain when sample

OF THE DISTRIBUTION OF M†

Lower value, P = 0·01

3·5	4·0	4·5	5·0	6·0	7·0	8·0	9·0	10·0
—	—	—	—	—	—	—	—	—
—	—	—	—	—	—	—	—	—
9·37	9·18	—	—	—	—	—	—	—
12·65	12·03	—	—	—	—	—	—	—
11·32	11·31	11·21	11·02	—	—	—	—	—
15·28	15·06	14·66	14·07	—	—	—	—	—
13·01	13·11	13·14	13·10	12·78	—	—	—	—
17·37	17·37	17·24	16·98	16·03	—	—	—	—
14·58	14·73	14·83	14·88	14·81	14·49	—	—	—
19·21	19·32	19·35	19·28	18·84	17·92	—	—	—
16·06	16·25	16·40	16·51	16·60	16·49	16·16	—	—
20·90	21·08	21·20	21·25	21·13	20·64	19·76	—	—
17·49	17·70	17·88	18·03	18·22	18·26	18·12	17·79	—
22·52	22·74	22·91	23·03	23·10	22·91	22·41	21·56	—
18·89	19·11	19·31	19·48	19·75	19·89	19·89	19·73	19·40
24·08	24·32	24·52	24·69	24·90	24·90	24·66	24·15	23·33
20·26	20·49	20·70	20·89	21·21	21·42	21·52	21·49	21·32
25·59	25·85	26·08	26·28	26·57	26·70	26·65	26·38	25·86
21·61	21·84	22·06	22·27	22·62	22·88	23·06	23·12	23·07
27·08	27·35	27·59	27·81	28·16	28·39	28·46	28·37	28·07
22·94	23·18	23·40	23·62	23·99	24·30	24·53	24·66	24·70
28·54	28·81	29·07	29·30	29·70	29·99	30·16	30·19	30·06
24·26	24·50	24·73	24·95	25·34	25·68	25·95	26·14	26·25
29·98	30·26	30·52	30·77	31·19	31·53	31·77	31·89	31·88
25·56	25·80	26·04	26·26	26·67	27·03	27·33	27·56	27·73
31·40	31·68	31·95	32·20	32·66	33·03	33·32	33·51	33·59

values. Table 32 of the *Biometrika Tables* (see Sources and References above) sizes are very unequal and for a method of interpolation.

TABLE V. CONTROL CHART FACTORS†

n	Limit factors for averages		Limit factors for ranges				
	$A'_{.025}$	$A'_{.001}$	$D'_{.001}$	$D'_{.025}$	$D'_{.975}$	$D'_{.999}$	d_2
2	1·229	1·937	0·00	0·039	2·810	4·12	1·128
3	0·668	1·054	0·04	0·178	2·174	2·99	1·693
4	0·476	0·750	0·10	0·288	1·933	2·58	2·059
5	0·377	0·594	0·16	0·365	1·806	2·36	2·326
6	0·316	0·498	0·21	0·419	1·720	2·22	2·534
7	0·274	0·432	0·26	0·462	1·660	2·12	2·704
8	0·244	0·384	0·29	0·495	1·620	2·04	2·847
9	0·220	0·347	0·32	0·522	1·583	1·99	2·970
10	0·202	0·317	0·35	0·544	1·556	1·94	3·078
11	0·186	0·294	0·38	0·562	1·532	1·90	3·173
12	0·174	0·274	0·40	0·578	1·510	1·87	3·258

† The A' factors are reproduced from B.S. 600R: 1942: *Quality Control Charts* (not current) by permission of the British Standards Institution. The D' factors have been computed from Tables 22 and 23 of the *Biometrika Tables* in which values of d_2, the mean of w_n/σ for samples drawn from a Gaussian distribution, are also to be found. $A'_{.025}$ and $A'_{.001}$ are $1{\cdot}96/d_2\sqrt{n}$ and $3{\cdot}09/d_2\sqrt{n}$ respectively. $D'_{.001}$ etc. are quantiles of the w_n/σ distribution divided by d_2.

TABLE VI. PERCENTAGE POINTS OF DIXON'S r†

n	Criterion	P 0·10	P 0·05	P 0·01
3	$r_{10} = \dfrac{x_2 - x_1}{x_n - x_1}$ or $\dfrac{x_n - x_{n-1}}{x_n - x_1}$	0·886	0·941	0·988
4		0·679	0·765	0·889
5		0·557	0·642	0·780
6		0·482	0·560	0·698
7		0·434	0·507	0·637
8	$r_{11} = \dfrac{x_2 - x_1}{x_{n-1} - x_1}$ or $\dfrac{x_n - x_{n-1}}{x_n - x_2}$	0·479	0·554	0·683
9		0·441	0·512	0·635
10		0·409	0·477	0·597
11	$r_{21} = \dfrac{x_3 - x_1}{x_{n-1} - x_1}$ or $\dfrac{x_n - x_{n-2}}{x_n - x_2}$	0·517	0·576	0·679
12		0·490	0·546	0·642
13		0·467	0·521	0·615
14	$r_{22} = \dfrac{x_3 - x_1}{x_{n-2} - x_1}$ or $\dfrac{x_n - x_{n-2}}{x_n - x_3}$	0·492	0·546	0·641
15		0·472	0·525	0·616
16		0·454	0·507	0·595
17		0·438	0·490	0·577
18		0·424	0·475	0·561
19		0·412	0·462	0·547
20		0·401	0·450	0·535
21		0·391	0·440	0·524
22		0·382	0·430	0·514
23		0·374	0·421	0·505
24		0·367	0·413	0·497
25		0·360	0·406	0·489

† Dixon (1953). The significance levels apply to tests on the observation at a predesignated end of the set. In testing an observation suspected after inspection of the values, the significance levels are double those quoted in the table, e.g. P = 0·05 becomes 0·10. The table is reprinted with the permission of Professor W. J. Dixon.

TABLE VII. PERCENTAGE POINTS OF THE DISTRIBUTION OF q†

f_p \ n	P = 0·1						
	3	4	5	6	7	8	9
10	1·68	1·92	2·09	2·23	2·33	2·42	2·50
11	1·66	1·90	2·07	2·20	2·30	2·39	2·46
12	1·65	1·88	2·05	2·17	2·28	2·36	2·44
13	1·63	1·86	2·03	2·16	2·26	2·34	2·41
14	1·62	1·85	2·01	2·14	2·24	2·32	2·39
15	1·61	1·84	2·00	2·12	2·22	2·31	2·38
16	1·61	1·83	1·99	2·11	2·21	2·29	2·36
17	1·60	1·82	1·98	2·10	2·20	2·28	2·35
18	1·59	1·82	1·97	2·09	2·19	2·27	2·34
19	1·59	1·81	1·96	2·08	2·18	2·26	2·33
20	1·58	1·80	1·96	2·08	2·17	2·25	2·32
24	1·57	1·78	1·94	2·05	2·15	2·22	2·29
30	1·55	1·77	1·92	2·03	2·12	2·20	2·26
40	1·54	1·75	1·90	2·10	2·10	2·17	2·23
60	1·52	1·73	1·87	1·98	2·07	2·14	2·20
120	1·51	1·71	1·85	1·96	2·05	2·12	2·18
∞	1·50	1·70	1·83	1·94	2·02	2·09	2·15

f_p \ n	P = 0·05						
	3	4	5	6	7	8	9
10	2·01	2·27	2·46	2·60	2·72	2·81	2·89
11	1·98	2·24	2·42	2·56	2·67	2·76	2·84
12	1·96	2·21	2·39	2·52	2·63	2·72	2·80
13	1·94	2·19	2·36	2·50	2·60	2·69	2·76
14	1·93	2·17	2·34	2·47	2·57	2·66	2·74
15	1·91	2·15	2·32	2·45	2·55	2·64	2·71
16	1·90	2·14	2·31	2·43	2·53	2·62	2·69
17	1·89	2·13	2·29	2·42	2·52	2·60	2·67
18	1·88	2·11	2·28	2·40	2·50	2·58	2·65
19	1·87	2·11	2·27	2·39	2·49	2·57	2·64
20	1·87	2·10	2·26	2·38	2·47	2·56	2·63
24	1·84	2·07	2·23	2·34	2·44	2·52	2·58
30	1·82	2·04	2·20	2·31	2·40	2·48	2·54
40	1·80	2·02	2·17	2·28	2·37	2·44	2·50
60	1·78	1·99	2·14	2·25	2·33	2·41	2·47
120	1·76	1·96	2·11	2·22	2·30	2·37	2·43
∞	1·74	1·94	2·08	2·18	2·27	2·33	2·39

TABLE VII *(cont.)*

f_p \ n	P = 0·01						
	3	4	5	6	7	8	9
10	2·78	3·10	3·32	3·48	3·62	3·73	3·82
11	2·72	3·02	3·24	3·39	3·52	3·63	3·72
12	2·67	2·96	3·17	3·32	3·45	3·55	3·64
13	2·63	2·92	3·12	3·27	3·38	3·48	3·57
14	2·60	2·88	3·07	3·22	3·33	3·43	3·51
15	2·57	2·84	3·03	3·17	3·29	3·38	3·46
16	2·54	2·81	3·00	3·14	3·25	3·34	3·42
17	2·52	2·79	2·97	3·11	3·22	3·31	3·38
18	2·50	2·77	2·95	3·08	3·19	3·28	3·35
19	2·49	2·75	2·93	3·06	3·16	3·25	3·33
20	2·47	2·73	2·91	3·04	3·14	3·23	3·29
24	2·42	2·68	2·84	2·97	3·07	3·16	3·22
30	2·38	2·62	2·79	2·91	3·01	3·08	3·15
40	2·34	2·57	2·73	2·85	2·94	3·02	3·08
60	2·29	2·52	2·68	2·79	2·88	2·95	3·01
120	2·25	2·48	2·62	2·73	2·82	2·89	2·95
∞	2·22	2·43	2·57	2·68	2·76	2·83	2·88

† The studentized extreme deviate, $(x_n - \bar{x})/s_p$ or $(\bar{x} - x_1)/s_p$, where n is the size of the sample containing the suspected observation and f_p is the number of degrees of freedom on which s_p is based. Nair (1952) with corrections by David (1956). (The latter should be consulted for other percentage points and for $n = 10$ and 12.)

TABLE VIII. TWO-SIDED TOLERANCE LIMIT FACTORS, l†

Degrees of freedom f	Confidence coefficient 1 − P = 0·90; Proportion of distribution P				Confidence coefficient 1 − P = 0·95; Proportion of distribution P				Confidence coefficient 1 − P = 0·99; Proportion of distribution P			
	0·90	0·95	0·99	0·999	0·90	0·95	0·99	0·999	0·90	0·95	0·99	0·999
4	3·51	4·18	5·49	7·02	4·29	5·11	6·72	8·58	6·64	7·92	10·40	13·29
5	3·14	3·74	4·92	6·28	3·72	4·44	5·83	7·45	5·35	6·38	8·38	10·71
6	2·91	3·47	4·55	5·82	3·38	4·02	5·29	6·75	4·62	5·51	7·24	9·25
7	2·75	3·27	4·30	5·50	3·14	3·74	4·92	6·28	4·15	4·95	6·51	8·31
8	2·63	3·13	4·12	5·26	2·97	3·54	4·65	5·94	3·83	4·56	5·98	7·66
9	2·54	3·02	3·97	5·08	2·84	3·39	4·45	5·69	3·59	4·27	5·62	7·17
10	2·47	2·94	3·86	4·93	2·74	3·26	4·29	5·48	3·40	4·05	5·32	6·80
12	2·36	2·81	3·69	4·72	2·59	3·08	4·05	5·18	3·13	3·73	4·90	6·26
14	2·28	2·72	3·57	4·56	2·49	2·96	3·89	4·96	2·95	3·52	4·61	5·89
16	2·22	2·65	3·48	4·44	2·40	2·86	3·76	4·80	2·81	3·35	4·40	5·62
18	2·17	2·59	3·40	4·35	2·34	2·79	3·66	4·68	2·70	3·22	4·23	5·41
20	2·14	2·54	3·34	4·27	2·29	2·72	3·58	4·57	2·62	3·12	4·10	5·24
25	2·07	2·46	3·23	4·13	2·19	2·61	3·43	4·39	2·47	2·94	3·87	4·94
30	2·02	2·40	3·16	4·04	2·13	2·54	3·33	4·26	2·37	2·82	3·71	4·74
40	1·95	2·33	3·06	3·91	2·05	2·44	3·20	4·09	2·24	2·67	3·50	4·47
50	1·91	2·28	3·00	3·83	1·99	2·37	3·12	3·99	2·16	2·57	3·38	4·31
60	1·89	2·25	2·95	3·77	1·96	2·33	3·06	3·91	2·10	2·50	3·29	4·20
70	1·86	2·22	2·92	3·73	1·93	2·30	3·02	3·85	2·06	2·45	3·22	4·11
80	1·85	2·20	2·89	3·69	1·91	2·27	2·98	3·81	2·02	2·41	3·17	4·05
90	1·83	2·18	2·87	3·67	1·89	2·25	2·96	3·78	2·00	2·38	3·13	3·99
100	1·82	2·17	2·85	3·64	1·87	2·23	2·93	3·75	1·98	2·35	3·09	3·95
200	1·76	2·10	2·76	3·53	1·80	2·14	2·82	3·60	1·87	2·22	2·92	3·73
300	1·74	2·07	2·73	3·48	1·77	2·11	2·77	3·54	1·82	2·17	2·85	3·64
400	1·73	2·06	2·70	3·45	1·75	2·08	2·74	3·50	1·79	2·14	2·81	3·59
500	1·72	2·05	2·69	3·43	1·74	2·07	2·72	3·48	1·78	2·12	2·78	3·55
600	1·71	2·04	2·68	3·42	1·73	2·06	2·71	3·46	1·76	2·10	2·76	3·53
800	1·70	2·03	2·66	3·40	1·72	2·05	2·69	3·43	1·75	2·08	2·74	3·50
1000	1·70	2·02	2·65	3·39	1·71	2·04	2·68	3·42	1·74	2·07	2·72	3·47
∞	1·64	1·96	2·58	3·29	1·64	1·96	2·58	3·29	1·64	1·96	2·58	3·29

† Hald, A. (1952). 1 − P denotes the probability that at least P of the normal distribution will be included between the tolerance limits $\bar{x} \pm sl$, where $\bar{x}$ and s are the mean and the standard deviation of the n observations and $f = n - 1$. The table is reproduced with the permission of John Wiley, New York.

TABLE IX. THE REDUCTION IN $\Sigma \hat{Y}^2$ FOR SIGNIFICANCE AT THE 5% LEVEL†

n	Percent reduction	n	Percent reduction
3	90·2	14	26·4
4	77·1	16	23·2
5	65·8	18	20·7
6	56·9	20	18·7
7	50·0	22	17·1
8	44·4	24	15·7
9	40·0	30	12·6
10	36·3	36	10·5
11	33·2	42	9·1
12	30·6	48	7·9

† Youden (1962). Reproduced by permission of Dr. W. J. Youden.

TABLE X. TRANSFORMATION OF r TO z AND z TO r

z		r	z		r	z		r	z	r	z	r
0·203	0·20	0·197	0·693	0·60	0·537	1·099	0·80	0·664	1·00	0·762	1·40	0·886
0·255	0·25	0·245	0·709	0·61	0·544	1·127	0·81	0·670	1·02	0·770	1·42	0·890
0·310	0·30	0·291	0·725	0·62	0·551	1·157	0·82	0·675	1·04	0·778	1·44	0·894
0·365	0·35	0·336	0·741	0·63	0·558	1·188	0·83	0·681	1·06	0·786	1·46	0·898
0·400	0·38	0·363	0·758	0·64	0·565	1·221	0·84	0·686	1·08	0·793	1·48	0·901
0·424	0·40	0·380	0·775	0·65	0·572	1·256	0·85	0·691	1·10	0·800	1·50	0·905
0·448	0·42	0·397	0·793	0·66	0·578	1·293	0·86	0·696	1·12	0·807	1·52	0·909
0·472	0·44	0·414	0·811	0·67	0·585	1·333	0·87	0·701	1·14	0·814	1·54	0·912
0·497	0·46	0·430	0·829	0·68	0·592	1·376	0·88	0·706	1·16	0·821	1·56	0·915
0·523	0·48	0·446	0·848	0·69	0·598	1·422	0·89	0·711	1·18	0·827	1·60	0·922
0·549	0·50	0·462	0·867	0·70	0·604	1·472	0·90	0·716	1·20	0·834	1·65	0·929
0·563	0·51	0·470	0·887	0·71	0·611	1·528	0·91	0·721	1·22	0·840	1·70	0·935
0·576	0·52	0·478	0·908	0·72	0·617	1·589	0·92	0·726	1·24	0·845	1·80	0·947
0·590	0·53	0·485	0·929	0·73	0·623	1·658	0·93	0·731	1·26	0·851	1·90	0·956
0·604	0·54	0·493	0·950	0·74	0·629	1·738	0·94	0·735	1·28	0·857	2·00	0·964
0·618	0·55	0·500	0·973	0·75	0·635	1·832	0·95	0·740	1·30	0·862	2·20	0·976
0·633	0·56	0·508	0·996	0·76	0·641	1·946	0·96	0·744	1·32	0·867	2·40	0·984
0·648	0·57	0·515	1·020	0·77	0·647	2·092	0·97	0·749	1·34	0·872	2·60	0·989
0·662	0·58	0·523	1·045	0·78	0·653	2·298	0·98	0·753	1·36	0·876	2·80	0·993
0·678	0·59	0·530	1·071	0·79	0·658	2·647	0·99	0·757	1·38	0·881	3·00	0·995

The first three sections of this table can be entered with either r or z as argument in the central column, the required value of z being read off to the left or of r to the right. Note that z is always numerically greater than r.

TABLE XI. CRITICAL VALUES OF Σd^2 FOR RANK CORRELATION†

n	P = 0·05	P = 0·01	*n*	P = 0·05	P = 0·01
5	0–40		23	1187–2861	960–3088
6	4–66	0–70	24	1370–3230	1115–3485
7	12–100	4–108	25	1570–3630	1287–3913
8	22–146	10–158	26	1789–4061	1475–4375
9	38–200	24–216	27	2028–4524	1681–4871
10	61–269	39–291	28	2287–5021	1906–5402
11	88–352	58–382	29	2569–5551	2149–5971
12	121–451	84–488	30	2873–6117	2414–6576
13	163–565	115–613	31	3199–6721	2700–7220
14	213–697	154–756	32	3550–7362	3008–7904
15	272–848	201–919	33	3926–8042	3338–8630
16	342–1018	257–1103	34	4328–8762	3693–9397
17	423–1209	322–1310	35	4757–9523	4073–10,207
18	515–1423	398–1540	36	5213–10,327	4476–11,064
19	621–1659	484–1796	37	5698–11,174	4908–11,964
20	740–1920	583–2077	38	6213–12,065	5366–12,912
21	873–2207	695–2385	39	6758–13,002	5853–13,907
22	1022–2520	820–2722	40	7334–13,986	6367–14,953

† Litchfield and Wilcoxon (1955). Reprinted from *Analytical Chemistry*, **27**, 299, 1955. Copyright 1955 by the American Chemical Society and reprinted by permission of the copyright owner. If the observed Σd^2 is within the tabulated limits the ranked values are not significantly associated.

TABLE XII. ORTHOGONAL POLYNOMIALS†

$n = 3$		$n = 4$			$n = 5$			$n = 6$			$n = 7$		
ξ_1	ξ_2	ξ_1	ξ_2	ξ_3	ξ_1	ξ_2	ξ_3	ξ_1	ξ_2	ξ_3	ξ_1	ξ_2	ξ_3
−1	+1	−3	+1	−1	−2	+2	−1	−5	+5	−5	−3	+5	−1
0	−2	−1	−1	+3	−1	−1	+2	−3	−1	+7	−2	0	+1
+1	+1	+1	−1	−3	0	−2	0	−1	−4	+4	−1	−3	+1
		+3	+1	+1	+1	−1	−2	+1	−4	−4	0	−4	0
					+2	+2	+1	+3	−1	−7	+1	−3	−1
								+5	+5	+5	+2	0	−1
											+3	+5	+1
2	6	20	4	20	10	14	10	70	84	180	28	84	6
$x - 2$		$2x - 5$			$x - 3$			$2x - 7$			$x - 4$		
$3x^2 - 12x + 10$		$x^2 - 5x + 5$			$x^2 - 6x + 7$			$^1/_2(3x^2 - 21x + 28)$			$x^2 - 8x + 12$		
		$^1/_3(10x^3 - 75x^2 + 167x - 105)$			$^1/_6(5x^3 - 45x^2 + 118x - 84)$			$^1/_6(10x^3 - 105x^2 + 317x - 252)$			$^1/_6(x^3 - 12x^2 + 41x - 36)$		

n = 8			n = 9			n = 10			n = 11			n = 12		
ξ_1	ξ_2	ξ_3	ξ_1	ξ_2	ξ_3	ξ_1	ξ_2	ξ_3	ξ_1	ξ_2	ξ_3	ξ_1	ξ_2	ξ_3
−7	+7	−7	−4	+28	−14	−9	+6	−42	−5	+15	−30	−11	+55	−33
−5	+1	+5	−3	+7	+7	−7	+2	+14	−4	+6	+6	−9	+25	+3
−3	−3	+7	−2	−8	+13	−5	−1	+35	−3	−1	+22	−7	+1	+21
−1	−5	+3	−1	−17	+9	−3	−3	+31	−2	−6	+23	−5	−17	+25
+1	−5	−3	0	−20	0	−1	−4	+12	−1	−9	+14	−3	−29	+19
+3	−3	−7	+1	−17	−9	+1	−4	−12	0	−10	0	−1	−35	+7
+5	+1	−5	+2	−8	−13	+3	−3	−31	+1	−9	−14	+1	−35	−7
+7	+7	+7	+3	+7	−7	+5	−1	−35	+2	−6	−23	+3	−29	−19
			+4	+28	+14	+7	+2	−14	+3	−1	−22	+5	−17	−25
						+9	+6	+42	+4	+6	−6	+7	+1	−21
									+5	+15	+30	+9	+25	−3
												+11	+55	+33
168	168	264	60	2772	990	330	132	8580	110	858	4290	572	12,012	5148
$2x - 9$			$x - 5$			$2x - 11$			$x - 6$			$2x - 13$		
$x^2 - 9x + 15$			$3x^2 - 30x + 55$			$\frac{1}{2}(x^2 - 11x + 22)$			$x^2 - 12x + 26$			$3x^2 - 39x + 91$		
$\frac{1}{3}(2x^3 - 27x^2 + 103x - 99)$			$\frac{1}{6}(5x^3 - 75x^2 + 316x - 330)$			$\frac{1}{6}(10x^3 - 165x^2 + 761x - 858)$			$\frac{1}{6}(5x^3 - 90x^2 + 451x - 546)$			$\frac{1}{3}(2x^3 - 39x^2 + 211x - 273)$		

† Owen (1962). Reproduced with the permission of the publishers, Addison-Wesley, Reading, Mass. The figure at the foot of each column is $\Sigma\xi^2$ for that column.

TABLE XIII. SUMMARY OF TESTS OF SIGNIFICANCE

Comparison of	H_0	H_a	Find	Reject H_0 if
(1) $\bar{x}$ with standard or given value, μ_0; σ_0 given (99).	$\mu = \mu_0$	$\mu \neq \mu_0$ $\mu \gtrless \mu_0$	$u = (\bar{x} - \mu_0)/(\sigma_0/\sqrt{n})$	$u > 1{\cdot}96$ $u \gtrless \pm 1{\cdot}64$
(2) $\bar{x}$ with standard or given value, μ_0; σ_0 not given (105).	$\mu = \mu_0$	$\mu \neq \mu_0$ $\mu \gtrless \mu_0$	$t = (\bar{x} - \mu_0)/(s/\sqrt{n}), \quad f = n - 1$	$t > t_{.05'', f}$ $t \gtrless \pm t_{.05', f}$
(3) s with standard or given value, σ_0; n large (104).	$\sigma = \sigma_0$	$\sigma \neq \sigma_0$ $\sigma \gtrless \sigma_0$	$u = (s - \sigma_0)/(\sigma_0/\sqrt{2n})$	$u > 1{\cdot}96$ $u \gtrless \pm 1{\cdot}64$
(4) s^2 with standard or given value, σ_0^2 (104).	$\sigma^2 = \sigma_0^2$	$\sigma^2 \neq \sigma_0^2$ $\sigma^2 \gtrless \sigma_0^2$	$F = s^2/\sigma_0^2$ or $\sigma_0^2/s^2, \quad f = n - 1$	$F > F_{.05'', f/\infty \text{ or } \infty/f}$ $F > F_{.05', f/\infty \text{ or } \infty/f}$
(5) two std. devs., s_1 and s_2; large samples.	$\sigma_1 = \sigma_2$	$\sigma_1 \neq \sigma_2$ $\sigma_1 \gtrless \sigma_2$	$t = (s_1 - s_2)/s_d, \quad s_d = \sqrt{(s_1^2/2n_1 + s_2^2/2n_2)}$	$t \doteq u > 1{\cdot}96$ $t \doteq u \gtrless \pm 1{\cdot}64$
(6) two variances, s_1^2 and s_2^2; $s_1^2 > s_2^2$, small samples (107).	$\sigma_1^2 = \sigma_2^2$	$\sigma_1^2 \neq \sigma_2^2$ $\sigma_1^2 > \sigma_2^2$	$F = s_1^2/s_2^2, \quad f_1 = n_1 - 1, \quad f_2 = n_2 - 1$	$F > F_{.05'', f_1/f_2}$ $F > F_{.05', f_1/f_2}$
(7) k variances, s_j^2; $j = 1, 2 \ldots k$ (112).	$\sigma_j^2 = \sigma^2$	$\sigma_j^2 \neq \sigma^2$	$M = \Sigma f_j \,.\, \ln s_p^2 - \Sigma(f_j \ln s_j^2)$, $s_p^2 = \Sigma f_j s_j^2/\Sigma f_j, \quad c = \Sigma(1/f_j) - 1/\Sigma f_j$	$M > M_{.05, k, c}$
(8) k ranges, w_j; $j = 1, 2 \ldots k$ (115).			w chart; $D'_{.025}\bar{w}, \quad D'_{.975}\bar{w}$	any w_j lies outside limits
(9) two means, $\bar{x}_1$ and $\bar{x}_2$; large samples.	$\mu_1 = \mu_2$	$\mu_1 \neq \mu_2$ $\mu_1 \gtrless \mu_2$	$t = (\bar{x}_1 - \bar{x}_2)/\sqrt{(s_1^2/n_1 + s_2^2/n_2)}$	$t \doteq u > 1{\cdot}96$ $t \doteq u \gtrless \pm 1{\cdot}64$
(10) two means, $\bar{x}_1$ and $\bar{x}_2$; $\sigma_1 = \sigma_2$, small samples (107).	$\mu_1 = \mu_2$	$\mu_1 \neq \mu_2$ $\mu_1 \gtrless \mu_2$	$t = (\bar{x}_1 - \bar{x}_2)\sqrt{(n_1 n_2)}/s_p\sqrt{(n_1 + n_2)}$ $s_p = \sqrt{[(\Sigma X_1^2 + \Sigma X_2^2)/f_p]}, \quad f_p = n_1 + n_2 - 2$	$t > t_{.05'', f_p}$ $t \gtrless \pm t_{.05', f_p}$
(11) two means, $\bar{x}_1$ and $\bar{x}_2$; $\sigma_1 \neq \sigma_2$, small samples (109).	$\mu_1 = \mu_2$	$\mu_1 \neq \mu_2$ $\mu_1 \gtrless \mu_2$	$t = (\bar{x}_1 - \bar{x}_2)/\sqrt{(s_1^2/n_1 + s_2^2/n_2)}$ $f = 1/[C^2/f_1 + (1 - C^2)/f_2]$ $C = (s_1^2/n_1)/(s_1^2/n_1 + s_2^2/n_2)$	$t > t_{.05'', f}$ $t \gtrless \pm t_{.05', f}$

(12) k means, $\bar{x}_j$; $j = 1, 2 \ldots k$, $\sigma_j = \sigma$, $n_j = n$.	$\mu_j = \mu$	$\mu_j \neq \mu$	$F = ns_{\bar{x}}^2/s_p^2$ $s_p^2 = \Sigma\Sigma X_j^2/f_p$, $f_p = k(n-1)$, $f_k = k - 1$	$F > F_{\cdot 05', f_k/f_p}$
Alternatively, (115). For unequal samples refer p. 112.			$\bar{x}$ chart; $\bar{\bar{x}} \pm A'_{\cdot 025}\bar{w}$	any $\bar{x}_j$ lies outside limits.
(13) pairs of values having n differences of mean $\bar{x}$ and std. dev. s (110).	$\mu = 0$	$\mu \neq 0$ $\mu \gtrless 0$	$t = \bar{x}/(s/\sqrt{n})$, $f = n - 1$	$t > t_{\cdot 05'', f}$ $t \gtrless \pm t_{\cdot 05', f}$
(14) p with standard or given value, P_0 (116).	$P = P_0$	$P \neq P_0$ $P \gtrless P_0$	$u = (p - P_0)/\sqrt{(P_0 Q_0/n)}$, $Q_0 = 1 - P_0$	$u > 1{\cdot}96$ $u \gtrless \pm 1{\cdot}64$
(15) two proportions, p_1 and p_2 (117).	$P_1 = P_2$	$P_1 \neq P_2$ $P_1 \gtrless P_2$	$u = (p_1 - p_2)/\sqrt{[\bar{p}\bar{q}(1/n_1 + 1/n_2)]}$ $\bar{p} = (n_1 p_1 + n_2 p_2)/(n_1 + n_2)$, $\bar{q} = 1 - \bar{p}$	$u > 1{\cdot}96$ $u \gtrless \pm 1{\cdot}64$
(16) k proportions, p_j; $j = 1, 2, \ldots k$, $n_j = n$.	$P_j = P$	$P_j \neq P$	p chart; $\bar{p} \pm 1{\cdot}96\sqrt{(\bar{p}\bar{q}/n)}$ $\bar{p} = \Sigma p/k$, $\bar{q} = 1 - \bar{p}$	any p_j lies outside limits
(17) observed, O, with expected, E, frequencies (119, 120).	Os do or do not differ randomly from Es.		$\chi^2 = \Sigma[(O - E)^2/E]$, k = no. of differences, c = constraints on E.	$\chi^2 > \chi^2_{\cdot 05, k-c}$
(18) g_1 with zero (test for normality, 105).	$\gamma_1 = 0$	$\gamma_1 \neq 0$	$u \doteqdot g_1/\sqrt{(6/n)}$	$u > 1{\cdot}96$
(19) g_2 with zero (test for normality, 105).	$\gamma_2 = 0$	$\gamma_2 \neq 0$	$u \doteqdot g_2/\sqrt{(24/n)}$	$u > 1{\cdot}96$
(20) r with zero (test of association, 191). For other tests refer p. 201. For partial and multiple correlation, refer p. 214.	$\varrho = 0$	$\varrho \neq 0$	$r = \Sigma XY/\sqrt{\Sigma X^2}\sqrt{\Sigma Y^2}$, $f = n - 2$	$r > r_{\cdot 05, f}$
(21) r with given value, ϱ_0, i.e. z with ζ_0 (191).	$\zeta = \zeta_0$	$\zeta \neq \zeta_0$ $\zeta \gtrless \zeta_0$	$u = (z - \zeta_0)/\sqrt{(n-3)}$	$u > 1{\cdot}96$ $u \gtrless \pm 1{\cdot}64$
(22) two correlation coefficients, r_1 and r_2, i.e. z_1 and z_2 (193).	$\zeta_1 = \zeta_2$	$\zeta_1 \neq \zeta_2$ $\zeta_1 \gtrless \zeta_2$	$u = (z_1 - z_2)/\sigma_{\Delta z}$ $\sigma_{\Delta z} = \sqrt{[1/(n_1 - 3) + 1/(n_2 - 3)]}$	$u > 1{\cdot}96$ $u \gtrless \pm 1{\cdot}64$

TABLE XIII *(cont.)*

Comparison of	H_0	H_a	Find	Reject H_0 if
(23) k correlation coefficients, r_j; $j = 1, 2 \ldots k$ (194).	$\varrho_j = \varrho$	$\varrho_j \neq \varrho$	$\chi^2 = \Sigma(n_j - 3) z_j^2 - (\Sigma(n_j - 3) z_j)^2/\Sigma(n_j - 3)$	$\chi^2 > \chi^2_{.05,\, k-1}$
(24) two pairs of means of correlated values, $\bar{x}_1, \bar{y}_1$ and $\bar{x}_2, \bar{y}_2$ (195).	$\mu_1 = \mu_2$ (for x and y)	$\mu_1 \neq \mu_2$	$F = (n_1 + n_2 - 3)\, T^2/2(n_1 + n_2 - 2)$ $T^2 = (t_x^2 - 2r_p t_x t_y + t_y^2)/(1 - r_p^2)$ $r_p = \Sigma\Sigma XY/\sqrt{\Sigma\Sigma X^2}\sqrt{\Sigma\Sigma Y^2}$, t_x, t_y as in (10).	$F > F_{.05',2/(n_1+n_2-3)}$
(25) b with given value, β_0 (148, 164). For partial regression coefficients refer pp. 212, 219.	$\beta = \beta_0$	$\beta \neq \beta_0$ $\beta \gtrless \beta_0$	$t = (b - \beta_0)/(s_e/\sqrt{\Sigma X^2})$ $s_e = \sqrt{[(\Sigma Y^2 - b\Sigma XY)/f]}$, $f = n - 2$ For a line constrained to pass through the origin, x and y replace X and Y and $f = n - 1$.	$t > t_{.05'',f}$ $t \gtrless \pm t_{.05',f}$
(26) a with given value, α_0 (148).	$\alpha = \alpha_0$	$\alpha \neq \alpha_0$ $\alpha \gtrless \alpha_0$	$t = (a - \alpha_0)/s_e\sqrt{(1/n + \bar{x}^2/\Sigma X^2)}$ $f = n - 2$	$t > t_{.05'',f}$ $t \gtrless \pm t_{.05',f}$
(27) a, b with given values, α_0, β_0. Refer pp. 151–2.				
(28) two regression lines. a) $s_{e_1}^2$ and $s_{e_2}^2$ (175) b) b_1 and b_2 (175)	 $\sigma_{e_1}^2 = \sigma_{e_2}^2$ $\beta_1 = \beta_2$	 $\sigma_{e_1}^2 \neq \sigma_{e_2}^2$ $\beta_1 \neq \beta_2$	 F as in test (6). $f_1 = n_1 - 2$, $f_2 = n_2 - 2$ $t = (b_1 - b_2)/s_{ep}\sqrt{(1/\Sigma X_1^2 + 1/\Sigma X_2^2)}$ $s_{ep} = \sqrt{(\Sigma\Sigma \hat{Y}^2/f_p)}$, $f_p = n_1 + n_2 - 4$	 $F > F_{.05'',f_1/f_2}$ $t > t_{.05'',f_p}$
c) a_1' and a_2' (for separation of lines) (176).	$\alpha_1' = \alpha_2'$	$\alpha_1' \neq \alpha_2'$	$t = (a_1' - a_2')/s_{\Delta a'}$ $s_{\Delta a'} = s_{ep}\sqrt{[1/n_1 + 1/n_2 + (\Delta\bar{x})^2/\Sigma\Sigma X^2]}$	$t > t_{.05'',f_p}$
(29) several regression lines, refer pp. 171–5.				

Table XIII is intended as a guide to tests which can be easily tabulated. Note that: (a) most assume the parent universe or universes to be of the normal form but only in the cases of (4), (6) and (7) is the effect of non-normality serious; (b) one- and two-sided values of test statistics are indicated (at the five per cent level for convenience) by single and double primes; (c) $t_{.05'}$ equals the tabulated $t_{.10''}$ (Table II) and $F_{.05''}$ may be taken as the tabulated $F_{.025'}$ (Table III); (d) in two-sided tests only the absolute value of the test statistic matters and differences in equation numerators may be interpreted accordingly, i.e. as $|\ldots|$; (e) to simplify tabulation some equations are given in a form differing slightly from that in the text. Page references to examples are given in parentheses in the first column.

EXERCISES

In this collection a second example is provided of most of the problems dealt with in the text. The order follows the text in the main but Table XIII of the Appendix may be found helpful as a quick guide to location and to the essential calculations in most cases.

1. A homogenized 1:1 oil/water emulsion was found to have the particle size distribution

Size (microns):	1·0	1·5	2·0	2·5	3·0	3·5	4·0	4·5	5·0	5·5	6·0	6·5
No. of particles:	2	24	34	36	43	26	19	13	9	3	2	1

where the sizes quoted are the mid-points of 0·5 micron intervals. (a) Calculate $\bar{x}$, m_g, m_q, m_c, m_l, m_s, m_v, s and s_g from the grouped data. (b) Find m_g and s_g from a plot of percentage cumulative frequency on logarithmic probability paper.

Ans. (a) $\bar{x} = 2{\cdot}974$, $m_g = 2{\cdot}784$, $m_q = 3{\cdot}162$, $m_c = 3{\cdot}344$, $m_l = 3{\cdot}362$, $m_s = 3{\cdot}738$, $m_v = 4{\cdot}089$, $s = 1{\cdot}074$, $s_g = 1{\cdot}446$. (b) Approximately 2·784 and 1·446.

2. Calculate $\hat{\mu}$, M, μ and μ_q from the following data, derived from the Maxwell distribution equation, for the speeds of oxygen molecules at 0°C. Compare the values obtained with those found by integrating Maxwell's equation.

Speed interval (m/sec):	<100	100–	200–	300–	400–	500–	600–	700–	800–	>900
No. of molecules/1000:	14	82	166	215	203	152	92	45	20	11

Ans. The correct values are given by $129\sqrt{(T/M)}$, $140\sqrt{(T/M)}$, $145\sqrt{(T/M)}$ and $158\sqrt{(T/M)}$ m/sec, respectively, where $T = 273{\cdot}2$°K and $M = 32$ in this instance. The values calculated from the grouped data are slightly greater.

3. The following are the angular variation percentage frequency distributions of Feret's and Martin's particle size diameters (profile length and length of area bisector in a given direction of measurement) for 1·5:1 rectangular and elliptical particle profiles, relative to the orientation invariant projected area diameter in each case. Describe and characterize the distributions.

Rectangle			Ellipse		
	(a)	(b)		(c)	(d)
d	f_{d_F}	f_{d_M}	d	f_{d_F}	f_{d_M}
0·724–	1·56	16·95			
0·75–	3·04	11·11			
0·80–	3·16	7·11			
			0·816–	16·67	24·33
0·85–	3·28	5·37	0·85–	10·56	13·84
0·90–	3·46	4·57	0·90–	8·49	9·94
0·95–	3·68	3·39	0·95–	7·88	8·24
1·00–	3·95	3·09	1·00–	7·87	7·55
1·05–	3·01	1·93	1·05–	8·36	7·24
1·085–	2·60	11·13			
1·10–	9.67	13·46	1·10–	9·64	7·60
1·15–	11·29	8·74	1·15–	13·26	9·56
1·20–	14·38	6.63	1·20–	17·27	11·70
			1·225		
1·25–	26·35	6·52 (1·25– and 1·30– combined)			
1·30–	10·57				
1·304					
$\Sigma f =$	100·00	100·00		100·00	100·00

Ans. (a) Negatively skewed, J-shaped. $\mu = 1{\cdot}152$, $\sigma = 0{\cdot}155$, $\gamma_1 = -1{\cdot}13$, $\gamma_2 = +0{\cdot}213$. (b) A compound distribution (the combination of two J-shaped distributions as, in fact, is (a)). $\mu = 0{\cdot}982$, $\sigma = 0{\cdot}186$, $\gamma_1 = -0{\cdot}042$, $\gamma_2 = -1{\cdot}51$. (c) U-shaped. $\mu = 1{\cdot}090$, $\sigma = 0{\cdot}140$, $\gamma_1 = -0{\cdot}103$, $\gamma_2 = -1{\cdot}51$. (d) U-shaped. $\mu = 0{\cdot}971$, $\sigma = 0{\cdot}139$, $\gamma_1 = +0{\cdot}332$, $\gamma_2 = -1{\cdot}40$.

4. Westgren (1916), investigating von Smoluchowski's theory of fluctuations, counted the number of particles of a gold sol to be seen in a 6·2 micron diameter field at 0·81 sec intervals. The frequency of occurrence of fields, f, characterized by number of particles per field, x, was found to be

x:	0	1	2	3	4	5	6	6+
f:	333	482	397	190	74	20	4	0

(a) Calculate the arithmetic mean and standard deviation of the distribution and compare them in the light of eqn. (3.18). (b) Calculate the frequencies to be expected were the distribution of the Poisson form (having the same mean) and compare them with the observed frequencies by χ^2 test.

Ans. (a) $\bar{x} = 1{\cdot}5107$, $s = 1{\cdot}207$. (b) 331·1, 500·2, 377·9, 190·3, 71·9, 21·7, 5·7, 1·2. $\chi^2 = 3{\cdot}05$ (with the last two classes combined) is not significant so the Poisson hypothesis is acceptable.

5. Baxter and Hale (1937) report the following observations obtained in their determination of the atomic weight of carbon—by pyrolysis of chrysene: 12·0080, 12·0095, 12·0101 (twice), 12·0105 (twice), 12·0106 (3 times), 12·0120, 12·0129; triphenylbenzene: 12·0095, 12·0097, 12·0101 (twice), 12·0102, 12·0103,

12·0106, 12·0107, 12·0116; anthracene: 12·0090 (twice), 12·0099, 12·0102 (twice), 12·0103, 12·0104, 12·0106, 12·0111, 12·0113, 12·0118. Compare the three sets of values for consistency by the methods of (a) Section 3·5(3), (b) Section 3·6(3), and, if there are no significant differences, find 95% confidence limits for the atomic weight of carbon.

Ans. (a) 12·01039 $\pm$ 0·00034, (b) 12·01036 $\pm$ 0·00027. The former result is to be preferred since there is no reason to suppose that the values of the first set are individually less reliable than the remainder.

6. The diffusion coefficient of a dye has been measured in six different cells with the following results (cm^2/day):

Cell 1	Cell 2	Cell 3	Cell 4	Cell 5	Cell 6
0·383	0·422	0·422	0·368	0·377	0·360
0·396	0·406	0·433	0·371	0·368	0·380
0·399	0·395	0·406	0·386	0·375	0·413

Compare the sets of values using a graphical test.

Ans. The "warning" limits are 0·004 to 0·054 for w and 0·375 to 0·409 for $\bar{x}$ so there are significant differences between the cells.

7. Some experimental results for the degree of dissociation, α, of HI at 629°K are 0·1914, 0·1953, 0·1968, 0·1956, 0·1937, 0·1938, 0·1949, 0·1948, 0·1954 and 0·1947. Check any value which may appear to be erroneous and find the 95% confidence interval for α and for the equilibrium constant of the reaction, $2HI = H_2 + I_2$, which is related to α by the expression $K=\alpha^2/4(1-\alpha)^2$.

Ans. The lowest value is suspect. Rejecting it, $\alpha = 0{\cdot}1950 \pm 0{\cdot}0007$, $K = 0{\cdot}01467 \pm 0{\cdot}00014$.

8. Some data for the equivalent conductance, Λ_c, of very dilute aqueous NaCl solutions at 25°C are

Λ_c (ohm^{-1} cm^2):	125·79	124·58	123·31	122·51	121·70
$10^2\sqrt{c}$ (equiv./litre):	0·771	2·066	3·667	4·661	5·645

(a) Assume a linear relationship with Λ_c, alone, subject to normally distributed errors and fit the equation $\Lambda_c = a + b\sqrt{c}$. (i) Find the 95% confidence limits for the true intercept, Λ_0. (ii) Alternatively, find the 95% interval for the true slope and compare this with the theoretical Debye–Hückel–Onsager value for aqueous solutions of NaCl at this temperature, —88·65. (iii) Alternatively, considering the joint unreliability of the estimates of slope and intercept, find limits for Λ_0 consistent with a slope of —88·65. (b) Assume that the theoretical slope does, in fact, apply to these data. Fit the constrained line $\Lambda_c = a - 88{\cdot}65\sqrt{c}$ and find the 95% confidence interval for Λ_0.

Ans. (a) (i) 126·38 $\pm$ 0·18; (ii) —83·45 $\pm$ 4·79; (iii) 126·55 $\pm$ 0·07. (b) 126·77 $\pm$ 0·04.

9. In an investigation of the effect of surfactant on the diffusion of dyes, diffusion coefficients of Anthralan Blue B were measured in the presence of varying relative amounts of the surfactant NP10 at 26°C.

Surfactant/dye ratio, S:	0	5	10	15
Diffusion coeff., D (cm²/day):	0·284	0·265	0·141	0·098
	0·303	0·275	0·147	0·106
	0·301		0·159	
	0·314			

Is D linearly related to S?

Ans. The fitted line is $\hat{D} = 0{\cdot}2573 - 0{\cdot}0062475S$ but the variation of the means about the line is very significantly greater than the variation within sets ($F = 194$), so either the relationship is non-linear or the D values do not reflect the error to be expected of truly independent values.

10. The retention index of 2- and 3-methyl alkanes in a gas chromatographic column with squalane as the stationary phase is linearly related to the number of carbon atoms in the main chain. The data, for 100°C, are

	Retention index, y_1		Retention index, y_2	Length of chain, x
2-methyl butane	476	3-methyl butane	—	4
2-methyl pentane	571	3-methyl pentane	586	5
2-methyl hexane	668	3-methyl hexane	678	6
2-methyl heptane	765	3-methyl heptane	774	7
2-methyl octane	—	3-methyl octane	870	8

Can a single linear equation be used to represent the points?

Ans. No, but the regression coefficients are comparable and may be pooled to give $\hat{y}_1 = 94{\cdot}2 + 95{\cdot}6x$, $\hat{y}_2 = 105{\cdot}6 + 95{\cdot}6x$, with $s_{ep} = 1{\cdot}225$. The 95% confidence interval for the true separation, $\Delta\alpha' = 11{\cdot}5 \pm 2{\cdot}6$.

11. If a positive biaxial crystal is mounted with its optic axial plane vertical in a series of liquids having refractive indices, n, between its own lowest and highest indices, α and γ, and the angle, θ, for a match of refractive index observed, and if $\cos^2\theta$ is plotted against $1/n^2$, then $1/\alpha^2$ is found against $\cos^2\theta = 0$ and $1/\gamma^2$ against $\cos^2\theta = 1$. Further, if the apparent optic axial angle, V′, is also measured then the intersection of the $\cos^2\theta$ and $\cos^2 V'$ versus $1/n^2$ lines gives $1/\beta^2$ where β is the intermediate crystal index. Assuming n is free of error and $\cos^2\theta$ and $\cos^2 V'$ may be regarded as normally distributed for given n, find 95% confidence intervals for α, β and γ from the following data for an organic acid.

$1/n^2$:	0·331	0·347	0·364	0·382	0·401	0·421
$\cos^2\theta$:	0·089	0·246	0·407	0·593	0·791	0·988
$\cos^2 V'$:		0·687	0·694	0·704	0·710	

Ans. $\alpha = 1{\cdot}540 \pm 0{\cdot}042$, $\gamma = 1{\cdot}764 \pm 0{\cdot}060$, $\beta = 1{\cdot}595 \pm 0{\cdot}001$. Note that several confidence intervals are being calculated from the same set of data in this problem.

12. The durability of moulds for steel ingots can be measured as consumption of mould material relative to the amount of steel produced. It should depend to some extent on the ratio of the weight of the mould to that of the ingot for which it is used (the M/I ratio) and might depend also on the actual weight of the mould. An investigation into the behaviour of small moulds in British foundries yielded the following data.

Consumption, y (lb/ton cast):	28	22	31	23	37	27	28	55	45	51
M/I ratio, x:	0·87	0·90	1·08	1·01	1·16	1·14	1·24	1·32	1·30	1·40
Weight, z (cwt):	100	68	67	52	84	49	57	66	46	38

(a) Find the correlation coefficient between x and y, test it for significance and calculate 95% confidence limits for the true value. Calculate the rank correlation coefficient as a check.

(b) From similar investigations in German and other foundries the values 0·74 (16 observations) and 0·42 (45 observations) have been obtained for the correlation coefficient. Compare these with the estimate from the data above and, if the three do not differ significantly, obtain an overall estimate and the corresponding confidence interval.

(c) Find the two regression lines relating x and y and draw them on a scatter diagram together with a confidence ellipse for the true point of intersection (μ_x, μ_y).

(d) Calculate the partial and multiple correlation coefficients, $r_{yx.z}$ and $r_{y.xz}$, and the multiple regression equation for y on x and z. Has z a significant influence on the variability of y?

Ans. (a) $r_{yx} = +0{\cdot}819$, 95% confidence limits of $\varrho = 0{\cdot}40$ to $0{\cdot}96$, $r_S = +0{\cdot}785$. (b) $\bar{r} = 0{\cdot}560$, 95% confidence limits of $\bar{\varrho} = 0{\cdot}37$ to $0{\cdot}71$. (c) $\hat{y} = -27{\cdot}05 + 54{\cdot}07x$, $\hat{x} = 0{\cdot}7113 + 0{\cdot}01241y$. (d) $r_{yx.z} = +0{\cdot}870$, $r_{y.xz} = 0{\cdot}880$, $\hat{y} = -33{\cdot}98 + 57{\cdot}375x + 0{\cdot}05029z$. The z term is not significant.

13. Bosworth (1956) has shown that in hindered sedimentation the ratio, y, of the volume of sediment at time t to the final volume (infinite time) is linearly related to the ratio, x, of the time t to that required to reach twice the final volume. From the data given, find a least squares relationship between y and x, the standard error of estimate and the 95% confidence interval for the slope. Note that the variability of y is approximately proportional to x, which can be assumed error free, and that the line must pass through the point 0, 1.

x:	0·5	1·0	1·5	2·0	2·5	3·0	3·5	4·0
y:	1·49	1·92	2·65	2·70	4·00	3·36	4·61	4·52

Ans. $\hat{y} = 1 + 0{\cdot}97x$, $s_e = 0{\cdot}137x$, 90% confidence limits of $\beta = 0{\cdot}97 \pm 0{\cdot}13$.

14. The concentration, C, of a complex formed in a particular system at equilibrium is given by K = C(S+A+B—C)/(A—C) (B—C) in which K is the equilibrium constant and A, B, and S are known concentrations of reactants and solvent. Neither K nor C are known but by taking trial values of K, finding the corresponding apparent values of C and fitting the latter to a known function, sums of squares of residuals (SSR) are obtained:

K taken:	0·40	0·45	0·50	0·55	0·60	0·65	0·70
SSR found:	1·1311	0·5431	0·2154	0·2117	0·2367	0·5406	1·0140

Find the best value of K, assuming this to be the one which minimizes the SSR.

Ans. K = 0·548 by Horton's (1955) method.

15. The following are some vapour pressure data for an alcohol:

T (°C):	20	30	40	50	60	70	80	90
p (mm):	30·90	56·23	102·33	173·8	288·4	446·7	676·1	1000
log p:	1·49	1·75	2·01	2·24	2·46	2·65	2·83	3·00

Fit to these data by the method of least squares (using the method of selected points as a check) equations of the form (a) $\log p = a + b/T$, (b) $\log p = a + b \log T + c/T$, (c) $\log p = a + b/(T + c)$, and (d) a polynomial in T of suitable degree (the degree to be determined by an examination of the item differences). Find the standard errors of estimate and compare the fitted equations for suitablility.

Ans. (a) $\log p = 9{\cdot}35718 - 2303{\cdot}63/T$ ($T = T°C + 273{\cdot}2$), $s_e = 0{\cdot}0124$.
(b) $\log p = 39{\cdot}93776 - 10{\cdot}3741 \log T - 3769{\cdot}70/T$ ($T = T°C + 273{\cdot}2$), $s_e = 0{\cdot}006$.
(c) $\log p = 7{\cdot}54536 - 1271{\cdot}543/(T + 189{\cdot}746)$ ($T = T°C$), $s_e = 0{\cdot}006$.
(d) $\log p = 0{\cdot}89922 + 3{\cdot}11847 \times 10^{-2}T - 8{\cdot}75012 \times 10^{-5}T^2$ ($T = T°C$), $s_e = 0{\cdot}005$.

The deviations of the observed values from those given by (a) vary systematically so s_e does not have its usual significance in this case.

16. In an investigation of the pyrolysis of cyclopentyl chloride at 359°C, the pressures observed at two minute intervals were 5·10, 5·40, 5·68, 5·94, 6·20, 6·41, 6·62, 6·82, 7·00, 7·16, 7·31, 7·45, 7·60 and 7·72 mm. The first of these was observed 2 minutes and the last 28 minutes after the commencement of the reaction. Assuming the reaction to be of the first order, find the initial and final pressures and the velocity constant.

Ans. $p_0 = 4{\cdot}76$ mm, $p_\infty = 9{\cdot}2$ mm, $k = 0{\cdot}039$ min^{-1}.

17. The percentage yields of a product obtained in a newly instituted batch process are

84·9, 84·5, 81·3, 86·0, 83·1, 85·9, 84·0, 83·3, 85·0, 85·5,
85·8, 86·4, 83·1, 82·6, 87·6, 84·5, 85·4, 86·6, 85·4, 84·4,
88·9, 81·3, 87·8, 81·9, 85·6, 90·9, 86·1, 86·9, 86·1, 84·9,
85·6, 82·0, 83·6, 84·6, 84·4, 85·1, 86·6, 83·7, 83·9, 89·9,
83·9, 84·1, 84·3, 80·5, 85·3, 87·8, 87·0, 88·0, 85·1, 85·9,

for 50 successive runs.

(a) Fit a Gaussian distribution to the sample data by the ordinate or area method and compare the observed and calculated frequencies by χ^2 test. (b) Plot the percentage cumulative frequencies on probability paper, draw approximate 95% limit lines for cumulative frequency as a test of normality, and find approximate 95% confidence intervals for the first quartile and for the cumulative frequency below 83% yield of the assumed parent distribution. (c) Find the standard deviation from the mean square successive differences and from the mean moving range and compare with the estimate obtained from the individual values. (d) Test the yield for trend by finding 95% limits for individual values using the mean moving range. (e) Within what limits can it be said that 95% of all batch yields should lie, with (i) 90%, (ii) 99% confidence? (f) The mean and standard deviation of the percentage yields of a very large number of batches made with previously installed equipment are 83 and 2·5, respectively. Can it be said that the new process has significantly increased the level and decreased the variability of the yield?

Ans. (a) $\chi^2 = 3{\cdot}5$ for a classification into 11 groups starting from 80 so the assumed parent universe can be regarded as Gaussian. (b) All points are within the limit lines. 95% confidence intervals are 83·1 to 84·3 for $Q_{.25}$ and 7 to 24 for $F_{<83}$. (c) $s' = 2{\cdot}25$, s from $\bar{w}' = 2{\cdot}15$, $s = 2{\cdot}11$. (d) There is no evidence of trend; 3 of the 50 lie outside the limits calculated but this is close to expectation. (e) 80·3 to 90, 79·7 to 90·6. (f) The average yield has increased significantly but there in no good reason to suppose that its variability has decreased.

18. The gas-liquid chromatogram of a mixture of two organic compounds consists of overlapping peaks. Readings of the height of the record curve, y, in chart scale units at millimetre intervals, x, are 0·30, 0·35, 0·39, 0·46, 0·58, 0·80, 1·07, 1·50, 2·06, 2·77, 3·52, 4·26, 4·98, 5·47, 5·72, 5·75, 5·53, 5·00, 4·58, 4·12, 3·90, 3·83, 4·06, 4·28, 4·60, 4·87, 4·80, 4·44, 3·75, 3·05, 2·32, 1·72, 1·18, 0·74, 0·55, 0·38, 0·32, 0·30. Taking $x = 0$ for the first of these and $x = 37$ for the last, resolve the curve into two Gaussian components and find their respective peak positions and their heights and areas above the base in the arbitrary units used.

Ans. 14·6, 5·48 and 60·5 for the first component and 25·6, 4·38 and 36·9 for the second.

19. The molar volumes, V_m, of the halogenated methanes, which exist in the liquid state at 15°C, are 63·67 (CH_2Cl_2), 79·66 ($CHCl_3$), 95·92 (CCl_4), 69·27 (CH_2Br_2), 87·06 ($CHBr_3$), 61·90 (CH_3I) and 80·33 (CH_2I_2), all in cm³/mole. Assuming these values are additive in terms of the atomic constitutions of the molecules (so that, for example, $V_m(CH_3I) = V_C + 3V_H + V_I$), find the most consistent set of atomic contributions and compare the calculated and observed molar volumes.

Ans. $V_C = 36{\cdot}742$, $V_H = 0{\cdot}001$, $V_{Cl} = 14{\cdot}460$, $V_{Br} = 16{\cdot}617$, $V_I = 22{\cdot}468$. The molar volumes calculated from these are 65·66, 80·12, 94·58, 69·97, 86·59, 59·21 and 80·33. The variance of the residuals is 7·87.

20. Thorpe and Roger's values for the viscosity of toluene (quoted by Timmermans (1950)) are

η (cP):	0·7655	0·6683	0·5900	0·5184	0·4667	0·4219
T (°C):	0·26	9·88	19·47	30·25	39·86	49·43

Smooth these data and construct a table of viscosity values at 10° intervals starting from 0°C.

Ans. 0·7684, 0·6661, 0·5856, 0·5203, 0·4657, 0·4195 approximately.

21. An old table of boiling points of the normal paraffins contains the following values for the series from ethane (C_2H_6) to n-hexadecane ($C_{16}H_{34}$): —88·5, —42·0, —0·5, 36·0, 69·0, 100·5, 125·5, 149·5, 173·0, 194·5, 214·5, 234·0, 252·5, 270·5 and 287·5 (all in °C). Smooth the values using the number of carbon atoms in the chain as argument.

(*Note*. Apart from the methods described in Chapter VI, a number of empirical formulae have been proposed for smoothing the values of properties of substances in homologous series. For example, T (°K) = aM^b (where M is the molecular weight and a and b adjustable constants) has been used for the boiling points above that of *n*-hexane. See also Tatevskii *et al.* (1961).)

22. The following table lists the vapour pressures of carbon dioxide at various temperatures.

p (mm):	10718	14781	19872	26142	33763	42959	54086
T (°C):	—30	—20	—10	0	10	20	30

For 5°C, find p, dp/dT and the latent heat of vaporization, L_v, from the relationship, $L_v = (dp/dT)\,(RT^2/p)$, in which R is the gas constant (1·987 cal/mole) and T is in °K.

Ans. 29777 mm, 760·6 mm/deg, 3960 cal/mole.

23. A sample of a plastic is tested by stretching a specimen of 1 cm^2 cross-section and 1 cm length until it breaks. The data obtained from the test are

f (kg/cm²):	0	120	122	94	72	74	76	77	79	80	82	83	84	85	86	88	89	91
l (cm):	1	2	3	4	5	6	7	8	9	10	11	12	13	14	15	16	17	17·4

where f is the force applied and l is the length of the sample between gauge marks. Calculate the work of rupture which is given by $W = \int f\,.\,dl$ between the limits of initial length and length of the specimen at rupture.

Ans. 1453 kg . cm/cm³.

24. A water-insoluble monolayer flows through a narrow canal due to a difference in surface tension, γ, maintained along it. The area of monolayer flowing per unit time, A, is a function of the width of the canal, w, the surface viscosity, η (having dimensions $[m]\,[t]^{-1}$), and the change in surface tension per unit length of the canal, $d\gamma/dl$. Obtain, dimensionally, a relationship between A and the other factors.

Ans. $A = (\text{constant})\,(w^3\,.\,d\gamma/dl/\eta)$.

25. (a) The volume, V, of a drop of liquid (in contact with its own vapour) falling from an orifice should depend on the surface tension, γ, density, ϱ, radius of orifice, r, and the gravitational constant, g. Obtain, dimensionally, a general relationship between these factors.

(b) The following data (all in absolute c.g.s. units) have been obtained in a drop volume determination. Using them and the dimensional relationship previously derived, obtain a complete empirical relationship between the factors.

V/r^3	γ	ϱ	r	g
3·00	72·80	0·998	0·5340	980·7
5·00	30·10	0·812	0·3865	
7·00	44·23	1·132	0·4016	
9·00	19·53	0·912	0·3002	
11.00	38·38	1·016	0·4106	

Ans. $V/r^3 = \varphi(\gamma/\varrho r^2 g)$, $V/r^3 = 632{\cdot}7 - 4{\cdot}654 \times 10^3(\gamma/\varrho r^2 g) + 8{\cdot}589 \times 10^3(\gamma/\varrho r^2 g)^2$.

26. 1000 gallons of a chemical solution is to be stirred for 10 hours in a lead-lined flat-bottomed cylindrical tank, 6 feet in diameter. The solution will attack the wall and base slowly but the manner of attack and conditions of stirring are such that a constant mass of metal is expected to pass into solution per unit time per unit area of exposed lining.

The reaction is to be checked in the laboratory in an inert glass container holding 0·1 gallon of solution. Assuming that the stirring and reaction conditions are the same as in the large reactor, what dimensioned cube of lead must be suspended in the laboratory vessel to give the same concentration of dissolved metal after 10 hours? The rate of attack is so slow that any change in the dimensions of the cube can be neglected.

Ans. 0·571 inch side.

REFERENCES

ACTON, F.S. (1959) *Analysis of Straight Line Data*, Wiley, New York.

AITCHISON, J. and BROWN, J.A.C. (1957) *The Lognormal Distribution*, Cambridge.

ANDERSON, R.L. and BANCROFT, T.A. (1952) *Statistical Theory in Research*, McGraw-Hill, New York.

ANGYAL, S.J. and MCHUGH, D.J. (1956) *Chemistry and Industry*, **1956,** 1147.

ASKOVITZ, S.I. (1957) *J. Amer. Stat. Assoc.*, **52,** 13.

ASKOVITZ, S.I. (1959) *J. Amer. Stat. Assoc.*, **54,** 668.

A.S.T.M. Manual (1951) *A.S.T.M. Manual on Quality Control of Materials*, American Society for Testing Materials, Philadelphia.

BAILEY, J.T., BEATTIE, W.H. and BOOTH, C. (1962) *J. Chem. Educ.*, **39,** 196.

BARTLETT, J.C. and SMITH, D.M. (1960) *Canadian J. Chem.*, **38,** 2057.

BASHFORTH, F. and ADAMS, J.C. (1883) *An Attempt to Test the Theory of Capillary Action*, Cambridge.

BAXTER, G.P. and HALE, A.H. (1937) *J. Amer. Chem. Soc.*, **59,** 506.

BERKSON, J. (1950) *J. Amer. Stat. Assoc.*, **45,** 164.

BIRGE, R.T. (1932) *Phys. Rev.*, **40,** 213.

BLAEDEL, W.J., MELOCHE, V.W. and RAMSAY, J.A. (1951) *J. Chem. Educ.*, **28,** 643.

BLISS, C.I. and JAMES, A.T. (1966) *Biometrics*, **22,** 573.

BOND, W.N. (1935) *Probability and Random Errors*, Arnold, London.

BOSWORTH, R.C.L. (1944) *Proc. Roy. Soc. (N.S.W.)* **78,** 81.

BOSWORTH, R.C.L. (1947a) *Proc. Roy. Soc. (N.S.W.)* **81,** 15.

BOSWORTH, R.C.L. (1947b) *Trans. Farad. Soc.* **43,** 399.

BOSWORTH, R.C.L. (1949) *Proc. Roy. Soc. (N.S.W.)* **83,** 8.

BOSWORTH, R.C.L. (1950) *Physics in the Chemical Industry*, Macmillan, London.

BOSWORTH, R.C.L. (1956) *J. Colloid Sci.*, **11,** 406.

BOWKER, A.H. and LIEBERMAN, G.J. (1959) *Engineering Statistics*, Prentice-Hall, New Jersey.

BROADBENT, S.R. (1953) *Some Uses of the Nomogram in Statistics*, Brit. Coal Utilisation Research Assoc. Information Circular No. 84.

BROWNLEE, K.A. (1960) *Statistical Theory and Methodology in Science and Engineering*, Wiley, New York.

BURTON, E.F. (1938) *The Physical Properties of Colloidal Solutions*, Longmans, London.

CAMPBELL, M. (1942) *Phil. Mag.*, **33,** 761.

CORNELISSEN, J., MEYER, H.W.J.H., KRUITHOF, A.M. and HAMAKER. H.C. (1962) *Advances in Glass Technology*, 488–510, Plenum, New York.

COWDEN, D.J. (1957) *Statistical Methods in Quality Control*, Prentice-Hall, New Jersey.

DAMKÖHLER, G. (1936) *Zeit. Electrochem.*, **42**, 846.

DAVID, H. A. (1956) *Biometrika*, **43**, 449.

DAVIES, O. L. (1954) *Design and Analysis of Industrial Experiments*, Oliver & Boyd, London.

DAVIES, O. L. (1957) *Statistical Methods in Research and Production*, 3rd Ed., Oliver & Boyd, London.

DAVIS, D. S. (1962) *Nomography and Empirical Equations*, 2nd Ed., Reinhold, New York.

DEMING, W. E. (1943) *Statistical Adjustment of Data*, Wiley, New York.

DINGLE, H. (1942) *Phil. Mag.*, **33**, 321, 692.

DIXON, W. J. (1950) *Ann. Math. Stat.*, **21**, 488.

DIXON, W. J. (1953) *Biometrics*, **9**, 89.

DIXON, W. J. and MASSEY, F. J. (1957) *Introduction to Statistical Analysis*, McGraw-Hill, New York.

DUDDING, B. P. and JENNETT, W. J. (1955) *British Standard 2564:1955*, British Standards Institution, London.

DURAND, D. (1954) *J. Amer. Stat. Assoc.*, **49**, 130.

DURBIN, J. (1953) *J. Amer. Stat. Assoc.*, **48**, 799.

DWYER, P. S. (1951) *Linear Computations*, Wiley, New York.

ELDERTON, W. P. (1938) *Frequency Curves and Correlation*, 3rd. Ed. Cambridge.

ESPENSCHEID, W. F., KERKER, M. and MATIJEVIC, E. (1964) *J. Phys. Chem.*, **68**, 3093.

FINNEY, D. J. (1947) *Probit Analysis*, Cambridge.

FISHER, R. A. (1963) *Statistical Methods for Research Workers*, 13th Ed., Oliver & Boyd, Edinburgh.

FISHER, R. A. (1955) *J. Roy. Stat. Soc.*, Series B, **17**, 69.

FISHER, R. A. and YATES, F. (1948) *Statistical Tables for Biological, Agricultural and Medical Research*, 3rd Ed., Oliver & Boyd, London.

FLEUNDY, M. A. D. and SMITH, E. B. (1962) *Quarterly Rev.*, **16**, 241.

FRENCH, D. (1950) *J. Amer. Chem. Soc.*, **72**, 4806.

FRENCH, D. and MCINTYRE, R. L. (1951) *J. Amer. Chem. Soc.*, **73**, 4541.

GIBSON, W. M. and JOWETT, G. H. (1957) *Applied Statistics*, **6**, 114, 189.

GUEST, P. G. (1961) *Numerical Methods of Fitting*, Cambridge.

GUGGENHEIM, E. A. (1926) *Phil. Mag.*, **2**, 538.

HALD, A. (1952) *Statistical Theory with Engineering Applications*, Wiley, New York.

HALD, A. (1952) *Statistical Tables and Formulas*, Wiley, New York.

HALPERIN, M. (1961) *J. Amer. Stat. Assoc.*, **56**, 567.

HAMMETT, L. P. (1952) *Introduction to the Study of Physical Chemistry*, McGraw-Hill, New York.

HARTLEY, H. O. (1948) *Biometrika*, **35**, 32.

HERDAN, G. (1948) *Quality Control by Statistical Methods*, Nelson, London.

HERDAN, G. (1960) *Small Particle Statistics*, 2nd Ed., Butterworth, London.

HEY, E. N. and HEY, M. H. (1960) *Biometrics*, **16**, 606.

HICKS, C. R. (1955) *Industrial Quality Control*, **9** (9), 1.

HILDEBRAND, J. H. (1962) *Regular Solutions*, Prentice-Hall, New Jersey.

HOGBEN, L. (1957) *Statistical Theory*, Allen & Unwin, London.

HORTON, W. S. (1955) *Analyt. Chem.*, **27**, 1190.

HUNTLEY, H. F. (1952) *Dimensional Analysis*, MacDonald, London.

HUYBERECHTS, S., HALLEUX, A. and KRUYS, P. (1955) *Bull. des Soc. Chim. Belges*, **64**, 203.

IRANI, R. R. and CALLIS, C. F. (1963) *Particle Size Measurement, Interpretation and Application*, Wiley, New York.

JENSON, V. G. and JEFFREYS, G. V. (1963) *Mathematical Methods in Chemical Engineering*, Academic Press, London and New York.

JOHNSTONE, R. E. (1939) *Trans. Inst. Chem. Engrs., London*, **17**, 129.

JOHNSTONE, R. E. and THRING, M. W. (1957) *Pilot Plants, Models* and *Scale-up Methods in Chemical Engineering*, McGraw-Hill, New York.

KASTENBAUM, M. A. (1959) *Biometrics*, **15**, 323.

KELLY, T. L. (1947) *Fundamentals of Statistics*, Harvard.

KENDALL, M. G. and BUCKLAND, W. R. (1957) *A Dictionary of Statistical Terms*, Oliver & Boyd, Edinburgh.

KENDALL, M. G. and STUART, A. (1958) *The Advanced Theory of Statistics*, Vol. 1, Griffin, London.

KULLBACK, S. (1959) *Information Theory and Statistics*, Wiley, New York.

KUPPERMAN, M. (1959) *Applied Statistics*, **9**, 37.

LAUPICHLER, F. G. (1938) *Ind. Eng. Chem.*, **30**, 578.

LEES, R. W. and LORD, F. M. (1962) *J. Amer. Stat. Assoc.*, **57**, 917.

LITCHFIELD, J. T. and WILCOXON, F. (1955) *Analyt. Chem.*, **27**, 299.

LORD, F. M. (1955) *J. Amer. Stat. Assoc.*, **50**, 1073.

LU, B. C.-Y. (1959) *Canadian J. Chem. Eng.*, **38**, 33.

LYLE, P. (1957) *Regression Analysis of Production Costs and Factory Operations*, 3rd Ed., Oliver & Boyd, London.

MCBRIDE, W. R. and VILLARS, D. S. (1954) *Analyt. Chem.*, **26**, 901.

MANDEL, J. (1947) *Analyt. Chem.*, **19**, 949.

MANDEL, J. (1957) *J. Amer. Stat. Assoc.*, **52**, 552.

MANDEL, J. and LINNIG, F. J. (1957) *Analyt. Chem.*, **29**, 743.

MERRINGTON, M. and THOMPSON, C. M. (1946) *Biometrika*, **33**, 296.

MILLER, D. G. (1963) *Ind. Eng. Chem. Fundamentals*, **2**, 68.

MODE, E. B. (1961) *Elements of Statistics*, 3rd Ed., Prentice-Hall, New Jersey.

MOSTELLER, F. and TUKEY, J. W. (1949) *J. Amer. Stat. Assoc.*, **44**, 174.

NAIR, K. R. (1952) *Biometrika*, **39**, 190.

NALIMOV, V. V. (1963) *The Application of Mathematical Statistics to Chemical Analysis*, Pergamon, Oxford.

OPFELL, J. B. and SAGE, B. H. (1958) *Ind. Eng. Chem.*, **50**, 803.

OWEN, D. B. (1962) *Handbook of Statistical Tables*, Pergamon, Oxford.

PARADINE, C. G. and RIVETT, B. H. P. (1953) *Statistics for Technologists*, E.U.P., London.

PARRATT, L. G. (1961) *Probability and Experimental Errors in Science*, Wiley, New York.

PASCHKIS, V. (1947) *Metal Progress*, **52**, 84.

PATTERSON, H. D. (1956) *Biometrics*, **12**, 323.

PAUL, M. A. (1951) *Principles of Chemical Thermodynamics*, McGraw-Hill, New York.

PEARSON, E. S. and HARTLEY, H. O. (1966) *Biometrika Tables for Statisticians*, Vol. 1, 3rd Ed. Cambridge.

PLUMMER, H. C. (1940) *Probability and Frequency*, Macmillan, London.

POWELL, R.E. (1953) in Frost and Pearson's *Kinetics and Mechanism*, Wiley, New York.
POWELL, R.W. (1940) *Trans. Inst. Chem. Engrs., London*, **18,** 36.
PROSCHAN, F. (1957) *Industrial Quality Control*, **13** (7), 2.
RUTH, B.F. (1935) *Ind. Eng. Chem.*, **27,** 708, 806.
RUTHERFORD, E. and GEIGER, H. (1910) *Phil. Mag.*, **20,** 698.
SCHEFFE, H. (1958) *J. Amer. Stat. Assoc.*, **53,** 106.
SCHMID, C.F. (1954) *Handbook of Graphic Presentation*, Ronald, New York.
SNEDECOR, G.W. (1956) *Statistical Methods*, Iowa State University Press, Ames, Iowa.
STEVENS, S.S. (1955) *Science*, **121,** 113.
STUBBINGS, G.W. (1948) *Dimensions in Engineering Theory*, Crosby Lockwood, London.
SVENSSON, H. (1959) *Science Tools*, **6,** 1 (LKB-Produkter Fabriksaktiebolag, Stockholm, Sweden).
SWAIN, C.G. (1944) *J. Amer. Chem. Soc.*, **66,** 1696.
SWINBOURNE, E.S. (1960) *J. Chem. Soc.*, **1960,** 2371.
SWINBOURNE, E.S. (1963) *Australian J. Chem.*, **16,** 170.
TATEVSKII, V.M., BENDERSKII, V.A. and YAROVOI, S.S. (1961) *Rules and Methods for Calculating the Physico-Chemical Properties of Paraffinic Hydrocarbons*, Pergamon, Oxford.
THOMSON, G.W. (1946) *Chem. Rev.*, **38,** 1.
TIMMERMANS, J. (1950) *Physico-Chemical Constants of Pure Organic Compounds*, Elsevier, Amsterdam.
TOPPING, J. (1955) *Errors of Observation and Their Treatment*, Institute of Physics, London.
VAN STRIK, R. (1960) *Sigma*, **6** (1), 10.
VOLK, W. (1958) *Applied Statistics for Engineers*, McGraw-Hill, New York.
WELCH, B.L. (1947) *Biometrika*, **34,** 28.
WELCH, B.L. (1949) *Biometrika*, **36,** 293.
WESTGREN, A. (1916) *Arkiv för Matematik, Stockholm*, **11,** No. 14.
WHITTAKER, E. and ROBINSON, G. (1944) *The Calculus of Observations*, 4th Ed., Blackie, London.
WILLIAMS, E.J. (1959) *Regression Analysis*, Wiley, New York.
WILLIAMS, W. (1892) *Phil. Mag.*, **34,** 234.
WILLSDON, B.H. (1934) *Suppl. J. Roy. Stat. Soc.*, **1,** 152.
WORTHING, A.G. and GEFFNER, J. (1943) *The Treatment of Experimental Data*, Wiley, New York.
YOUDEN, W.J. (1947) *Analyt. Chem.*, **19,** 946.
YOUDEN, W.J. (1962) *J. Assoc. Official Agricultural Chemists*, **45,** 169.
YULE, G.V. and KENDALL, M.G. (1950) *An Introduction to the Theory of Statistics*, 14th Ed., Griffin, London.

GENERAL BIBLIOGRAPHY

The following is a brief selection of books and articles for general reference and further reading. The categories overlap to some extent.

NUMERICAL AND GRAPHICAL METHODS

BUTLER, R. and KERR, E. (1962) *Introduction to Numerical Methods*, Pitman, London.

WHITTAKER, E. and ROBINSON, G. (1944) *The Calculus of Observations*, 4th Ed., Blackie, London.

WORTHING, A.G. and GEFFNER, J. (1943) *The Treatment of Experimental Data*, Wiley, New York.

STATISTICAL METHODS—GENERAL

DIXON, W.J. and MASSEY, F.J. (1957) *Introduction to Statistical Analysis*, 2nd Ed., McGraw-Hill, New York.

GOODMAN, R. (1957) *Teach Yourself Statistics*, E.U.P., London.

SNEDECOR, G.W. (1956) *Statistical Methods*, 5th Ed., Iowa State University Press, Ames, Iowa.

YULE, G.U. and KENDALL, M.G. (1950) *An Introduction to the Theory of Statistics*, 14th Ed., Griffin, London.

STATISTICAL METHODS—CHEMICAL, ENGINEERING ETC. APPLICATIONS

BENNETT, C.A. and FRANKLIN, N.L. (1954) *Statistical Analysis in Chemistry and the Chemical Industry*, Wiley, New York.

BROWNLEE, K.A. (1960) *Statistical Theory and Methodology in Science and Engineering*, Wiley, New York.

DAVIES, O.L. (Editor) (1957) *Statistical Methods in Research and Production*, 3rd Ed., Oliver & Boyd, Edinburgh.

HALD, A. (1952) *Statistical Theory with Engineering Applications*, Wiley, New York.

MANDEL, J. (1964) *The Statistical Analysis of Experimental Data*, Interscience, New York.

NALIMOV, V.V. (1963) *The Application of Mathematical Statistics to Chemical Analysis*, Pergamon, Oxford.

PARADINE, C.G. and RIVETT, B.H. (1953) *Statistics for Technologists*, E.U.P., London.

PARRATT, L.G. (1961) *Probability and Experimental Errors in Science*, Wiley, New York.

VOLK, W. (1958) *Applied Statistics for Engineers*, McGraw-Hill, New York.

WINE, R.L. (1964) *Statistics for Scientists and Engineers*, Prentice-Hall, New Jersey.

YOUDEN, W.J. (1951) *Statistical Methods for Chemists*, Wiley, New York.

Curve Fitting, Graphical Methods and Regression Analysis

DAVIS, D.S. (1962) *Nomography and Empirical Equations*, 2nd Ed., Reinhold, New York.

GUEST, P.G. (1961) *Numerical Methods of Fitting*, Cambridge.

HERDAN, G. (1948) *Quality Control by Statistical Methods*, Nelson, London.

LYLE, P. (1957) *Regression Analysis of Production Costs and Factory Operations*, 3rd Ed., Oliver & Boyd, London.

Dimensional Analysis

HVISTENDAHL, H.S. (1964) *Engineering Units*, Macmillan, London.

JOHNSTONE, R.E. and THRING, M.W. (1957) *Pilot Plants, Models and Scale-up Methods in Chemical Engineering*, McGraw-Hill, New York.

LANGHAAR, H.L. (1951) *Dimensional Analysis and the Theory of Models*, Wiley, New York.

Bibliographical Articles

WERNIMONT, G. (1949) *Analyt. Chem.*, **21**, 115.

HADER, R.J. and YOUDEN, W.J. (1952) *Analyt. Chem.*, **24**, 120.

MANDEL, J. and LINNIG, F.J. (1956) *Analyt. Chem.*, **28**, 770.

NELSON, B.N. (1960) *Analyt. Chem.*, **32**, 161R.

NELSON, B.N. (1962) *Analyt. Chem.*, **34**, 294R.

NELSON, B.N. (1964) *Analyt. Chem.*, **36**, 344R.

HILL, H. M. and BROWN, R. H. (1968) *Analyt. Chem.*, **40**, 376R.

INDEX

MADE IN GREAT BRITAIN